MW00837051

STATISTICAL ANALYSIS OF GEOGRAPHIC INFORMATION WITH ArcView GIS® AND ArcGIS®

STATISTICAL ANALYSIS OF GEOGRAPHIC INFORMATION WITH ArcView GIS® AND ArcGIS®

DAVID W. S. WONG

JAY LEE

JOHN WILEY & SONS, INC.

This book is printed on acid-free paper. ♾

Published by John Wiley & Sons, Inc., Hoboken, New Jersey.
Published simultaneously in Canada.

For general information on our other products and services please contact our Customer Care Department within the United States at (800) 762-2974, outside the United States at (317) 572-3993 or fax (317) 572-4002.

Wiley also publishes its books in a variety of electronic formats. Some content that appears in print may not be available in electronic books. For more information about Wiley products, visit our web site at *www.wiley.com.*

Library of Congress Cataloging-in-Publication Data:

Wong, David W. S.
Statistical analysis of geographic information with ArcView GIS and ArcGIS / David Wong, Jay Lee.
p. cm.
Rev. ed. of: Statistical analysis with ArcView GIS, 2001.
Includes bibliographical references and index.
ISBN-13: 978-0-471-46899-8 (cloth)
ISBN-10: 0-471-46899-1 (cloth)
1. Geographic information systems. 2. Spatial analysis (Statistics) 3. ArcView. I. Wong, David W. S. (David Wing-Shun) II. Title: Statistical analysis with ArcView GIS. III. Title.
G70.212L43 2005
910′.285—dc22

2005005178

Printed in the United States of America

10 9 8 7 6 5 4 3 2 1

We dedicate this book to our wives and daughters:
Regina Lau and Melody Wong
Lin Lee and Rebecca Lee
Without their support and much sacrifice of family time,
we would not be able to finish this project.

CONTENTS

PART II SPATIAL STATISTICS 183

PREFACE

While spatial analysis and spatial statistics are more than half a century old, GIS, regardless of whether the "S" stands for system or science, has a relatively short history. Logically, the relatively recent development and maturation of GIS should have been closely tied to the advances in spatial analysis and spatial statistics. Despite the claim that a major strength of GIS is its spatial analytical capabilities (Goodchild, 1987; Fotheringham and Rogerson, 1994), the development of GIS has not provided adequate support parallel to the use of spatial analysis and spatial statistics. Though spatial statistical packages were available (such as SpaceStat by Luc Anselin), no commercial off-the-shelf GIS packages before the year 2000 could support spatial analytical procedures and statistics beyond rudimentary GIS procedures such as buffering, overlay, and simple distance calculation.

To our knowledge, our effort starting in 1999 and the resultant book published in 2001, *Statistical Analysis with ArcView GIS* (Wiley), was the first serious attempt to integrate spatial analytical and statistical functions with GIS. In that project, procedures and functions were developed, using Avenue script in ArcView 3.X, for spatial statistical analyses. The scripts were made available as embedded functions in individual project files (.apr). Since then, we have received overwhelmingly positive responses and eager encouragement from readers and users, including those from disciplines other than geography or GIS. Because of their responses, we knew that we were on the right track.

Our first book and its accompanying GIS tools seemed to have achieved the pedagogical purposes of encouraging broader and better uses of spatial statistics. They were successful in introducing spatial statistics and analysis to both geographers and nongeographers. However, readers let us know that the book would have had a broader impact and would have contributed more if it had been a stand-alone text so that both students and researchers could use it. These urges and encouragements were our primary motivation to develop the current book with added coverage of basic statistical concepts. We believe that this book is now a self-contained text that can be used in a quantitative methods class to introduce both basic statistical concepts and spatial statistical issues to students at various levels. In its current format, we believe the book is not merely a textbook. It can also be used to introduce spatial analysis and statistics to professionals, practitioners, and policy makers when the spatial dimension is important in their work.

Since the introduction of our first book, the need for spatial analytical and statistical techniques in GIS has been partly recognized by commercial vendors. In the development of the newest flagship GIS package, ArcGIS 9, the Environmental Systems Research Institute (ESRI, Redlands, California) has incorporated many of the techniques we discussed in our first book. These techniques are now accessible in ArcGIS 9.X's Spatial Statistics wizards. These tools allow users to calculate basic centrographic measures and both local and global measures of spatial autocorrelation, as well as perform hot/cold spot analysis. Even thought ArcGIS 9.X's implementation of spatial statistics covers only a portion of what we discuss in this book, we are delighted to see the beginning of this trickle-down process. We sincerely hope that both ArcGIS and other GIS software packages incorporate more spatial statistical tools with better implementations.

It is now time for a new, comprehensive, and expanded book to better serve readers. We are pleased to be able to offer this book and to have the support from the publisher, John Wiley & Sons. We wish to note that our approach to developing statistical tools is slightly different from that of the vendors. In this book, we still use ArcView 3.X and the associated Avenue scripts as our development platforms. It may be true that the newer ArcGIS is more functionally rich in many aspects, especially in cartographic design and map rendering. However, it seems relatively slow and demanding in terms of hardware support, making it costly (in memory and computation time) to implement and use analytical tools for spatial statistics. In addition, Visual Basic for Applications (VBA), used in ArcGIS, performs much more slowly than Avenue scripts in ArcView 3.X, as tested by one of the authors. Because some spatial analytical and statistical procedures are computationally intensive, using ArcView 3.X with Avenue scripts has clear advantages.

This book offers several improvements over our first book:

- In contrast to the first book, in which the tools were disseminated through ArcView project files (.apr), the tools developed for the current book are organized into different ArcView extensions (.avx). Each extension is developed for a chapter. Users can load multiple extensions to the same project file to access the various suites of tools.

- Though the tools have been developed in ArcView 3.X with Avenue scripts, we provide an overview of compatible functions in ArcGIS 9.X in this book's Appendix. Tools in our extensions provide broader support for general statistical analyses and spatial analytical-statistical techniques. Readers may want to review the Appendix to execute some of the procedures, parallel to those in this book, that are available in ArcGIS 9.X.
- ArcView examples have been added in all chapters to demonstrate how spatial statistics may be used and how to use the accompanying ArcView extensions.
- Traditional statistics textbooks usually include probability tables of different statistical distributions, such as the standard normal distribution (z-score), t-distribution, F-distribution, and chi-square. In our book, such tables are not included. Instead, various ArcView extensions have these probabilities calculated whenever they are needed.
- All the programs (ArcView 3.X extensions) and datasets are available on the accompanying CD-ROM. It has a README.TXT file that provides detailed instructions for installing the extensions, related files, and data sets.

To our knowledge, this book is the only one available that (1) comes with ready-to-use tools and programs in spatial analysis and statistics that are integrated with a highly accessible GIS package; (2) allows readers to verify the spatial analytical and statistical concepts discussed; and (3) provides tools and data so that readers can experiment with the statistical analyses demonstrated in the book in order to gain experience that can be applied to research or put to practical use.

As for our next project, we are currently developing new sets of tools, also in the format of wizards, for all of the spatial statistical tools discussed in this book. Please send inquiries to us or visit our sites at *http://mason.gmu.edu/~dwong2* or *http://gis.geog.kent.edu/AVStat* for updates and new developments.

The spatial statistical tools developed for this book are not intended to be a commercial software package. Our main purpose is to provide educational tools and to share our experience. Though we can program GIS tools, we are by no means professional software programmers. Our programs may not have been optimized for performance, but we are certain that they work correctly and are relatively efficient. We have included, whenever possible, error trappings in our programs. Of course, there is no way for us to develop procedures that trap all potential errors created either by users or by the data. For this reason, we ask for your help in pointing out to us any problems, mistakes, or room for improvement. We are always eager to communicate with readers, even for collaborative research.

Some of the procedures have a limit in terms of how many records they can handle. This limit is likely 256. This is partly due to the column limitation of the dbf format. For some of the tools, we have developed alternatives to get around the case limitation. If readers need such expanded tools, they should contact us.

REFERENCES

Fotheringham, A. S. and P. A. Rogerson (eds.). 1994. *Spatial Analysis and GIS*. Bristol, PA: Taylor and Francis.

Goodchild, M. F. 1987. A spatial analytical perspective on geographical information systems. *International Journal of Geographical Information Systems* 7(1): 3–19.

DAVID WONG AND JAY LEE
October 2005

ACKNOWLEDGMENTS

We wish to thank Yingqi Tang, a graduate student in the Earth Systems and GeoInformation Sciences Program in the School of Computational Sciences at George Mason University. He developed the DLL programs for the probability calculations of several statistical distributions incorporated in several extensions. We also wish to express our appreciation to our respective departments, the Department of Geography at Kent State University and the Earth Systems and GeoInformation Sciences Program in the School of Computational Sciences at George Mason University, for their support and for the laboratory equipment used in the development of this book and its accompanying tools, as well as for tolerating two chairs taking time away from their administrative work to finish this book. We must also thank Jim Harper of Wiley, who has been very patient with this second project, and has provided assistance and encouragement.

Finally, we are thankful to our parents, who supported our postgraduate education in the United States, and eventually allowed us to freely pursue our interests and dreams.

CHAPTER 1

INTRODUCTION

1.1 WHY STATISTICS AND SAMPLING?

Attempts to understand, explain, estimate, or predict events or phenomena occurring around us often start by simplifying the information we have about them. In many cases, statistics have been devised and used to digest large quantities of information and to provide streamlined and concise impressions of the events or phenomena that we are trying to comprehend. For example, population counts of the 164 cities in Ohio would provide little meaning to us unless we know the largest, smallest, or average size among these cities or the range within which these city population sizes vary. In this case, the maximum, minimum, average, and the range of population counts are among the summary information that is known as statistics because they help to describe how values in a set of numeric information, or data, are distributed.

With this understanding, we can state that, given a set of numeric data, **statistics** are quantitative measures derived from the data to describe various aspects of the data. If they are classified by their functions, we have descriptive statistics and inferential statistics. **Descriptive statistics** are calculated from a set of data to describe how the values are distributed within the set of data. For example, the maximum, minimum, range, and average of a set of data are all in this category. **Inferential statistics** are calculated from sample data for the purpose of making an inference to a population or for making comparisons between sets of data. Depending on the areas of application, **classical statistics,** or conventional statistics, are generally used in different application areas, such as sociology, political science, medicine, and engineering. But these statistics have been modified and extended to accommodate specific application areas. In this book, we will include a great deal of statistics known as **spatial statistics.** These statistics are strongly based upon classical statistics but have been extended to work with data that are spatially referenced. Other statistics that are extensions of classical statistics for various application areas include *econometrics, psychometrics, biostatistics, geostatistics,* and several others. Certain statistics discussed in this book are sometimes classified as **geostatistics,** which originated in geoscience.

With statistics, an analysis can be performed to understand how data values concentrate or disperse around certain values, how they are compared with each other or with another set of data, or whether they are just subsets of a larger set of data. When analyzing data statistically, each observation should be independent so that its values or data are not dependent on, or tied to, values of other observations in the same data set. This independence assumption is one of the most fundamental assumptions in statistical analysis. Unfortunately, it is often violated for data collected to describe events or phenomena that are spatially referenced. This is because, in many geographic events or phenomena, what happened in a location is highly correlated to what happened in its surroundings. Because of this characteristic of spatially referenced data, much of our discussion in this book will focus on how statistics and associated methods can be modified to analyze spatially referenced data.

When one attempts to answer a scientific question, one will rarely draw conclusions based on just one or a few observations. For instance, if there was one case or a few cases of malaria in a community, can we say that there is a real epidemic or should we treat those occurrences as accidents or events occurring by chance? To take another example, can we conclude that the soil in a farm has lost its fertility if the farmer harvested much less crop this year than last year? Could the decline in yield be a one-time event or a short-term fluctuation? Will this happen again next year? Is soil fertility the only factor determining the amount of crop yield? Before any conclusions can be drawn, we need to understand the nature of these events or occurrences. In other words, when a certain phenomenon occurs, it may be due to a **random process** or a **systematic process.** We have to determine if the process is random or systematic. If an event or a phenomenon is triggered by a random process, there may not be much that we can do to identify its underlying cause in order to explain why the event or phenomenon happens the way that it does. But if it is part of a systematic process, the numeric or spatial patterns will be interesting to study and explore. As the first step in understanding these processes, statistical analysis is usually the tool used to help us decide if the events are random or not.

Using the soil example again, if we suspect that the soil fertility of a farm is low, and if this suspicion is based on some observations made in this farm, we are essentially formulating a **hypothesis.** This hypothesis can be tested to see if it should be rejected. To test this hypothesis, we would need to gather more information or data about the soil. Instead of just focusing on a small plot in the farm, we may want to examine different plots around the field for soil fertility levels. For a more rigorous study, we may want to drill holes at various locations in the field to collect soil samples to conduct a soil chemical analysis in a laboratory. By selecting different locations in the field to drill holes, we are essentially collecting a **sample** of soil for further examination instead of examining the entire **population,**[1] which will require us to examine every location in the field. Each examined location can be regarded as an **observation** or a case in the sample, and the number of observations selected is known as the **sample size.** Similarly, by examining the same location over time and treating each examination of that location as an observation, we are collecting these observations from a population along the temporal dimension. Finally, the measured value from an observation is normally referred to as a data value. When there is a set of such values, they are referred to as a dataset.

After a sample of soil is assembled from different locations, a chemical analysis can be conducted to evaluate the levels of different chemicals, such as phosphorus, nitrogen, and potassium, in the sample. A measurement of each chemical can be derived by examining all observations in the sample, such as on average 30 mg of nitrogen per 1 kg of soil. This measurement is then a **statistic,** because it is derived from all the observations in the sample. If the data-gathering process covers the entire population, a similar measurement is derived from that process. This measurement is then known as a **parameter.** For instance, in the U.S. decennial census, certain questions were asked of all individuals in the United States in principle (we know that some people were missed due to the difficulty of reaching them—so-called undercounting). The measurements derived from those questions are parameters.

When analyzing a sample, a logical question to ask is, why should we examine a sample but not the entire population? Isn't it more accurate to enumerate the entire population? Of course, we would prefer to survey the entire population if we could. But often it is impossible and/or impractical for one or more of the following reasons:

1. The population is too large to be enumerated completely.
2. The cost of enumerating the entire population may be prohibitive.
3. The study requires a quick turnaround time, and studying the entire population may take too long.
4. If the enumeration process requires destroying the observations, such as in certain processes of quality control, then a full enumeration will destroy the entire population.

Using the soil study example again, it is impossible to evaluate the fertility level of every cubic foot of soil in the field for a complete examination. It is also very expensive and will take much too long to get the full result. In

[1]The term **population** is used here to represent the entire collection of objects of interest from which sample can be drawn. It is easily confused with another use of this term referring to the number of people in a given region at a given time.

addition, if a hole is drilled for every location to gather the soil, there will be no soil left in the field. Therefore, sampling is often used instead of examining the entire population. For this reason, studying statistics becomes necessary.

The statistics on chemical levels that are generated from the soil sample may offer descriptive information about the condition of the soil in the field, including a numerical distribution of the chemical levels. Therefore, these statistics are regarded as descriptive statistics. How accurate the statistics are in describing the distribution of chemical levels in the entire field or in describing the population is dependent upon many factors. We know that these statistics will never be 100% accurate (since they are not from a complete survey of every inch of soil in the farm) and that the level of accuracy is dependent upon how representative the sample is of the population.

Fortunately, procedures have been developed, based on random processes, to allow us to draw conclusions on whether a sample is a reliable representative of the population or not. This process of drawing a conclusion about a population based on information derived from a sample is known as **inference.** The process of drawing an inference normally includes

1. formulating one or more hypotheses,
2. collecting relevant data by making observations,
3. computing descriptive or test statistics, and
4. deciding if the hypothesis should be rejected based on the computed statistics.

If sampling is desirable or preferred because an exhaustive survey of the population is not possible, then the sampling process should be carefully considered. But how should one select sample observations from the population? There are two general sampling schemes one may adopt: random sampling and systematic sampling. **Random sampling** is the process of selecting observations randomly from the population without any specific predefined structure or rules. Often, random numbers are used to assist the selection process. For example, items in an ordered set of objects are selected as samples if their positions correspond to those assigned by the random numbers. Alternatively, all objects in the set can be mixed up randomly before selection.

In contrast to random sampling, **systematic sampling** is the process of selecting observations based on certain rules developed according to certain principles. These principles are based on the objective(s) of the studies. Often one would like to adopt a sampling principle to cover the entire spectrum of the population. For instance, one may select every fifth observation from an ordered list of objects or select the households at the northwest corner of every street block in the city.

But sometimes a study may want to emphasize a specific segment or segments of the population, such as minority groups in the general population. For this purpose, sampling can be set up so that a particular minority group is sampled more than other groups. However, this should be done only with careful consideration of what the sample may represent and how it may affect the results because it is possible that those segments may be oversampled.

Within the two general sampling schemes, additional variations of thc sampling process have been developed. For instance, observations sharing certain common characteristics may be grouped into different strata. With objects in different strata or groups, either random or systematic sampling can be performed within each stratum or group. This is called **stratified sampling.**

For example, selecting 30 cities from the 164 cities in Ohio may be performed in several ways. In random sampling, all 164 cities may be ordered or ranked by their population sizes. Next, we can select cities if their ordered positions match the first 20 random numbers from a random number table. Or we can select every 8th city until we have selected 20 from the list of 164 Ohio cities to perform the systematic sampling. Finally, we can use stratified sampling by first dividing the 164 cities into four groups based on their locations in northeast, northwest, southeast, or southwest Ohio and then selecting either randomly or systematically, 5 cities from each of the four groups to ensure that the sampled cities provide a good representation of Ohio cities over the entire state.

If the sampling of observations involves objects that have geographic references, more variations are needed to accommodate the geographic dimension. The sampling scheme that is designed to accommodate the sampling of observations in the geographic space is called **spatial sampling.** A good summary is available for further reading in Griffith and Amrhein (1991, pp. 215).

In the spatial sampling framework, locations are randomly selected to perform random sampling. When this process is implemented in a computer environment or with a Geographic Information System (GIS), the random locations are usually defined by the *x*-*y* coordinates taken from two sets of random numbers, as shown in Figure 1.1a. If the *x*-coordinates and *y*-coordinates are randomly determined, the resulting points defined by these *x*-*y* pairs are thought to be randomly distributed. In its simplest form, systematic sampling selects regularly spaced locations to ensure complete coverage of the entire study area, such as the structure shown in Figure 1.1b. Note that the distances between adjacent points are kept the same or approximately the same along the *x*- and *y*-directions only, not along the diagonals. If one prefers a spatial systematic sampling framework with observations regularly spaced, but with equal distances to their nearest neighbors, then the structure will be a triangular lattice, which resembles a hexagonal structure.

With these two general schemes of spatial sampling, we can create more variations. For example, we can combine random sampling with systematic sampling so that the geographical space is divided systematically but sampling

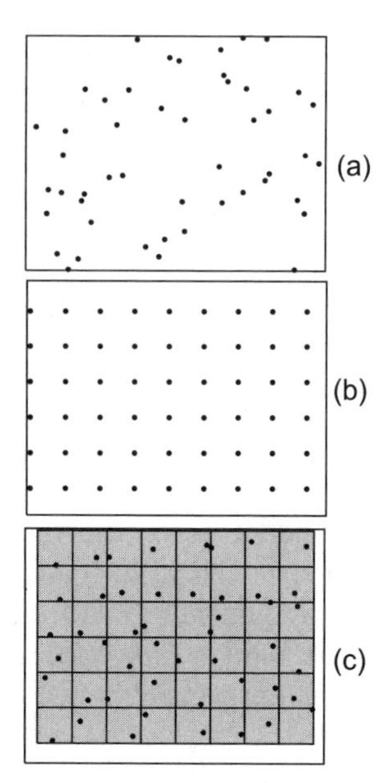

Figure 1.1 Selected spatial sampling schemes. (a) Random points; (b) regular points; (c) a random point within each regular grid cell.

is done randomly within each partitioned region. Of course, the partition of the geographical space should be mutually exclusive and collectively exhaustive. Figure 1.1c combines the systematic and random sampling frameworks by first dividing the entire region into subregions and then randomly selecting a point within each subregion.

One final note about spatial sampling is that our sampling unit so far is limited to locations in space, or points. There are, in fact, alternative sampling units. For example, Griffith and Amrhein (1991, p. 215) reviewed two other types of sampling units: linear units or traverses and areas. Sampling by areas will be discussed in Chapter 6, which deals with point pattern analysis. When Quadrat Analysis is used to analyze point patterns, the sampling areal units are known as **quadrats.**

1.2 WHAT ARE SPECIAL ABOUT SPATIAL DATA?

Techniques for statistical analysis have been very well developed and are widely used in many research fields and practical applications. However, most of the statistical techniques and models were developed not for observations with explicit geographic referencing information, but rather for data most likely compiled by selecting sample observations randomly from the population. When conventional statistical methods are used to analyze data derived from these observations, it is assumed that these observations and associated data can be considered independent. But for spatial data gathered from nearby observations or within the study region, these data tend to be related to each other. Thus, we cannot assume that observations are independent of each other. For this reason, using conventional statistical methods to analyze spatial data derived from these observations may cause problems.

One way to describe geographic phenomena or events is by representing real-world objects with three geometric primitives: **points, lines** or **arcs,** and **polygons.** Points are used for features or events with no or very little spatial extent at a given geographic scale of representation. Lines are used for features or phenomena with a linear extent or directional movements. Polygons are used to represent areas delineated by the spatial extent of the studied phenomena. Putting line features aside, point and polygon data often represent individual observations or aggregates of individual observations. In other words, spatial data are subject to spatial aggregation, and they may need to be represented differently at different scales or with different resolutions.

In working with geographic objects at different scales or with different resolutions, two conceptual issues need to be further explored because they are critical in handling and analyzing spatial data in a meaningful way. These two issues are the modifiable areal unit problem and spatial autocorrelation.

1.2.1 MAUP—The Modifiable Areal Unit Problem

The first critical issue regarding spatial data is the **modifiable areal unit problem (MAUP).** A region can be partitioned in many ways. It can be divided into smaller areal units based on any given criteria. For example, the United States is divided into states based on state boundaries. Similarly, a subunit can be further subdivided into smaller areal units. Using the U.S. example, each state is divided into a number of counties, and each county is further divided into smaller census areal units or administrative units. The continued subdividing of a region into smaller, or multiple levels of, subregions will create a nested hierarchical spatial partitioning system.

The census geography used in the United States partitions the country into four census regions. These four regions are further subdivided into nine divisions. Under the nine divisions, the entire country is covered by 51 states. Within the states are counties, and within the counties are census tracts, then census block groups, and finally census blocks. Together, they make up what is known as census geography (Figure 1.2).

In creating the hierarchical spatial partitioning system, larger units are divided into smaller units. But a larger region can be subdivided into the same or a similar number of subunits by using different zonal configurations. For example, the U.S. Bureau of the Census divides the United States into nine divisions, but the U.S. Environmental Protection Agency (EPA) divides the

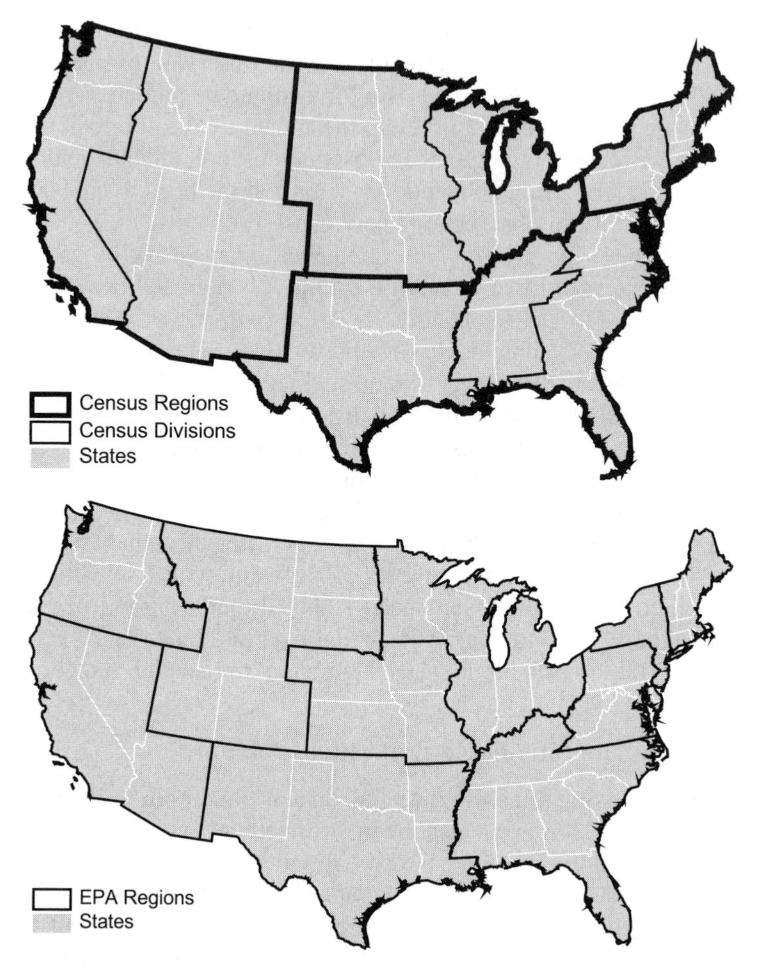

Figure 1.2 Different partitioning systems adopted in the United States: 4 census regions, 9 census divisions, 10 EPA regions, and 51 states.

country into 10 regions (Figure 1.2). In this case, the two partitioning systems are not nested within each other hierarchically—they coalesce with each other.

When data tabulated at multiple levels of spatial resolution or multiple geographic scales in a nested hierarchy are analyzed, they may provide results that are not consistent across these levels of spatial resolution or scales. This inconsistency is known as the **scale effect.** When analyzing data derived from different zonal systems with approximately similar numbers of areal units, we may expect the results to be inconsistent. This is known as the **zoning effect.** The scale and zoning effects together are known as the **modifiable areal unit**

problem (MAUP). This is because both of these subproblems are related to the changing definition of areal units (Openshaw, 1984; Wong 2004).

It is well documented that the MAUP effects are pervasive among almost all statistical techniques. The most significant effect is the impact on the correlation[2] between variables (Openshaw and Taylor, 1979). Because the relationship between variables is the basis of most classical statistical techniques, including most modeling techniques, MAUP subsequently affects many classical statistical techniques (Fotheringham and Wong, 1991; Wong and Amrhein, 1996). Often, relationships between variables of larger geographic units (such as counties in the United States) are stronger than the relationships between the same variables of smaller geographic units (such as census tracts or block groups). As a result, outcomes of statistical analysis of data from different scales or spatial resolution levels do not yield the same result.

Relationships between variables also vary when data derived from different zoning systems of similar scale or resolution levels are used in the analysis. Because MAUP is a problem with spatial data analysis, it has significant implications for various techniques and tools handling and analyzing spatial data, including those used in GIS (Tate and Atkinson, 2001) and remote sensing (Quattrochi and Goodchild, 1997).

A large portion of the MAUP literature has focused on documenting its effects. Different approaches have been proposed to handle MAUP. A simple but somewhat passive approach is to acknowledge the presence of MAUP, and thus multiscale analysis is suggested to demonstrate the range of results one may obtain (Fotheringham, 1989). Another approach is to identify or introduce spatial analytical techniques that are relatively scale insensitive. This is one of the directions taken by MAUP research in the past decade in attempting to identify solutions to the problem. So far, this approach has achieved only limited success. Very few techniques have demonstrated the property of scale invariant, and they are application-specific (e.g., Tobler, 1989; Wong, 2001). Some are computationally intractable (e.g., Holt et al., 1996). An accounting approach was proposed to deal with the scale effect. It acknowledges the presence of scale effects and tries to identify their sources from different locations at different scales. An example of its application can be seen in measuring spatial segregation (Wong, 2003).

1.2.2 Spatial Autocorrelation

Another major conceptual issue in the analysis of spatial data is the presence of spatial autocorrelation in the data. Spatial autocorrelation is essentially the

[2] **Correlation** is a statistic that summarizes the strength and direction of the relationship between two variables.

nature of geography and, consequently, will almost always be present in spatial data. Tobler (1970) refers to it as the First Law of geography: *all things are related, but closer things are more related.*

It goes without saying that things on the surface of the Earth are related to each other, though perhaps to various extents. It is interesting to note that a few recent studies in chaos theory and global circulation models (Glick, 1987) have popularized some notions that may contradict the second part of Tobler's First Law of geography. For example, one of these notions is the butterfly effect. One version of the butterfly effect is that a butterfly flapping its wings over the Great Wall of China may cause a hurricane landfall in the Middle Atlantic states on the East Coast of the United States due to spatial propagation of air disturbances. Even if we could not prove this type of effect, it is still true in most cases that events or phenomena affect each other more when they are closer together than when they are farther apart. As such, the second part of the first law of geography is not disputable. The implication of this law in analyzing spatial data is that we should not assume that the observations in space are unrelated to or independent of each other, even if this violates the fundamental assumption adopted in most traditional or classical statistical analyses.

The study of spatial autocorrelation has a relatively long history in geographical research. One of the most famous and influential events was the introduction of Moran's I, an index that measures the level of spatial autocorrelation. But what is the significance of spatial autocorrelation in statistical data? Griffith (1987) has offered a detailed analysis of the impacts of spatial autocorrelation. Overall, it does not affect the accuracy of estimating parameters in inferential statistics. However, it may cause incorrect conclusions regarding whether the relationships between variables or factors are true or not, that is, the statistical significance level of the parameters describing the relationships. This is because, if observations are dependent on each other, that is essentially the same as having duplicate observations. If a large number of observations indicate the relationship, that will lead us to believe that there is a strong relationship between the variables (a significant parameter). However, if we have duplicate or autocorrelated observations, the actual relationship should be weaker. That is, the **effective sample size** is smaller than the sample size because of autocorrelated observations (Rogerson, 2001, p. 98).

To determine if spatial autocorrelation should be a concern in statistical analysis, we can evaluate the level of spatial autocorrelation using simple indices (Cliff and Ord, 1981). This book will discuss some of the statistics that explicitly evaluate the level of spatial autocorrelation. If this level is significant, then inferential statistics should take into account the presence of spatial autocorrelation. Readers interested in more advanced topics on this issue can find discussions of spatial regression models in Griffith (1988) and spatial econometric models in Anselin (1988).

1.3 SPATIAL DATA AND THE NEED FOR SPATIAL ANALYSIS/STATISTICS

In geographical analysis, each object or feature occupies a location in space. This location may be depicted as an **absolute location,** such as by latitude-longitude coordinates, or as a **relative location** in reference to other spatial features or objects. For example, the City of Kent in northeast Ohio may be defined by latitude and longitude: (41°N, 81°W). Alternatively, its relative location can be defined as 42 miles southeast of Cleveland, Ohio, or as 12 miles east of Akron, Ohio, and so on. When defining absolute locations, all we need is the coordinates. When defining relative locations, we need another location or spatial feature as a reference point.

In classical statistics, location information is often not included in the analysis. These statistics often focus only on the nonspatial characteristics or attributes of the observations. In the two simple hypothetical landscapes in Figure 1.3, it is easy to conclude from a spatial perspective that the landscape in Figure 1.3a has a clustering pattern, while the one in Figure 1.3b is rather random. When one resorts to classical statistics, such as using average elevations to represent these landscapes, the two landscapes are not discernible. In fact, regardless of how many classical statistics are used, if no location information is included in the analysis, we will not be able to tell the difference between these two landscapes. In this case, classical statistics are clearly insufficient to handle spatial data because they fail to capture all pertinent information, especially location information. Therefore, spatial statistics should be used because they deal explicitly with the spatial characteristics of the observations.

Using GIS terminology (Dueker, 1979), geographic objects can be described by two types of data:

(a)

1	3	5	8
2	6	10	12
4	9	13	15
7	11	14	16

(b)

1	3	10	8
2	16	5	12
14	9	6	15
7	11	4	13

Figure 1.3 Two hypothetical landscapes indicating a clustering pattern (a) and a random pattern (b).

1. **Spatial data** define the locations and geometric forms of the objects. For example, a point is defined by a pair of coordinates. Many points may be connected to form a line segment, and a set of closed line segments makes up a polygon.
2. **Attribute data** describe various characteristics of the geographic objects (points, lines, or polygons). They help distinguish one observation from another, even if all of them are points, in addition to using location information.

For example, a dataset of power poles in a city has a set of coordinate pairs in which each pair of coordinates defines precisely the location of a power pole. The other part of this spatial data set is the attribute data associated with these power poles. As shown in Figure 1.4a, each point is defined by a pair of coordinates along the vertical and horizontal axes. At the same time, each point is associated with a record in the attribute data table. Two parts of this dataset are linked by a set of unique identifiers. In this example, the attribute data table (Figure 1.4b) contains the identifiers, coordinates, and two additional attributes showing when each power pole was installed and what material was used.

In a dataset such as that described in Figure 1.4, spatial data capture the spatial characteristics of the objects, while attribute data describe various characteristics of these objects. Spatial data sometimes are also called **feature data** or **cartographic data,** and aspatial data are sometimes called **attribute data.** As mentioned above, geographic objects can be abstractly represented as points, lines, or polygons.

In general, three types of analytical frameworks can be used in analyzing data describing geographic features:

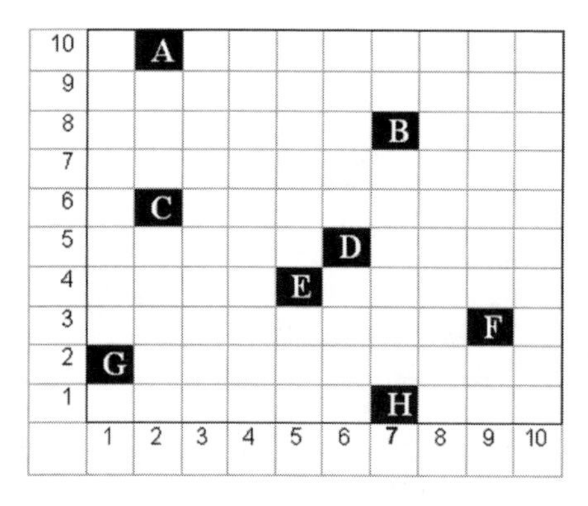

(a)

ID	X	Y	Date Installed	Type
A	2	10	8/24/1978	Wood
B	7	8	11/11/1959	Wood
C	2	6	2/5/1963	Cement
D	6	5	7/7/1988	Wood
E	5	4	11/1/1947	Cement
F	9	3	9/17/1959	Cement
G	1	2	2/5/1960	Cement
H	7	1	1/5/1988	Wood

(b)

Figure 1.4 (a) Locations of power poles in a city as defined by coordinates. (b) Attribute data table.

1. If only attribute data are analyzed, then traditional or classical statistics will be adequate since no spatial information is involved or available.
2. If only spatial data are analyzed, then only spatial information on the geographic objects will be used but the analysis will make no distinctions among geographic objects beyond their locations. In other words, point a and point b will be no different except in their locations.
3. Finally, if both spatial data and attribute data are analyzed, then statistical techniques have to incorporate both the spatial or location information on the observations and the attribute data describing them.

1.4 FUNDAMENTALS OF SPATIAL ANALYSIS AND STATISTICS

Before we discuss various statistical topics in detail, we will address some fundamental concepts and common notations used in statistical analysis. In addition, we will review some basic issues related to the representation or display of spatial data.

1.4.1 Scales of Measurement

Attribute information on geographic objects can be measured and presented at different scales of measurement, depending on the purpose of data collection and/or the analysis to be performed. For example, a set of cities can be recognized as a set of points without any additional information. In this case, we only know that different points represent different cities. We do not know which city is represented by which point.

Also, when looking at a map that shows a number of streets with the same line symbol, we may be able to identify different streets by their names, but we do not know which street has heavier traffic than the others or which street is a highway or a local street. Similarly, if we distinguish the 88 Ohio counties only as different counties, then our understanding of these counties would stop at knowing their names and their boundaries as depicted on the map. All these limits are due to the different scales of measurement applied to the data collected. These limits also exist because the information we have about these geographic objects is measured at a **nominal scale.**

At a nominal scale, data may exist in numeric form, but they may not be compared meaningfully, like real numeric variables. For example, five-digit zip codes may look like numbers, but they are actually nominal data. Zip codes do not work with any mathematical operations such as addition, subtraction, multiplication, or division. In fact, zip codes cannot even be compared to say, for example, that one zip code is greater than another. In this manner, zip codes function like names of objects that only allow the objects to be distinguished from one another. Using another example, land parcels in a city may be zoned for different uses. Unless a criterion is defined as the basis for comparison, different land use types are just different categories. They cannot be ordered or compared mathematically in any way.

Now consider data derived from remote sensing; pixels of an image can be classified into land cover categories such as urban, vegetation, water, barren, and others. Different types of land cover are another example of data measured at the nominal scale. Again, with no predefined criterion to determine the meanings of these categories, it would not be appropriate to order these categories or to perform mathematical operations on them.

If there are only two values or outcomes in a nominal scale variable, the data in that variable are known to have a **binary** classification scheme or simply to be binary data. Classification schemes with more than two categories are known as **polychotomous** classifications, and the data are known to be polychotomous data. Data measured at the nominal scale reflect the observation memberships in the categories. It is possible to distinguish observations among categories, but it is not possible to distinguish observations within the same category.

Another scale of measurement that provides stronger power to discriminate observations is the **ordinal scale.** Observations on ordinal scale data are ranked according to a predefined criterion. Ordinal data may be **weak ordered** or **strong ordered.** If observations are grouped into a number of ordered categories, the data are said to be in weak ordering. For instance, when the average family income values of the 88 Ohio counties are lumped into high, medium, and low income groups, we know that those counties in the "high" group have higher average family income than those counties in the "medium" or "low" group. However, we do not know which county in the "high" group has an average family income higher than that of other counties in the same group. In other words, weak-ordered data allow us to differentiate observations between groups but not within groups.

In a dataset, if we can assign a rank to each observation according to a specific criterion, then these data are regarded as strong ordered data. These data allow us to distinguish individual observations from each other, but only by their ranks. Note, however, that the dataset could have tied observations in its ranks. In that case, it would not be possible to distinguish them from each other. Examples of strong ordered data are everywhere. The 164 cities in Ohio can be ranked according to their population sizes. Once they are so ranked, we can compare any two cities to see which is larger than the other, but we can only tell their difference by their ranks, not their absolute difference in population size. Similarly, all fast food restaurants in a city may be individually ranked according to their monthly sale volumes, their floor areas, and so on.

In general, it is not appropriate to apply mathematical operations to ordinal data. Ranks cannot be added, subtracted, multiplied, or divided to derive other ranks in a meaningful manner. For example, the top-ranked observation cannot be multiplied by the second-ranked observation to derive observations of another rank. One characteristic to be noted here, however, is that ordinal data are transitive. If A is ranked higher than B and B is ranked higher than C, then A will be ranked higher than C.

Based on the discussion so far, neither nominal nor ordinal data allow us to distinguish data values individually. For that purpose, observations need to be measured at either **interval or ratio scale,** which are more precise measurement scales.

Many researchers often lump interval and ratio scales together because they both work with mathematical operations such as addition, subtraction, multiplication, and division. But these scales are different because data measured at interval scale do not have a true or meaningful zero or minimum value, while data measured at ratio scale do.

For example, temperature is in ratio scale because when the temperature on the Kevin scale is zero, there is absolutely no energy, and that is the theoretical minimum for temperature. Other examples of ratio scale data include population density (we cannot have negative population density) and weight (no negative weight). On the other hand, income level can be regarded as interval data, though normally one prefers to have an infinite amount of money and does not prefer to be in debt—negative income. But we may not be able to determine the level of one's minimum (negative) income or one's maximum amount of debt.

We should also be careful about the meaning of differences calculated from interval data. For example, the difference between 80°F and 90°F is 10°F. The difference between 30°F and 40°F is also 10°F. But do the two 10°F differences mean the same thing in both situations? The first 10°F difference makes weather go from comfortable to warm, while the second 10°F difference makes it go from freezing to cold. They are definitely different, even though they are the same numerically.

As for data measured at ratio scale, the preciseness of the data is the same as that of interval scale data, but there is an absolute lower bound in the value. Take population density as an example. This measure is derived by dividing the number of people in a region by the region's area. As ratio data, the number of people can be 0, meaning no people at all, or any other integer. Similarly, area can be 0, meaning no area at all or any other areal extent. After the number of people is divided by the area, the resulting ratio is the population density. The ratio between two values can be compared with other ratios.

In short, the four scales of measurement have increasing levels of information details or preciseness, from nominal data (the lowest), to ordinal data, and then to interval/ratio data (the highest). Similarly, the degree of mathematical complexity of the data measured at these scales goes from nominal data, which allow only recognition; to ordinal data, which allow recognition and comparison; to interval/ratio data, which allow recognition, comparison, and differentiation.

In general, interval/ratio data can be converted to ordinal data, and ordinal data can be converted to nominal data. Data lose their detailed information when they are converted to lower scales of measurement.

Consider population density again. The subfigures in Figure 1.5 show the population densities of the 13-county region of northeast Ohio in different ways. Each subfigure corresponds to one of the scales of measurement. When displayed as ratio data, individual population densities are shown in Figure 1.5a, where comparisons can be made between any densities, and the measure is presented at the highest level of detail.

In Figure 1.5b, the same population densities are shown as interval data. The original ratio data have been converted to interval data, using Mahoning County's 636 persons per square mile as a reference point, though this is not the usual interval scale at which we will present population density. The resulting numbers show how the population densities of other counties deviate from that of Mahoning County. Note that the population density of Mahoning County was chosen arbitrarily as the reference point. Any other population density can be used similarly and will have the same effect.

In Figure 1.5c, the same population densities are shown in terms of their rankings. In this case, we may know which county has a higher or lower density than others, but no further information is available. Because each density value is ranked individually, the values are displayed at the scale of ordinal data with strong ordering. In Figure 1.5d, however, the population densities are grouped into five ordered intervals. Therefore, the data are in ordinal scale with weak ordering. Those counties displayed by the same gray shade cannot be distinguished from each other, but counties displayed by different shades show their different population densities.

Finally, Figure 1.5e shows each county in its own color shade without an order. The result still allows us to distinguish different counties from each other. However, we only know that there are 13 different counties.

A related issue concerning scales of measurement is how to group observations into categories. Such grouping is necessary for many practical reasons. When there are too many individual data values to display on a map, we often group them into a small number of categories first. Each of these categories is then assigned a color or a shading pattern to display its associated data values on the map. A number of cartographic studies suggest that people usually cannot distinguish categories if too many categories are shown on a map. In general, six or fewer categories are suggested as acceptable and efficient for displaying data values on a map, while some suggest 5 to 12 and others suggest 7 with ± 2 categories. From a statistical perspective, too few categories will fail to reveal the distribution characteristics of the data, as observations with very different values are lumped together. On the other hand, too many categories will provide highly detailed distribution information on the data, but may defeat the original purpose of avoiding the need to handle a large number of individual observations.

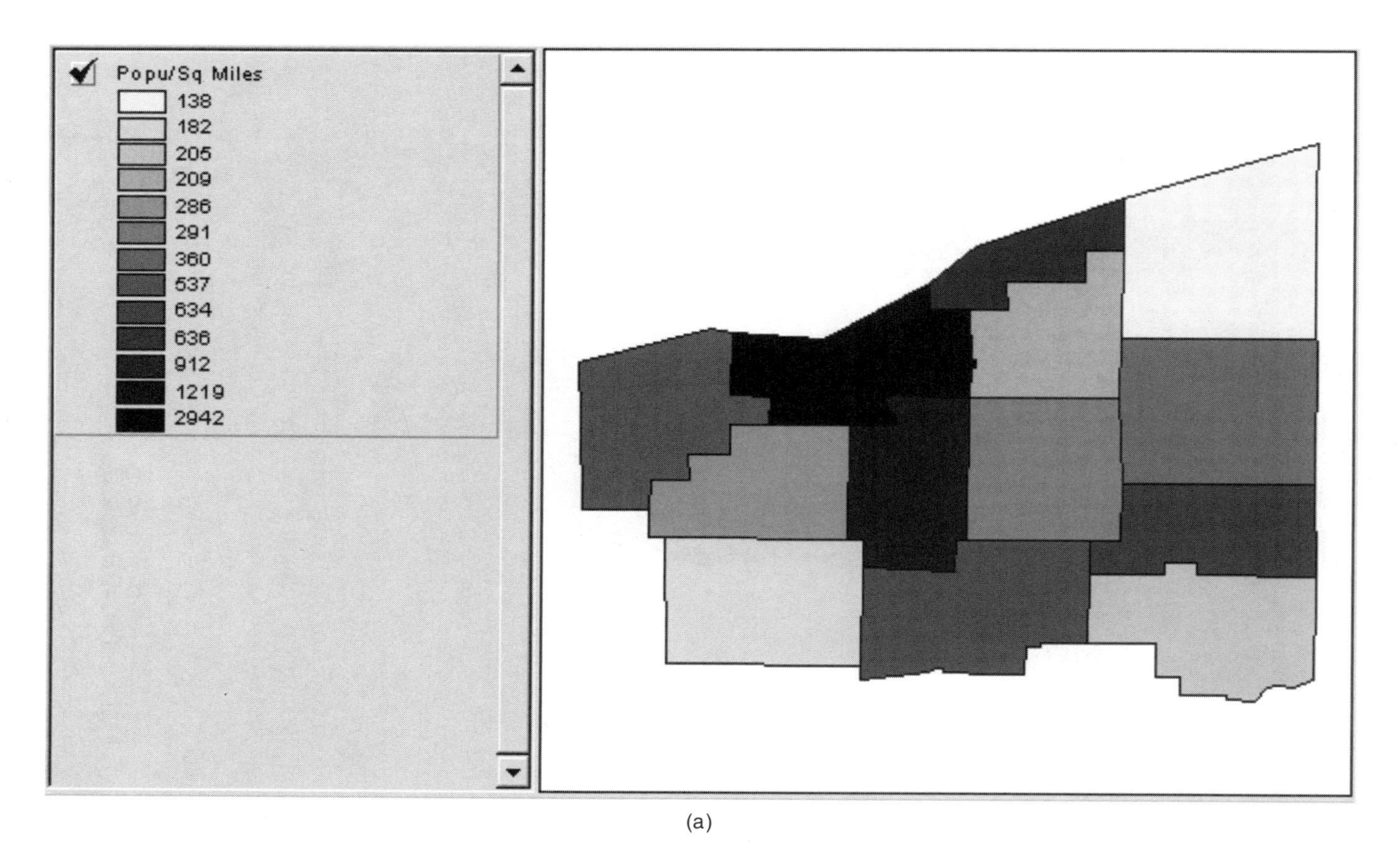

(a)

Figure 1.5 (a–e) Population densities displayed at different scales of measurement. (a) Population densities as ratio data.

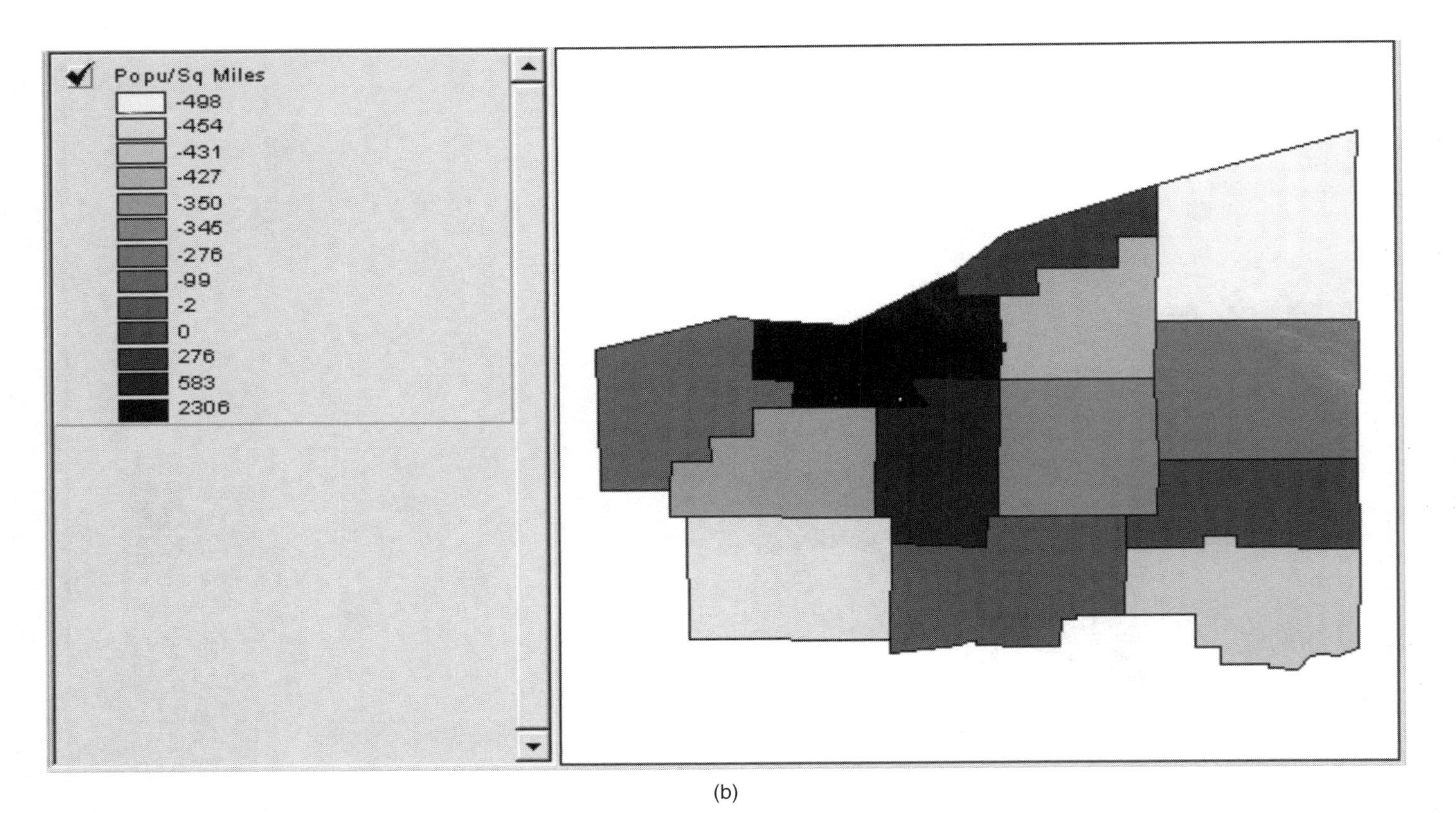

(b)

Figure 1.5 (*Continued*). (b) Population densities as interval data.

(c)

Figure 1.5 (*Continued*). (c) Population densities as ordinal data (strong ordering).

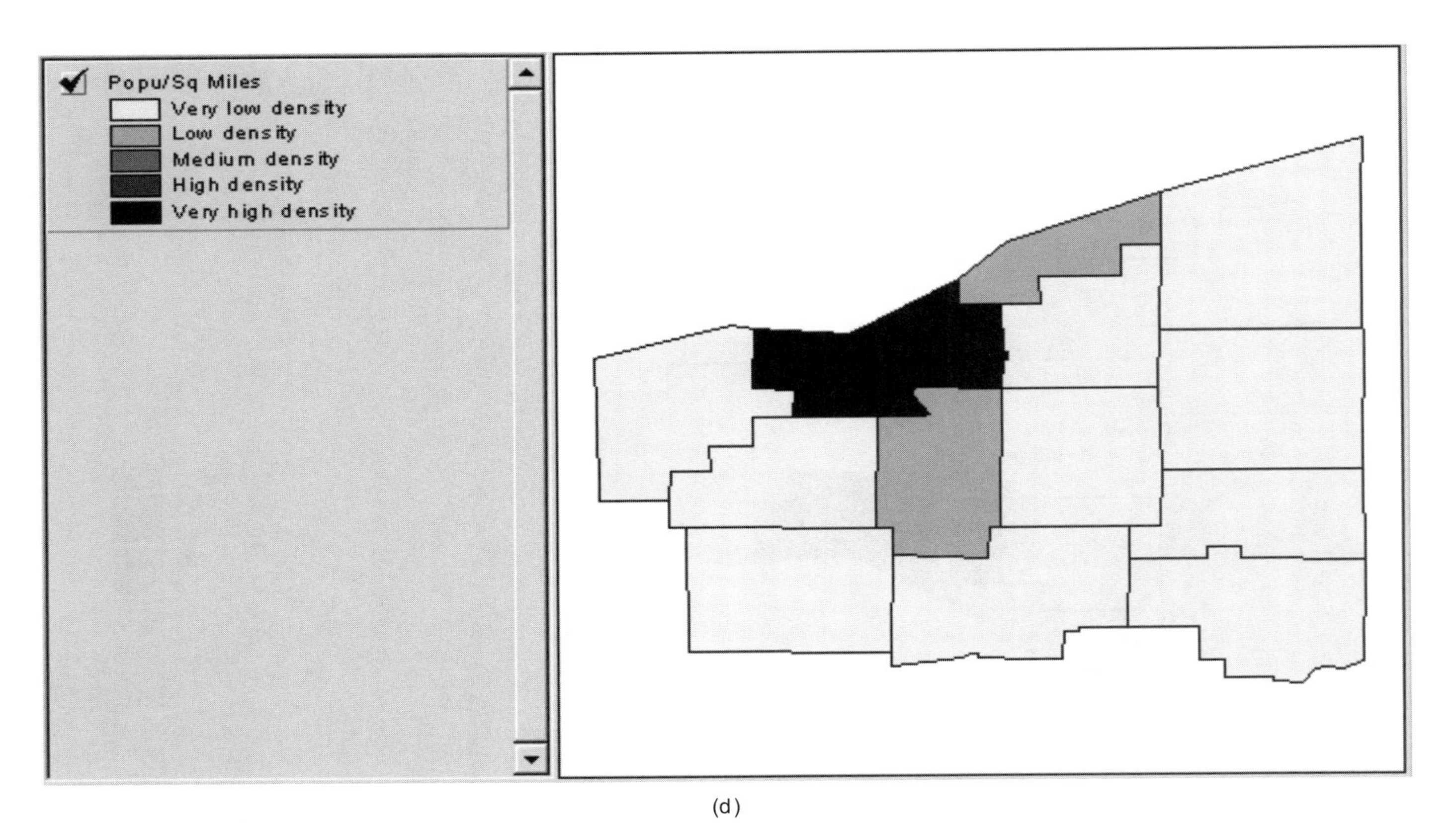

(d)

Figure 1.5 (*Continued*). (d) Population densities as ordinal data (weak ordering).

(e)

Figure 1.5 (*Continued*). (e) Nominal data.

Two principles should be kept in mind when classifying data values into categories. First, the categories should be *mutually exclusive.* This ensures that any data value is assigned to one and only one category. Second, the categories should be *collectively exhaustive.* This means that all categories together should provide complete coverage of all possible data values to ensure that no data values are left unclassified.

These two principles have significant implications, especially in relation to today's widespread use of GIS in mapping quantitative data. The map in Figure 1.6 illustrates how the use of GIS can complicate compliance with the two principles. In this figure, a map of U.S. population densities by county was created using the *quantile* classification method. The quantile method, which is one of several common methods for classifying data in GIS packages, classifies data values so that each category includes approximately the same number of observations. But sometimes classes constructed by GIS do not satisfy the principle of forming categories that are mutually exclusive and collectively exhaustive.

As shown in the legend of Figure 1.6, while population density is a continuous variable, there are gaps between the categories on this map. Therefore, GIS users should be aware that even standard procedures and tools available in off-the-shelf GIS packages do not always handle spatial data according to statistical principles, though newer systems seem to be more intelligent in constructing classes. Thus, knowledge of spatial data handling and statistics is essential to ensure the appropriate use of GIS to analyze and display spatial data.

1.4.2 Mathematical Notations

This book will introduce many statistics, both classical and spatial. To present and explain these statistics efficiently and effectively, we have to use certain mathematical notations. These notations are simply condensed presentations of simple algebraic operations so that the formulas will be more compact. Several basic notations are related to summation and multiplication. They are:

Factorial: $n! = n \times (n - 1) \times (n - 2) \times (n - 3) \times \ldots \times 3 \times 2 \times 1$

Summation: $\Sigma_{i=1}^{n} x_i = x_1 + x_2 + x_3 + \ldots + x_n$

Product: $\Pi_{i=1}^{n} x_i = x_1 \times x_2 \times x_3 \times \ldots \times x_n$

The "!" notation refers to *factorial,* which is the product of multiplying values with successively smaller numbers until the last number ends with 1. For example, $4! = 4 \times 3 \times 2 \times 1$.

The notation $\Sigma_{i=1}^{n} x_i$ simply represents the sum of a series of values named $x_1, x_2, \ldots,$ to x_n. For example, if there are four values, represented as x_1, x_2,

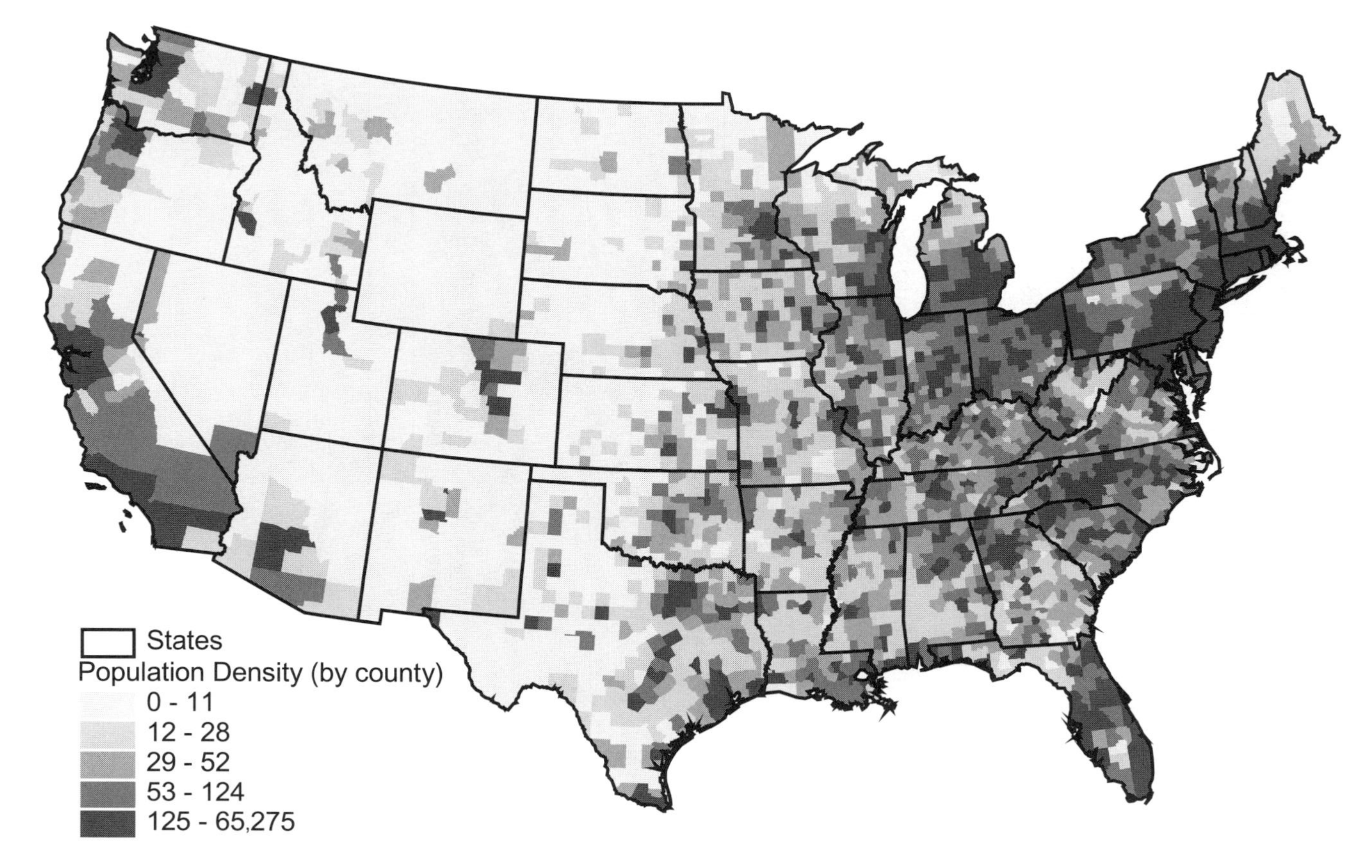

Figure 1.6 A map generated from GIS fails to conform to standard cartographic principles in the creation of a legend.

x_3, and x_4, to be summed, n will be four and the sum will be $x_1 + x_2 + x_3 + x_4$. In its general form, i functions as a counter, or an index, that provides references to individual data values. The notation sigma (Σ), represents the sum of all the values of that variable. For simplification, $\Sigma_{i=1}^{n} x_i$ is sometimes simplified to $\Sigma_i x_i$ or just Σx.

The notation $\Pi_{i=1}^{n} x_i$ represents the product of multiplying values of x_1, x_2, x_3, . . . , to x_n. It, too, is sometimes simply written as $\Pi_i x_i$ or Πx. These notations are sometimes lengthy, but they simplify algebraic operations in a compact manner.

The above notations involve only a variable identified by one dimension. For variables with two dimensions, we use two subscripts: i and j. Their use can be best depicted by a data matrix such as the one in Table 1.1. Notice that the elements in the matrix are x_{ij} values in the ith row and the jth column.

In Table 1.1, $\Sigma_{i=1}^{m} \Sigma_{j=1}^{n} x_{ij}$ is the grand total of all values in the matrix. It is derived by accumulating all x_{ij} by both the i and j subscripts. To write it out in steps, it is the same as

$$\begin{aligned}\sum_{i=1}^{m}\sum_{j=1}^{n} x_{ij} &= \sum_{i=1}^{m} (x_{i1} + x_{i2} + x_{i3} + \ldots + x_{in}) \\ &= (x_{11} + x_{12} + x_{13} + \ldots + x_{1n}) + (x_{21} + x_{22} + x_{23} + \ldots + x_{2n}) \\ &\quad + (x_{31} + x_{32} + x_{33} + \ldots + x_{3n}) + \ldots \\ &\quad + (x_{m1} + x_{m2} + x_{m3} + \ldots + x_{mn})\end{aligned}$$

TABLE 1.1 A Two-Dimensional Data Matrix

	$j = 1$	$j = 1$	$j = 3$	. . .	$j = n$	Row Sum
$i = 1$	x_{11}	x_{12}	x_{13}	. . .	x_{1n}	$\sum_{j=1}^{n} x_{1j}$
$i = 2$	x_{21}	x_{22}	x_{23}	. . .	x_{2n}	$\sum_{j=1}^{n} x_{2j}$
$i = 3$	x_{31}	x_{32}	x_{33}	. . .	x_{3n}	$\sum_{j=1}^{n} x_{3j}$
. . .	. . .	. . .	. . .	x_{ij}	. . .	. . .
$i = m$	x_{m1}	x_{m2}	x_{m3}	. . .	x_{mn}	$\sum_{j=1}^{n} x_{mj}$
Column Sum	$\sum_{i=1}^{m} x_{i1}$	$\sum_{i=1}^{m} x_{i2}$	$\sum_{i=1}^{m} x_{i3}$	. . .	$\sum_{i=1}^{m} x_{in}$	$\sum_{i=1}^{m}\sum_{j=1}^{n} x_{ij}$

Several common algebraic expressions using these notations are also useful to know in order to understand the computation of some elementary statistics.

First, if a is a constant, then $\Sigma_{i=1}^{n} a = a + a + \ldots + a = n \times a$ represents adding the value of a n times to derive na.

Next, $\Sigma_{i=1}^{n} x_i y_i = (x_1 \times y_1) + (x_2 \times y_2) + (x_3 \times y_3) + \ldots + (x_n \times y_n)$ represents the sum of paired products of x_i and y_i, where $i = 1, 2, \ldots, n$. Note that the parentheses are optional. They are used merely to make the expression more readable.

Some notations are quite confusing. For example, $\Sigma_{i=1}^{n} x_i^m = x_1^m + x_2^m + x_3^m + \ldots + x_n^m$ is not the same as $(\Sigma_{i=1}^{n} x_i)^m = (x_1 + x_2 + x_3 + \ldots + x_n)^m$.

Sometimes, constants can be moved or factorized out of sigma as $\Sigma_{i=1}^{n} ax_i = a \Sigma_{i=1}^{n} x_i$, and sometimes expressions can be expanded as $\Sigma_{i=1}^{n} (ax_i + b) = a \Sigma_{i=1}^{n} x_i + n \times b$.

As for multiplication, we have, $\Pi_{i=1}^{n} a = a \times a \times a \times \ldots \times a = a^n$, $\Pi_{i=1}^{n} x_i y_i = (x_1 \times y_1) \times (x_2 \times y_2) \times (x_3 \times y_3) \times \ldots \times (x_n \times y_n)$, and $\Pi_{i=1}^{n} x_i^m = x_1^m \times x_2^m \times x_3^m \times \ldots \times x_n^m = (x_1 \times x_2 \times x_3 \times \ldots \times x_n)^m$.

1.4.3 Scale, Extent, and Projection

When working with geographic data, we should be concerned not only with the scale of measurement that is appropriate for the attribute data in hand but also with the geographic scale that is appropriate for the associated spatial data. Because all geographic data have a spatial dimension (*0-dimension* for points, *1-dimension* for lines, and *2-dimension* for polygons), it is particularly important to choose a geographic scale so that spatial data can be properly analyzed.

On printed maps, the size of the map dictates the degree to which details of geographic objects can be shown. Consequently, the scale of a map determines in how much detail the geographic objects are displayed. For example, the City of Cleveland is represented as only a point occupying almost no space on a world map. However, Cleveland would occupy the entire map sheet if the map shows all the streets in that city. A section of a river may be represented as a thin line at one scale but as a thick line or even a linear-shaped polygon at another scale. Other similar examples can be found easily.

For instance, on the national map in Figure 1.7a, cities, including Washington, D.C., in a portion of the Mid-Atlantic area are represented as points given the spatial scale. But when only the Mid-Atlantic area is presented in Figure 1.7b, some of the cities are shown as polygons.

In a computerized environment, map displays can be dynamically zoomed in or out to show more or less detail of the same set of geographic objects. On the same monitor screen, geographic objects are shown in more detail at a larger geographic scale (e.g., 1:2000), while less detail of the same geographic objects is shown at a smaller geographic scale (e.g., 1:2,000,000). Furthermore, between 1:2000 and 1:2,000,000 on the same monitor screen,

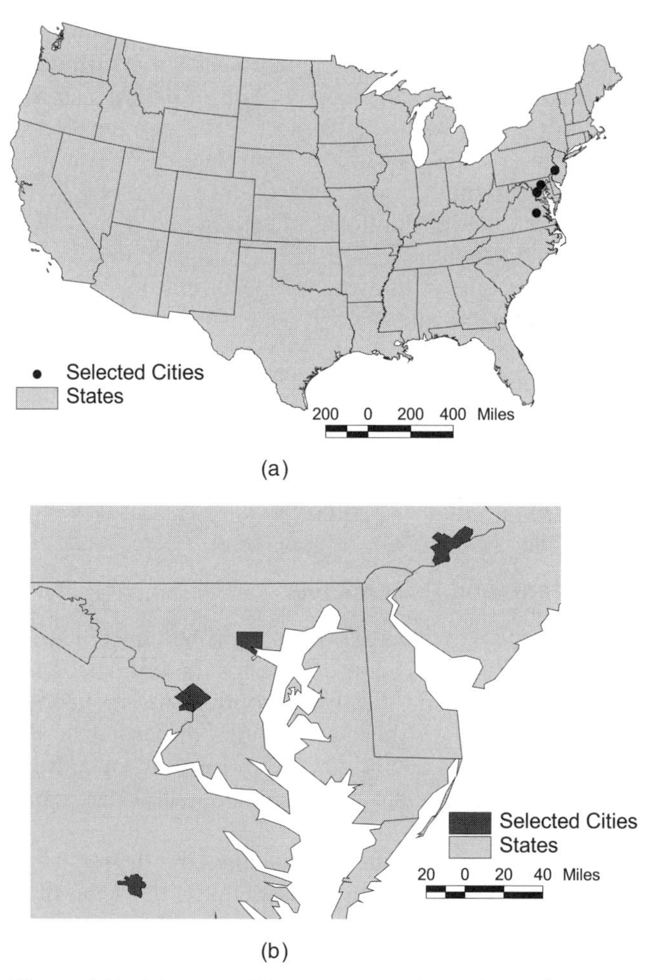

Figure 1.7 Maps at different geographic scales and extents.

the former scale shows less geographic extent (a smaller region) than the latter scale.

In cartography, the scale of a map is the ratio between the size of a feature shown on the map and the actual size of the same feature in the real world. In other words, a geographic scale can be calculated as

$$\text{Scale} = \frac{\text{Distance between two points on the map}}{\text{Distance between the two points on the ground}}$$

Therefore, a geographic scale is often represented by a ratio, such as 1:24,000, that is, 1 unit on the map represents 24,000 units in the real world.

Another popular way to show the scale of a map is to use a scale bar. The maps in Figure 1.7 have scale bars indicating how 1 unit on either map is equivalent to the real-world distance. Finally, a geographic scale can be given with a verbal description. For example, a scale may be 1 inch to 2000 meters. This means that 1 inch on the map equals 2000 meters on the ground, the literal version of the ratio scale.

Maps with relatively large ratios, such as 1:2000, or numerically 1/2000, are considered large-scale maps when compared with maps with small ratios, such as 1:1,000,000 or 1/1,000,000, which are classified as small-scale maps. In general, a small-scale map covers a large geographic area or extent, such as the map in Figure 1.7a, covering the entire United States. A large-scale map, such as the one in Figure 1.7b, can show details of the landscape or features but covers only a relatively small area.

Note that, due to changes in geographic scales (or map scales), what appears to be a clustered pattern at one scale may become a dispersed pattern at another scale. For example, the same cities identified in Figures 1.7a and 1.7b show just that. They appear to be clustered together when their distribution is viewed against the entire United States. However, they seem to be rather dispersed when we zoom in on a sufficiently large map scale. For this reason, we should always keep in mind that no matter what statistics we use, geographic events and phenomena are highly dependent on the geographic scale at which they are observed or described.

Finally, we must consider map projections when displaying the distribution of geographic objects. Because the Earth is a three-dimensional sphere, while a map is only two-dimensional, the display of the Earth on any map involves projecting the geographic features or phenomena from the three-dimensional sphere onto the two-dimensional piece of paper. Distortions of the spatial relationship between features cannot be avoided. There are different ways to translate the spatial/geometric relationships among features from the globe to a map. Different ways can be regarded as different projections, and different projections distort the relationships differently.

Four possible distortions may be caused by different projections: distortions in *area, shape, direction* and *distance.* Among many different projections that can be applied in mapping, there is no single projection that can perfectly transform geographic locations from where they are on the globe (a three-dimensional space) to a map (a two-dimensional plane) without any distortion. Thus, we will need to be sensitive to the focus of the study at hand. Careful consideration of the purposes, scales, projection, and accuracy of data is needed for a successful study. As in spatial analysis and statistics in general, areas of geographic objects and distances between geographic objects are used in almost all studies that involve spatial data. Choosing an appropriate projection to display them or to quantify them is critical for meaningful and useful analysis. How can one choose the appropriate projection given the existence of so many different types of spatial data at different scales? Monmonier (1993) provides a set of simple but useful principles.

Not surprisingly, the larger the study area, the more significant the impact of different projections will be. If the study area is small, such as a residential neighborhood or a small village, using different projections may not have much effect on areal and distance measurements. However, studies that encompass the entire United States or the world must be careful to select proper projections. This is evident in the maps shown in Figure 1.8.

In Figure 1.8, the 48 contiguous states of the United States are displayed in different coordinate systems. These maps clearly show shape changes, and in turn obviously would affect direction. Similar changes in shape and direction are also evident in the three displays of Ohio. Since these maps are at the same scale, 1:45,300,000 for the 48 U.S. states and 1:3,340,000 for Ohio, the different sizes of the displays suggest that the distances and areas would also change.

1.5 ARCVIEW NOTES—DATA MODEL AND EXAMPLES

We have developed a suite of computer codes for calculating various statistics and for performing statistical tests. These computer codes were written in Avenue scripts. They are structured as extensions to ArcView 3.X, one extension per chapter in this book. Each extension can be loaded and executed in ArcView 3.X to ensure full compatibility. If the reader has followed the instructions in the readme file on the CD, these extensions should have been put in the EXT32 folder under ArcView. The datasets that are used in the examples and exercises in different chapters should have been copied to C:\Temp\Data by chapters. Please contact either author if further information is needed. David Wong can be contacted at *dwong2@gmu.edu,* and Jay Lee can be contacted at *jlee@kent.edu.*

As available from the vendor, the Environmental System Research Institute (ESRI, Redlands, California), ArcView 3.3 is the latest version of the relatively lightweight first-generation desktop GIS. ESRI's newer GIS packages, ArcGIS 8 and, most recently, ArcGIS 9, also have an ArcView component, but it is a completely different product and is incompatible with ArcView 3.X. ArcView 3.X uses Avenue scripting language as the programming platform, but ArcGIS 8/9 uses Visual Basics for Applications (VBA, Microsoft, Redmond, Washington) to support customization and additional development.

In this book, we decided to keep using Avenue scripting language in developing the routines for various statistics and techniques discussed here. This is because ArcView 3.3 costs less and demands fewer computing resources. ArcView 3.3 also performs faster than the newer generation of GIS packages on the same computation tasks. We will continue using ArcView 3.3 to illustrate examples and to provide step-by-step instructions for implementing many analytical techniques for spatial statistics discussed in this book. In the Appendix, we will discuss similar tasks in ArcGIS 9 as additional references. Developing tools in ArcGIS 9 parallel to those covered in this book is under

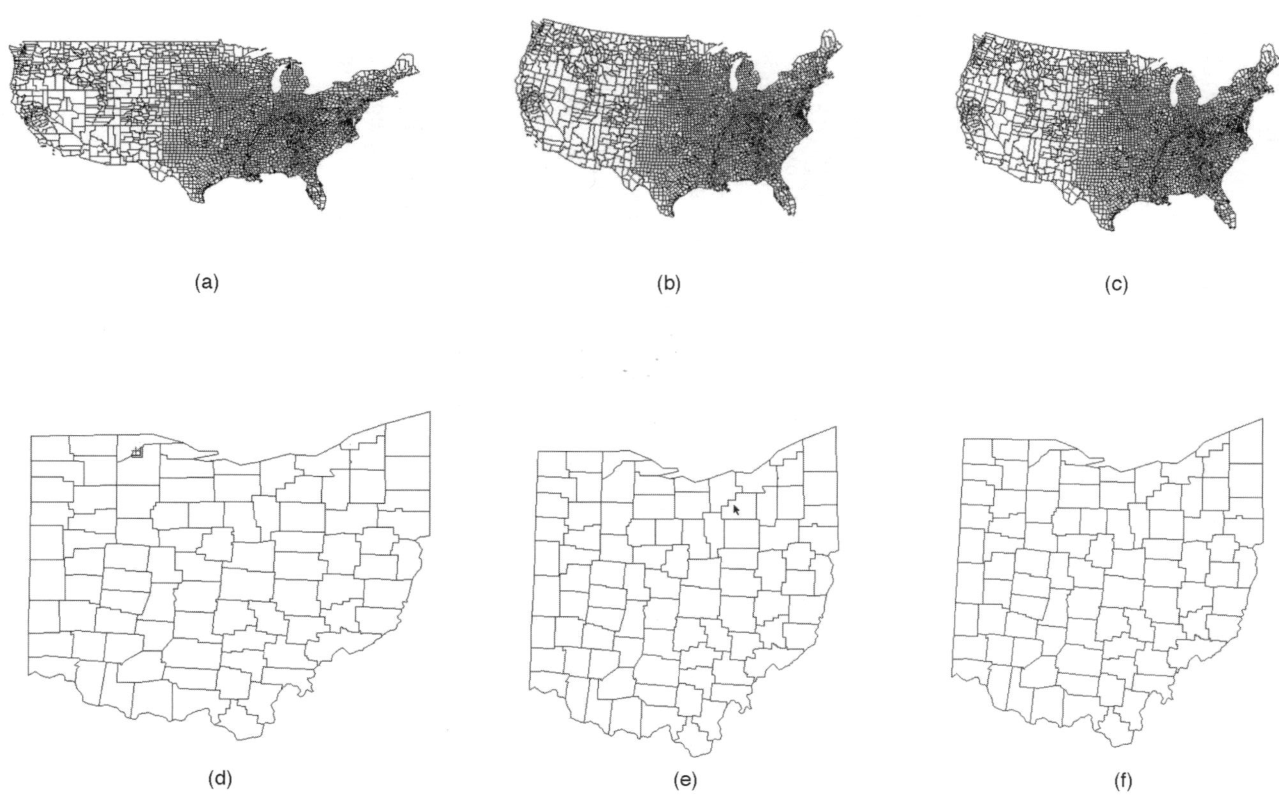

Figure 1.8 The United States and the State of Ohio in various projections. (a) Forty-eight U.S. states displayed in latitudes and longitudes. (b) Albers Equal Area projection. (c) Lambert Conformal Conic projection. (d) Ohio in latitudes and longitudes. (e) State Plane projection, NAD83, North Ohio zone. (f) UTM projection, NAD83, Zone 17.

consideration. Interested users may contact the authors or visit the book's website through the link at *http://mason.gmu.edu/~dwong2* for development updates.

We assume that the readers are familiar with basic operations in ArcView 3.X. Therefore, these operations will not be repeated in this book. Beginners in GIS may wish to consult a number of online training courses available at ESRI's online campus or free web-based tutorials from academic institutions to gain the necessary experience before using the computing routines for this book.

1.5.1 Data Model Used in ArcView GIS

As discussed earlier, a GIS dataset or theme includes both spatial and attribute data. Attribute data are often structured in a matrix or tabular form in which information for each geographic object or observation is organized into a **record.** Combining all records produces an **attribute data table** that contains information on all geographic objects in that GIS layer. Within each record, there are **attribute fields.** Each attribute field has an **attribute item** that holds the value of that attribute field in a given record.

GIS packages handle the linkages between spatial data and attribute data internally. These linkages are usually transparent to GIS users. Typically, each geographic object is assigned a feature identifier that is also included in the corresponding record in the attribute table. For example, Figure 1.9 shows that selecting a polygon would also mark the corresponding record in the attribute table as a selected record.

1.5.2 Random Sampling—the Generic Way

Like many events in life, there are many ways to do the same thing. In this section, we will discuss steps for carrying out random sampling of geographic objects in a GIS database using a standard installation of ArcView. Here is a list of tasks to be carried out:

1. Bring in a data layer.
2. Open the attribute table of the data layer.
3. Make the attribute table editable.
4. Add an attribute field to hold random numbers.
5. Generate random numbers and store them in the newly added field.
6. Sort random numbers in ascending order.
7. Select the desirable number of records.

In this process, we will use an ArcView request called MakeRandom. Its format is MakeRandom (min, max), where min and max define the range of the generated random numbers. In addition, we will use tools and functions

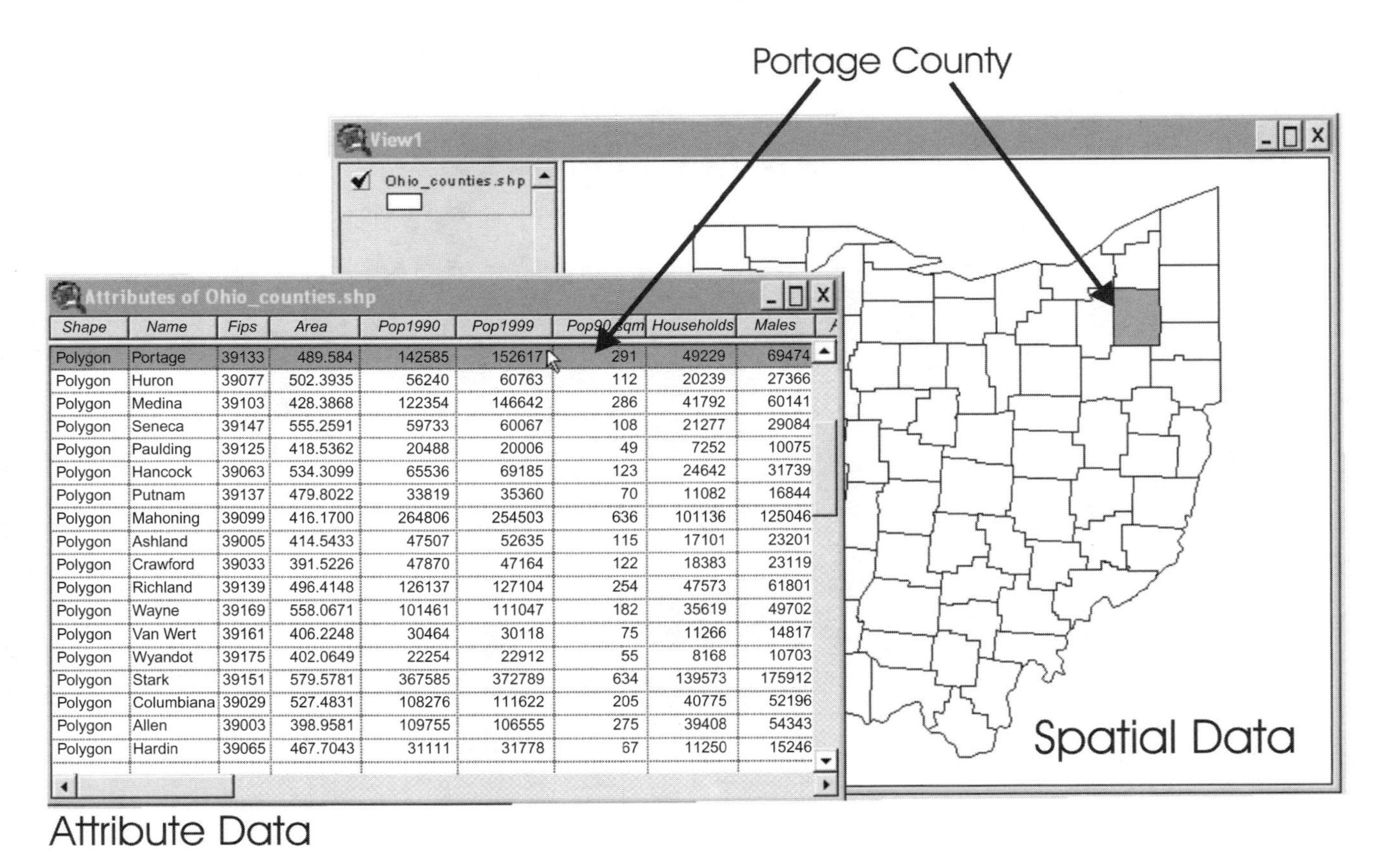

Shape	Name	Fips	Area	Pop1990	Pop1999	Pop90 sqm	Households	Males
Polygon	Portage	39133	489.584	142585	152617	291	49229	69474
Polygon	Huron	39077	502.3935	56240	60763	112	20239	27366
Polygon	Medina	39103	428.3868	122354	146642	286	41792	60141
Polygon	Seneca	39147	555.2591	59733	60067	108	21277	29084
Polygon	Paulding	39125	418.5362	20488	20006	49	7252	10075
Polygon	Hancock	39063	534.3099	65536	69185	123	24642	31739
Polygon	Putnam	39137	479.8022	33819	35360	70	11082	16844
Polygon	Mahoning	39099	416.1700	264806	254503	636	101136	125046
Polygon	Ashland	39005	414.5433	47507	52635	115	17101	23201
Polygon	Crawford	39033	391.5226	47870	47164	122	18383	23119
Polygon	Richland	39139	496.4148	126137	127104	254	47573	61801
Polygon	Wayne	39169	558.0671	101461	111047	182	35619	49702
Polygon	Van Wert	39161	406.2248	30464	30118	75	11266	14817
Polygon	Wyandot	39175	402.0649	22254	22912	55	8168	10703
Polygon	Stark	39151	579.5781	367585	372789	634	139573	175912
Polygon	Columbiana	39029	527.4831	108276	111622	205	40775	52196
Polygon	Allen	39003	398.9581	109755	106555	275	39408	54343
Polygon	Hardin	39065	467.7043	31111	31778	67	11250	15246

Figure 1.9 Spatial data and attribute data table.

that are available in the Table document, such as Field Calculator, sorting, and the Select tool. The above steps can be applied to point data layers, line data layers, or polygon data layers.

Let's assume that we need to select 20 cities from the 164 cities in Ohio for an analysis. We can follow the following procedures to obtain a random selection of 20 cities.

ArcView Example 1.1: Random Sampling

Step 1 Start ArcView GIS and start with a new View document

Assume the data are under C:\Temp\Data. If not, use the new path and folder.

Step 2 Add data layers into ArcView

- Use the **Add Theme** button, to bring in

  `C:\Temp\Data\Ch1_data\ohio_cities.shp`,
  `C:\Temp\Data\Ch1_data\ohio_roads.shp`, and
  `C:\Temp\Data\Ch1_data\ohio_counties.shp`.

- Once completed, make sure `ohio_cities.shp` is placed above `ohio_roads.shp` and `ohio_roads.shp` is placed above `ohio_counties.shp` in the Table of Contents of the View document in ArcView. Make sure that all layers are checked for display and that Ohio_cities.shp is the active layer.

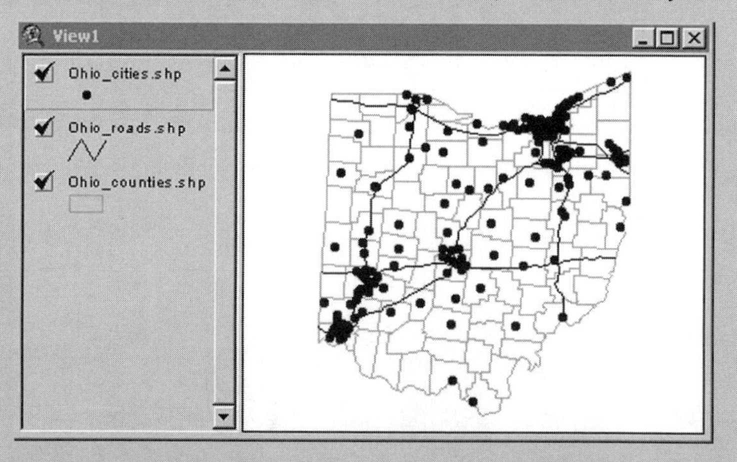

Step 3 Open attribute data table and set it editable

- Click the **Open Theme Table** button, , to open Attribute of Ohio_cities.shp. From the menu, choose **Table,** then **Start Editing.** This sets the table to be modifiable.

Step 4 Add a new field for random numbers

- Choose from the **Edit** menu the item for **Add Field** . . . Name the new field `RanNum` and set it to be of the `Number` type with a width of `4` and `0` decimal places.

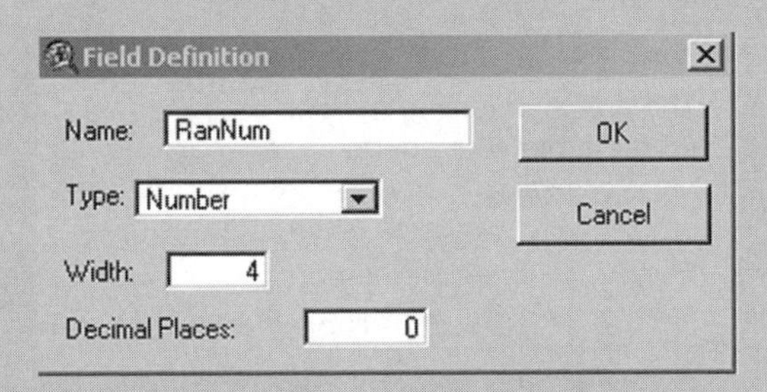

- The new field is added with its content empty.

Step 5 Calculate random numbers

- Click the **Calculate** button, [calculator icon], to invoke the **Field Calculator.**
- Notice that the left side of the equation has already been defined as "[RanNum] =" because the new field, **RanNum,** was selected after it was created.
- From the list of **Requests,** select `MakeRandom` and type in the range for random numbers to be from `1` to `1000`.
- The **Field Calculator** should look similar to this:

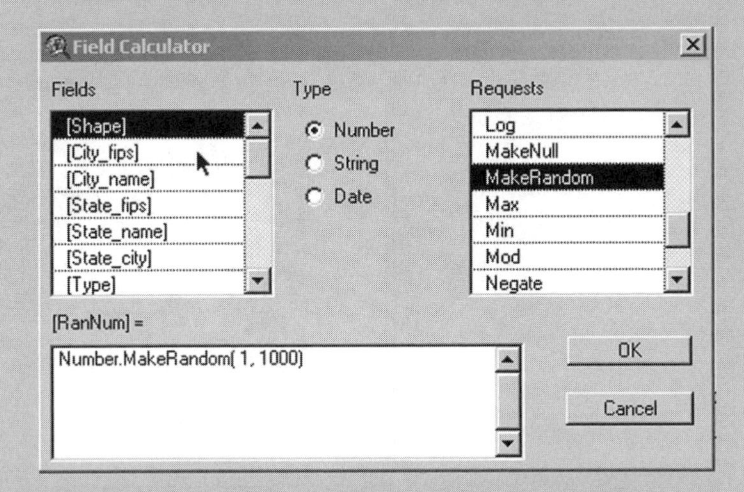

- Click **OK** to fill in random numbers. Notice that, for the **RanNum** field, random numbers with values from 1 to 1,000 have been generated.

Step 6 Sort random numbers

- With the **RanNum** field highlighted, click the **Sort Ascending** button, , to sort the records according to the random numbers in ascending order.

Step 7 Select records as sample

- The actual sampling is done by selecting from the top of the attribute table as many records as are required or desired to sample. For example, selecting the first 20 records will provide a sample with a size of 20 records.
- The actual selection can be carried out by using the **Select** button, , to select multiple records by holding the Shift key while highlighting records to select them.

Attributes of Ohio_cities.shp

Units2	Units3_9	Units10_49	Units50_up	Mobilehome	RanNum
1287	2752	1245	393	538	70
283	925	937	528	2	71
5306	2852	4627	4631	8	75
449	583	297	0	367	77
454	164	346	247	1	77
254	822	265	163	90	105
436	286	311	0	272	124
488	808	628	58	8	125
481	485	130	241	67	136
14	337	433	117	2	136
2637	2246	2539	714	297	137

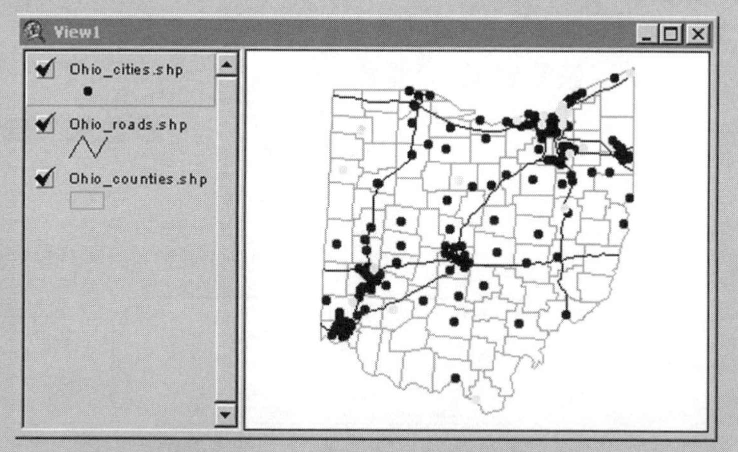

- Selected records can be used in subsequent tasks for mapping, analysis, or charting. Alternatively, selected records can be exported as a DBF file for use with other software or exported to another shapefile (under the View document interface, choose the **Theme** menu and select the **Convert to Shapefile** item. This is essentially subsetting the original theme or clipping).

Note that this approach allows us to select as many records/objects as are needed for any sample size up to the entire data set. It also allows similar processes to be performed for a line data layer or a polygon data layer. While it is possible to use this approach to do random sampling, it is rather tedious. In the next section, we will describe a greatly simplified way to do not only random sampling but also systematic sampling.

1.5.3 Random and Systematic Point Sampling—an Extended Function

While the steps outlined in Section 1.5.2 work for almost all types of data themes, including point themes, line themes, and polygon themes, going through these steps is somewhat tedious. To speed up the process of sampling, a set of sampling tools have been developed as an extension for ArcView GIS. These tools include:

1. Sampling locations: random points
   - A specified number of point locations are randomly generated and saved in a new shapefile. The created shapefile can be used in subsequent analytical steps.
2. Sampling locations: systematic
   - A number of regularly spaced point locations are generated so that they cover the extent of the data theme or the extent of the View document. These point locations can be saved in a shapefile for further use.
3. Sampling locations: systematic-random
   - The extent of a selected theme or the extent of the View document is first divided evenly as specified. A point location is then randomly selected within each partitioned area.
4. Sampling features: random
   - In the selected theme, the specified number of features are randomly selected. Data themes can be point, line, or polygon themes.
5. Sampling features: regular/systematic
   - The features in the selected data theme are sampled systematically such as every 5 records.

6. Sampling features: stratified
   - Features in a data theme are sampled based on their categorical memberships. A specified number of observations will be selected in each category.

These tools have been assembled into an extension to ArcView GIS. They are accessible as menu items once the extension is loaded. Note the following limitations when using the ArcView extensions:

1. When using extensions developed for this book, the data theme at the top of the Table of Contents (TOC) is normally the data theme with which the extension works unless specified otherwise. It is assumed that users will perform statistical analysis on one data theme at a time.
2. The extensions developed for this book are limited in terms of the numbers of features or data records they accommodate. When the size of your data sets exceeds these limits, please contact the authors for an expanded version of these extensions.

In general, ArcView extensions for this book should be copied to the EXT32 folder inside the installation folder of ArcView GIS—for example, C:\ESRI\AV_GIS30\ARCVIEW\EXT32.

Below is an application in which a number of cities are selected from all cities in Ohio for further study. Instead of using all cities, a smaller set of cities allows site visits or on-site surveys to be practical.

For a study of consumer preferences, the available budget allows on-site surveys to be conducted in up to 20 cities. Among the 164 cities, 20 can be selected in various ways. For example, we can randomly select 20 of the 164 cities. Alternative, we can systematically select 20 cities based on certain criteria. Or we can create 20 random points and select one city that is closest to each random point.

ArcView Example 1.2: Sampling Cities

Step 1 Load Ch1.avx extension

- For this example, **Ch1.avx** will be used.
- Start ArcView GIS and choose to start a blank project.
- From the **File** menu, choose **Extensions . . .** to bring up the Extensions dialog box.
- Check the box besides **Ch1 Extension** and then click the **OK** button.

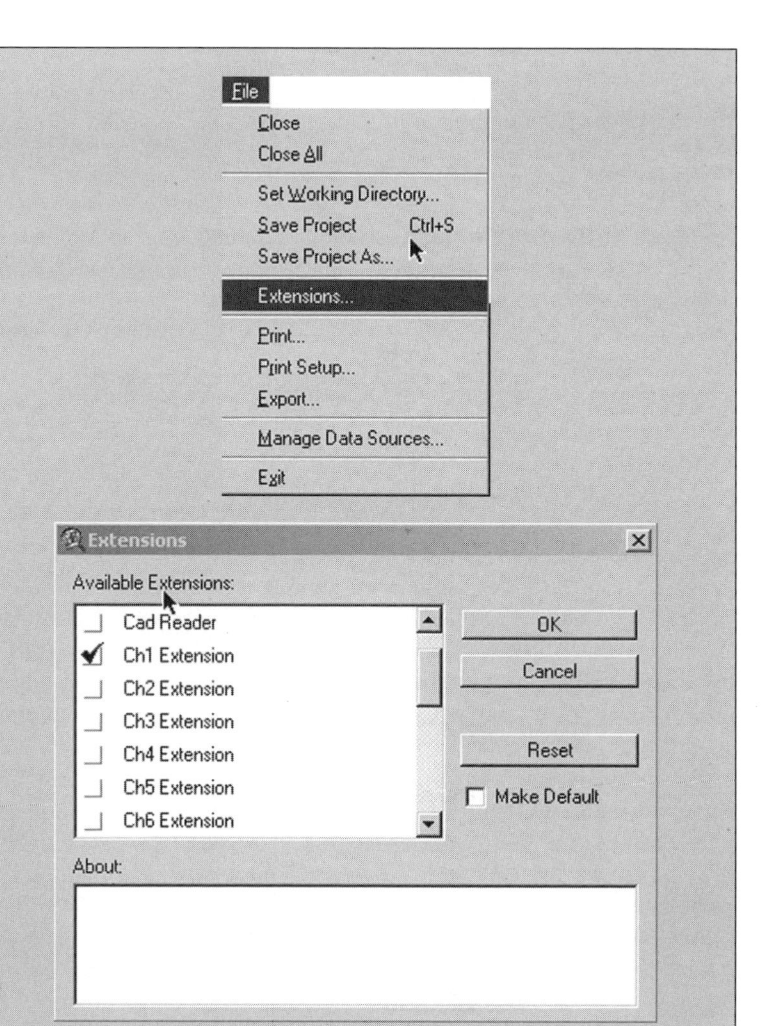

Step 2 Start ArcView GIS and add data themes

- In the project window (Untitled), click the **View** icon, [Views icon], and then the **New** button to create a new View document. Notice that a **Ch1** menu item has been added because of the loaded Ch1 extension.
- While the View document is opened, use the **Add Theme** button, [Add Theme icon], to bring in

  `C:\Temp\Data\Ch1_data\ohio_cities.shp`
  `C:\Temp\Data\Ch1_data\ohio_roads.shp`, and
  `C:\Temp\Data\Ch1_data\ohio_counties.shp`.
- We will use `ohio_cities.shp` with `ohio_roads.shp` and `ohio_counties.shp` providing background display.

- Make sure that ohio_cities.shp is placed above ohio_roads.shp and ohio_roads.shp is placed above ohio_counties.shp in the Table of Contents of the View document.

Step 3 Random location sampling

- From the **Ch1** menu, select Sampling Locations: random points, as

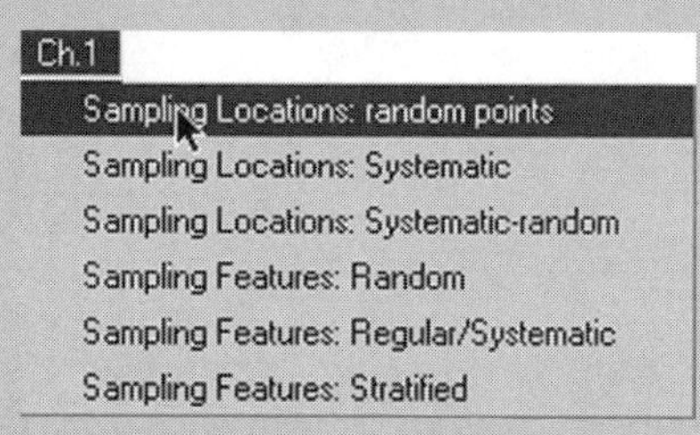

- In the **Random point sampling** dialog box, click Ohio_counties.shp to use its theme extent for sampling so that the sampled locations are kept within the county boundaries.
- Enter 20 as the number of points, as

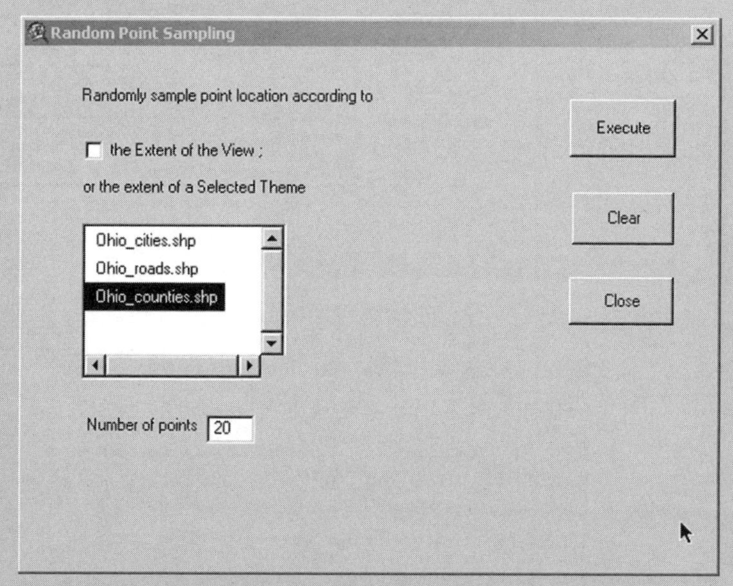

- Click the **Execute** button to proceed.
- In the **Random Point File** dialog box, navigate to C:\Temp or any preferred folder and click **OK** to save the random points as a dbf (dBase) file, as

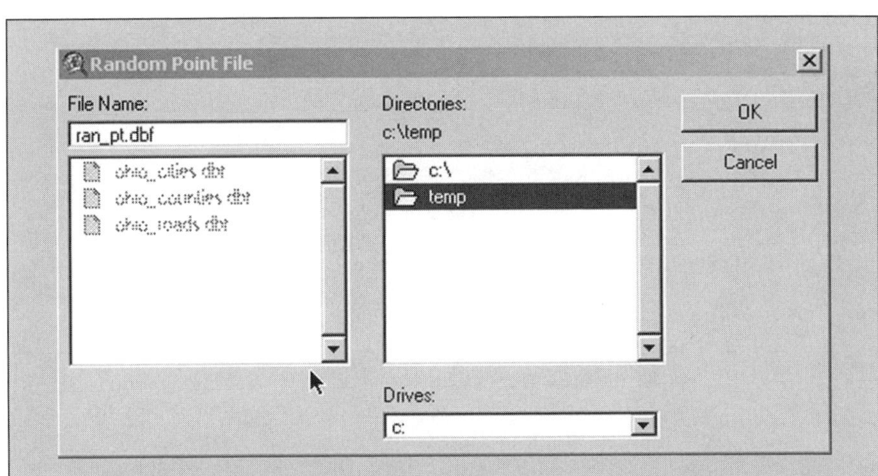

- Notice that 20 random points have been drawn and marked by the "*" symbol in the View document, as

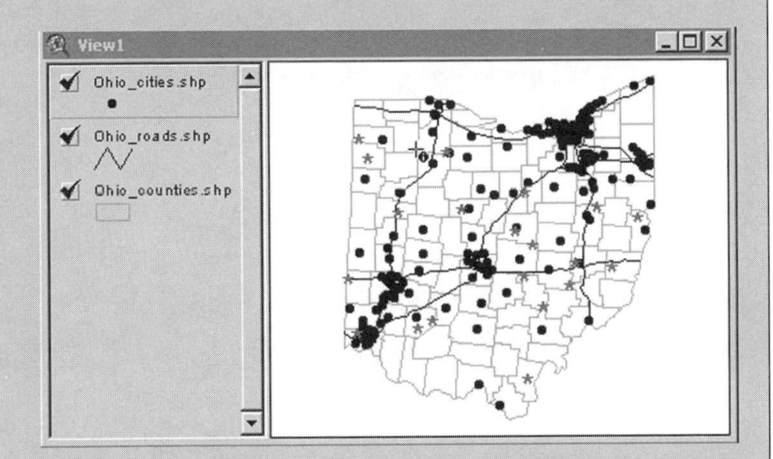

- Add the random points theme to the View document.
- Using the **Select Feature,** [tool icon], tool, points that are close to the "*" symbols can be selected as sampled cities.
- Note that the process may take a long time if there are many polygons in the theme.

Step 4 Randomly select features as the sample

- From the **Ch1** menu, select `Sampling Features: Random`, as

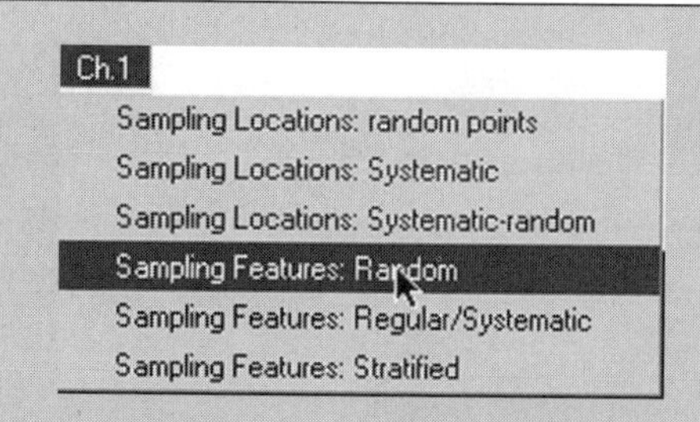

- This will randomly select a specified number of features (points in this case).
- In the **Sample Features: Random** dialog box, click `Ohio_cities.shp` to select it and type `20` for the number of sampled features. Click the **Execute** button to proceed, as

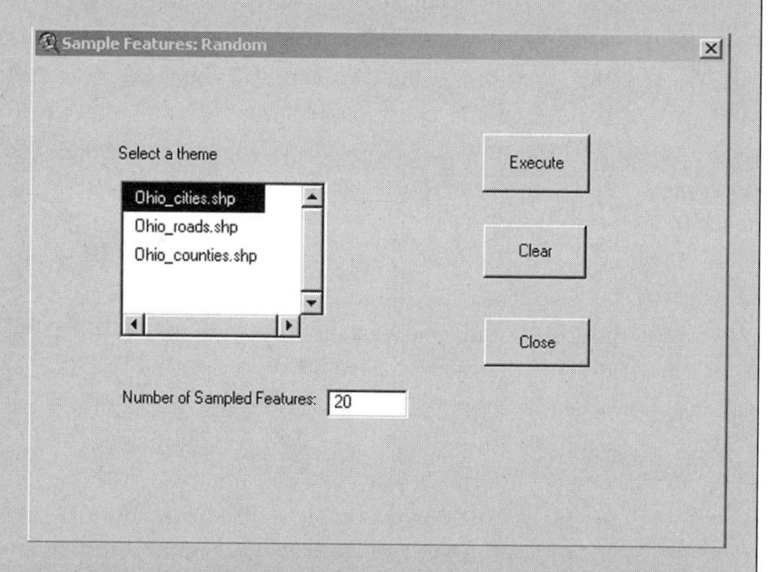

- Once executed, the extension creates a new field, called **sampled,** and adds it to the attribute table. (If this step is repeated, it prompts to replace the older sampled field.)
- For features that have been selected as a sample, a value of 1 is entered in the **Sampled** attribute field; otherwise, a value of 0 is entered.

Step 5 Identify sampled features

- Make sure that the sampled data theme is the active theme.
- Use the **Query Builder,** [icon], to invoke a dialog box similar to this:

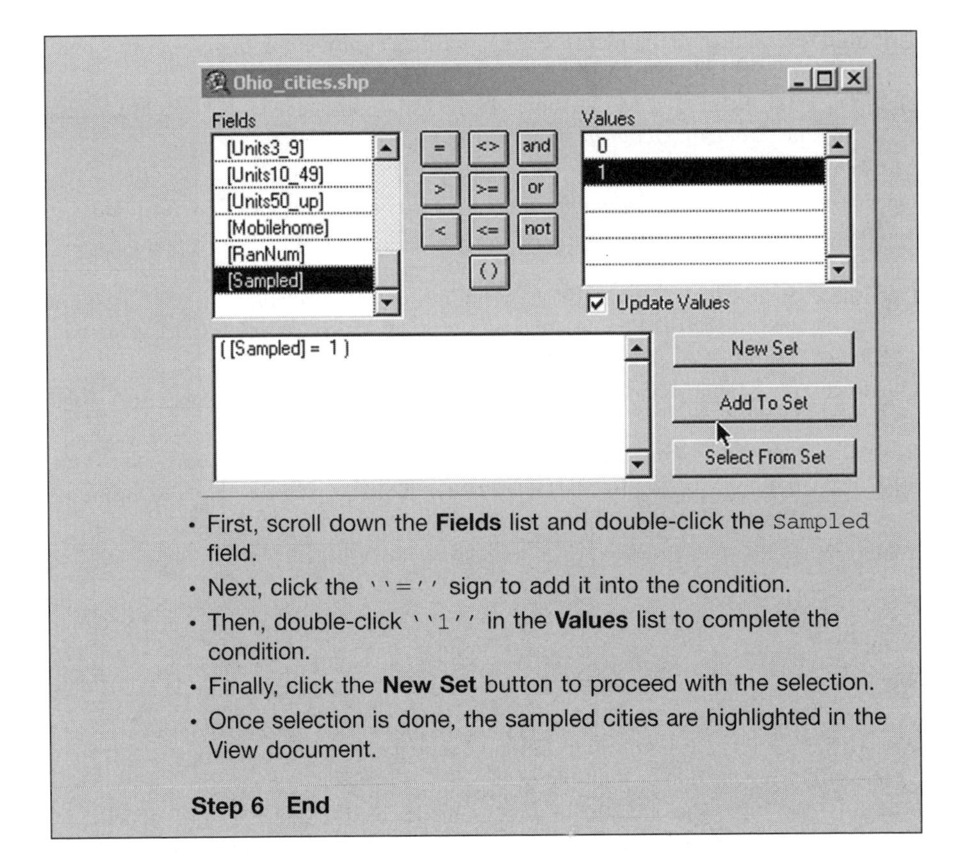

REFERENCES

Anselin, L. 1988. *Spatial Econometrics: Methods and Models.* Dordrecht: Martinus Nijhoff.

Cliff, A. and J. K. Ord. 1981. *Spatial Processes: Models and Applications.* London: Pion.

Dueker, K. J. 1979. Land resource information systems: a review of fifteen years' experience." *Geo-processing* 1(2): 105–128.

Fotheringham, A. S. 1989. Scale-independent spatial analysis. In M. F. Goodchild and S. Gopal (eds.), *Accuracy of Spatial Databases.* London: Taylor and Francis, pp. 221–228.

Fotheringham, A. S. and D. W. S. Wong. 1991. The modifiable areal unit problem in multivariate statistical analysis. *Environment and Planning A* 23: 1025–1044.

Glick, J. 1987. *Chaos: Making a New Science.* New York: Penguin Books.

Griffith, D. A. 1987. Toward a theory of spatial statistics: another step forward. *Geographical Analysis* 19: 69–82.

Griffith, D. A. 1988. *Advanced Spatial Statistics.* Dordrecht: Martinus Nijhoff.

Griffith, D. A. and C. G. Amrhein. 1991. *Statistical Analysis for Geographers.* Englewood Cliff, NJ: Prentice Hall.

Holt, D., D. G. Steel, and M. Tranmer. 1996. Area homogeneity and the Modifiable Areal Unit Problem. *Geographical Systems* 3(2–3): 181–200.

Monmonier, M. 1993. *Mapping It Out.* Chicago: University of Chicago Press.

Openshaw, S. 1984. *The Modifiable Areal Unit Problem. Concepts and Techniques in Modern Geography (CATMOG) No. 38.* Norwich: Geo Books, University of East Angolia.

Openshaw, S. and P. J. Taylor. 1979. A million or so correlation coefficients: three experiments on the modifiable areal unit problem. In N. Wrigley (ed.), *Statistical Applications in the Spatial Sciences.* London: Pion, pp. 127–144.

Quattrochi, D. A. and M. F. Goodchild. 1997. *Scale in Remote Sensing and GIS.* Boca Raton, FL: CRC Press.

Rogerson, P. A. 2001. *Statistical Methods for Geography.* London: Sage Publications.

Tate, N. and P. M. Atkinson. 2001. *Modelling Scale in Geographical Information Science.* New York: John Wiley & Sons.

Tobler, W. 1970. A computer movie simulating urban growth in the Detroit region. *Economic Geography* 46(Supplement): 234–240.

Tobler, W. 1989. Frame independent spatial analysis. In M. F. Goodchild and S. Gopal (eds.), *Accuracy of Spatial Databases.* London: Taylor and Francis, pp. 115–122.

Wong, D. W. 2001. Location-specific Cumulative Distribution Function (LSCDF): an alternative to spatial correlation analysis. *Geographical Analysis* 33(1): 76–93.

Wong, D. W. 2003. Spatial decomposition of segregation indices: a framework toward measuring segregation at multiple levels. *Geographical Analysis* 35(3): 179–194.

Wong, D. W. 2004 The Modifiable Areal Unit Problem (MAUP). In D. Janelle, B. Warf, and K. Hansen (eds.), *WorldMinds: Geographical Perspectives on 100 Problems.* Cambridge, MA: Kluwer Academic Publishers, No. 93, pp. 571–575.

Wong, D. W. and C. G. Amrhein, 1996. Research on the MAUP: old wine in a new bottle or real breakthrough? (An editorial note). *Geographical Systems* 3: 73–76.

EXERCISES

1. The table below reports the average population density by census block group in 10 of the largest metropolitan areas in the United States.

Metro Areas	Population Density
New York City	607.9
Los Angeles	179.0
Chicago	380.4
Washington, D.C.	280.0

Metro Areas	Population Density
San Francisco	308.9
Philadelphia	349.4
Boston	319.1
Detroit	298.9
Dallas	212.9
Houston	205.4

a. At what measurement scale are the population density data? Does the variable population density have an "absolute zero"?

b. Convert the population density data into the strong ordering scale.

c. Using a three-category scheme, derive a new variable in the weak ordering scale based upon the population density data.

d. Using a density level of 300, derive a new binary nominal scale variable based upon the population density data.

2. Let $x = \{1, 3, 5, 7\}$, $y = \{2, 4, 6, 8\}$, $i = \{1, 2, 3, 4\}$, $a = 3$, and $b = 4$. Evaluate the following:

a. $a!\ b!$

b. $\Sigma_{i=1}^{4} x_i$

c. $\Sigma_{i=2}^{3} y_i$

d. $\Sigma_{i=1}^{4} x_i y_i$

e. $a\ \Sigma_{i-1}^{4} x_i$

f. $b\ \Sigma_{i=1}^{4} y_i^a$

g. Does $a\ \Sigma_{i=1}^{4} x_i = \Sigma_{i=1}^{4} ax_i$?

h. Simplify $\Sigma(b + y_i)$

i. Πx_i

j. $\Pi b y_i$

3. Let x_{ij} denotes the *i-j* element in the matrix below; evaluate the following:

		j		
	i/*j*	1	2	3
i	1	5	2	7
	2	3	9	4
	3	12	1	8

a. $\Sigma_j x_{1j}$ or $x_{1\cdot}$

b. $\Sigma_i x_{i2}$ or $x_{\cdot 2}$

c. $\Sigma_i x_{ii}$

d. $\Sigma_i \Sigma_j x_{ij}$

4. Define and differentiate the following terms:
   Descriptive statistics and inferential statistics
   Classical statistics and spatial statistics
   Random sampling and systematic sampling
   Population and sample
   Absolute location and relative location
   Spatial data and attribute data
   Binary variable and polychotomous variable

5. Using the CensusTrcts.shp in C:\Temp\Data\Ch1_data,
   a. randomly select 100 census tracts
   b. systematically select 100 census tracts

   How are the two sets of selected census tracts different?

PART I

CLASSICAL STATISTICS

While the focus of this book is on analyzing spatial data, it is necessary to start our discussion with classical statistics. A great variety of spatial statistics and analytical methods have been developed to handle spatial and attribute data. Many if not all of them originated in classical statistics.

Spatial data are unique in that observations describing geographic phenomena or events may have spatial dependency embedded in them, as described in the First Law of geography (Tobler, 1970; Sui, 2004). Because of the presence of spatial dependency, classical statistics may not be useful in analyzing spatial data, as a fundamental assumption in classical statistics is that data values are derived from independent observations. Therefore, classical statistics and related analytical methods may not be appropriate to handle spatial data. Consequently, classical statistics or analytical methods have to be modified in order to address the spatial dependency issue. In addition, spatial data include location or geographic information, which requires methods different from those of classical statistics to analyze them meaningfully. Some of these methods, including those used in analyzing point patterns, are unique to spatial data analysis, though some of the conceptual underpinnings originated in classical statistics.

Beginning in Chapter 5, various concepts and analytical methods handling spatial data will be discussed in detail. Chapters 2 to 4 will cover concepts and techniques used in classical statistics to provide a foundation for subsequent discussions on spatial data analysis.

In Chapters 2 and 3, we will focus on descriptive statistics. These statistics are used to describe the statistical properties of data we analyze. As defined earlier, **data values** are observations made by measuring various properties of geographic phenomena or events. Since the set of data may capture various characteristics or properties of the observations, each property or characteristic represented by data values is termed a **variable** in statistics and an **attribute** in GIS. As an extension of this, statistics and methods for handling data of a single variable are called **univariate statistics.** Statistics and methods for handling data of two variables are called **bivariate statistics,** and those for data with more than two variables are known as **multivariate statistics.**

In this book, only univariate and bivariate statistics will be discussed. This limited scope is partly a programmatic issue because currently GIS, especially vector-based GIS, rarely embraces multivariate methods. In addition, multivariate statistics require a solid understanding of matrix algebra, which is generally covered by more advanced textbooks.

In Chapter 2, we will review univariate descriptive statistics. In Chapter 3, we will focus on bivariate statistics, that is, the relationship between two variables. In Chapter 4, we will review the basic concepts of probability theory and hypothesis testing. In statistics, we often compute statistics from the observations. These statistics capture certain distributional characteristics of the set of observations. Then we formulate a null hypothesis, usually in a format in which a certain characteristic of the observed data is not significantly different from the characteristic of a known theoretical distribution or another observed distribution. Then we compare the statistic derived from the observations with the measurement of a known distribution or the statistic from another set of observations. Depending on the difference from the comparison, we decide to reject or not to reject the null hypothesis.

If we know the distribution characteristics of the observations or if we make an assumption about the distribution, the statistics used are called **parametric statistics** because the distributions can be described by parameters. This is the case when the data are measured in interval or ratio scales. When the data are measured in nominal or ordinal scales, their distributional characteristics cannot be summarized by statistical parameters meaningfully. In those cases, we must use **nonparametric statistics.** In Chapter 2, we will discuss only parametric statistics. When comparisons of two variables are addressed in Chapter 3, some nonparametric statistics will be employed.

REFERENCES

Sui, D. Z. 2004. Tobler's First Law of Geography: a big idea for a small world. *Annals of the Association of American Geographers* 94(2): 269–277.

Tobler, W. 1970. A computer movie simulating urban growth in the Detroit region. *Economic Geography* 46: 234–240.

CHAPTER 2

DISTRIBUTION DESCRIPTORS: ONE VARIABLE (UNIVARIATE)

Every scientific inquiry of a phenomenon starts by collecting information to describe the phenomenon. There are many ways to describe a phenomenon, such as verbal descriptions or graphic drawings. Using a quantitative or statistical description is probably one of the most effective approaches because it provides concise and objective information that helps us understand the phenomenon or make decisions about how to deal with it.

For instance, when studying the economic status of people in a poor inner-city neighborhood, we may want to find a number of representative households in that neighborhood as the subjects to study. These selected households are our observations. For the purpose of discussion here, let's use n to represent the number of selected households. If we measure their well-being by using annual household income as an indicator, then that income will be the variable and the n figures representing their levels of annual household income will be our data. To study these data, we normally start by calculating the descriptive statistics that will help us understand how these income figures are distributed. We will measure the central tendency of the income levels and find out how clustered or dispersed the income figures are.

2.1 MEASURES OF CENTRAL TENDENCY

Income is a quantity that is measured at interval/ratio scale, as a household could have negative income if it owes money, though we normally will not tabulate such a statistic in a report. Therefore, a variable representing income figures is on a continuous scale. Given the set of income figures from the households in a poor neighborhood, x_i, $i = 1, 2, \ldots, n$ we may wish to obtain an impression of the overall level of income in that neighborhood. Determining the overall level may allow us to compare the condition of this neighborhood with that of another neighborhood. Often this overall, or general, condition can be reflected by the "central location" of the set of values along a numeric value line from negative infinity to positive infinity—or the measures of **central tendency.** The set of central tendency measures can provide overall representative values of the variable.

For example, the average household income of the poor neighborhood gives us an idea of the economic status of households in the area. Using the average household income to represent all income figures in that neighborhood allows us to quickly obtain an overall impression of its economic status. Similarly, we can compute the average household income of another neighborhood and compare it with the original neighborhood to assess the difference in their economic conditions.

In surveying students at George Mason University to determine their means of commuting to school, we found that most of them drive. These specific commuting choice data, that is, different commuting means and the numbers of students selecting these means of transportation, are in nominal scale. The most common or frequent value is "driving," which is the mode of the data. When one attempts to develop an overall impression of housing conditions in a neighborhood, one may choose to use the middle price of all houses as the indicator. Choosing the middle value can avoid the pitfall of using an indicator that will be affected by extreme housing values, such as the value of a mansion or a shack in a middle-class neighborhood.

The concept of central tendency is applied in everyday life. People use average Scholastic Aptitude Test (SAT) scores of freshman classes to compare how well a college does over the years or to compare it with other colleges. We use phrases such as the *typical temperature* of a region or the *typical speed* on a specific road or highway to provide an overall impression of the situation. The description of the situation often relies on measures of central tendency from data describing the situation. Several popular measures of central tendency exist. Each of them possesses different characteristics, and some are preferable to others given the characteristics of the data. When the statistical properties of the data are not known, it is desirable to use multiple measures of central tendency to obtain a comprehensive understanding of the data.

2.1.1 Mode

Mode is the simplest measure of central tendency. It is the value that occurs most frequently in a set of data. Consequently, that value is also known as the **modal value.** For categorical or nominal data, the category that has the most observations or the highest frequency is regarded as the **mode.** If two or more categories have the highest frequency, then we have a **bimodal** or

multimodal distribution. When working with ordinal data, the mode is usually the rank shared by two or more observations. If no rank has more than one observation, then there is no mode in the data. Multimodal situations may also appear in a set of ordinal data.

Measured at interval or ratio scale, the data may not have a modal value because no one value may occur more than once. Alternatively, researchers may choose to degrade or simplify interval or ratio data into ordinal (weak or strong ordering) or even nominal scale by assigning individual data values to different ranks or different categories that are defined by ranges of the original interval/ratio data values. In Table 2.1, we selected 17 countries from South America and listed only their total population counts in 1990, their area, and their population density (number of people per square mile). We also classified these countries into three categories—high, medium, and low—according to their population density levels. This assignment process essentially degraded the population density ratio variable into an ordinal variable. As a result, the downgraded data are in weak ordering because the three categories are in the orders of high, medium, and low population densities. The population density variable is a ratio variable; therefore, a negative population density value is not interpretable.

In Table 2.1, none of the population density levels occurs twice or more. But when we examine the "Category" column, the mode of the category variable is "Low," as this category appears 10 times. The other two categories, "Medium" and "High," appear three and four times, respectively.

TABLE 2.1 Population, Area, and Density of 17 Countries in South America

Country	Population (1990)	Area (sq mi)	Population Density	Category
Argentina	33,796,870	1,073,749	31.48	Low
Aruba	67,074	71	949.68	High
Bolivia	7,648,315	420,985	18.17	Low
Brazil	151,525,400	3,284,602	46.13	Low
Chile	13,772,710	286,601	48.06	Low
Colombia	34,414,590	440,912	78.05	Medium
Ecuador	10,541,820	99,201	106.27	Medium
French Guiana	130,219	32,359	4.02	Low
Grenada	95,608	142	675.00	High
Guyana	754,931	81,560	9.26	Low
Netherlands Antilles	191,572	311	616.84	High
Paraguay	4,773,464	154,475	30.90	Low
Peru	24,496,400	500,738	48,92	Low
Suriname	428,026	56,177	7.62	Low
Trinidad and Tobago	1,292,000	1,989	649.56	High
Uruguay	3,084,641	68,780	44.85	Low
Venezuela	19,857,850	353,884	56.11	Medium

Figure 2.1 is a horizontal bar graph of the population densities of the 17 countries in descending order. Note that this figure is a **bar graph,** not a **histogram.** Both histograms and bar graphs can graphically show the numeric distribution of data. The first difference between them is that bar graphs are more appropriate for nominal or ordinal data, while histograms are better for interval or ratio data. The second difference is that bar graphs can be used to show the values of individual observations (like Figure 2.1) or the frequencies of observations in categories, while histograms only show frequencies of observations in interval or ratio scale in different data intervals. In Figure 2.1, it is obvious that most of the selected countries have relatively low population densities, along with a few countries with relatively high population densities.

2.1.2 Median

The median is another popular measure of central tendency. In a set of data, the **median** is the middle value after all values are sorted in ascending or descending order. Another way to describe the median is that when the data

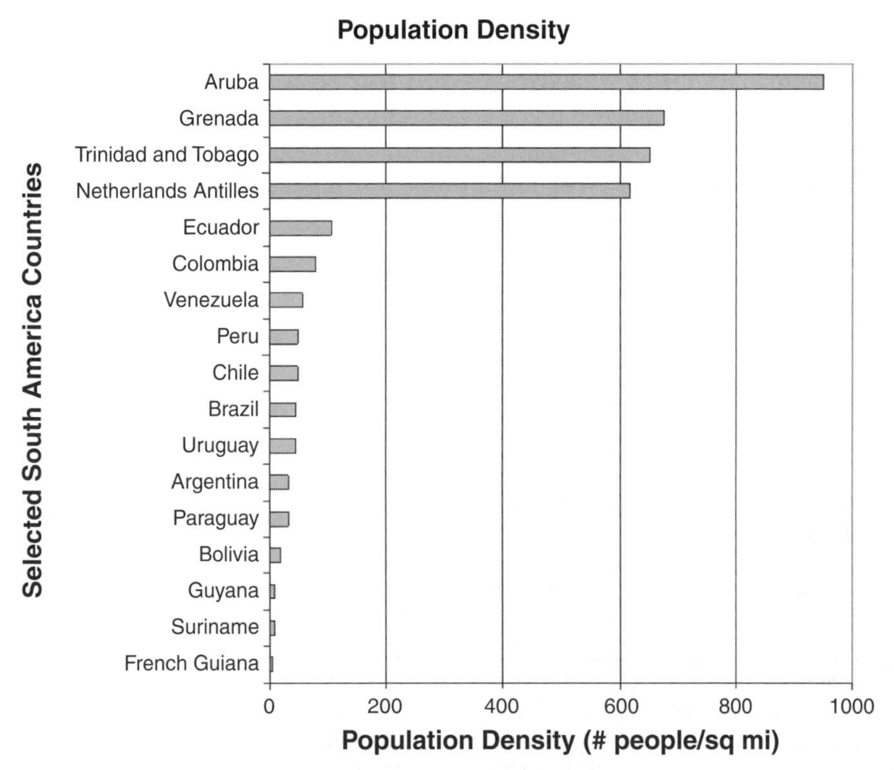

Figure 2.1 A bar graph of population density of 17 countries in South America.

values are ordered (either in ascending or descending order), half of them will be smaller than the median and half will be larger than the median.

Let us use x_i to represent the ith rank value of x after the variable is sorted and n to represent the total number of observations. The median can then be defined in two ways according to the value of n. If n is an odd number, then the median will be the $x_{(n+1)/2}$ or the $[(n + 1)/2]$-th value in the ordered data or the value in the middle of the ordered value list. If n is even, then we do not really have a value in the middle of the list. The two values closest to the middle are $x_{n/2}$ and $x_{(n/2)+1}$. Then we can define the median as the average of these two values closest to the middle, or $(x_{(n/2)} + x_{(n/2+1)})/2$.

To find the median of the population densities variable in Table 2.1 with 17 countries, we first sort the table by population densities to derive Table 2.2. Because we have an odd number of observations (17), the median will be the ninth value, which is $(17 + 1)/2 = 9$. In Table 2.2, the ninth-ranked population density is that of Chile, with 48.06 persons per square mile. In this case, eight countries have population densities lower and eight countries have population densities higher than that of Chile. If one more country is included in the region with a relatively low population density, then we will have 18 observations. To find the median in this hypothetical case, we must first identify the two values closest to the middle. They are the 9th and 10th observations, which are 46.13 for Brazil and 48.06 for Chile. The median is then the average of these two values, which is $(46.13 + 48.06)/2 = 47.09$.

In general, a median can be found in a set of data measured at the interval or ratio scale. One very desirable property of the median is that it is not

TABLE 2.2 Sorted Population Densities of 17 Countries in South America

Country	Population (1990)	Area (sq mi)	Population Density	Category
French Guiana	130,219	32,359	4.02	Low
Suriname	428,026	56,177	7.62	Low
Guyana	754,931	81,560	9.26	Low
Bolivia	7,648,315	420,985	18.17	Low
Paraguay	4,773,464	154,475	30.90	Low
Argentina	33,796,870	1,073,749	31.48	Low
Uruguay	3,084,641	68,780	44.85	Low
Brazil	151,525,400	3,284,602	46.13	Low
Chile	13,772,710	286,601	48.06	Low
Peru	24,496,400	500,738	48.92	Low
Venezuela	19,857,850	353,884	56.11	Medium
Colombia	34,414,590	440,912	78.05	Medium
Ecuador	10,541,820	99,201	106.27	Medium
Netherlands Antilles	191,572	311	616.84	High
Trinidad and Tobago	1,292,000	1,989	649.56	High
Grenada	95,608	142	675.00	High
Aruba	67,074	71	949.68	High

severely affected by extremely large or extremely small values, the so-called outliers, in the data set. This is because the median is defined by its relative position in the ordered sequence of data values. The extreme values will not affect the location of the middle value along the value line. For instance, the first value line in Figure 2.2 has nine numbers, and therefore the median is the number in the (9 + 1)/2-th or fifth position, that is, 5. The second value line also has nine numbers, but the last one, 18, can be regarded as an outlier because it is very different or far from the rest. But even with this much larger value, the median of the second value line is still 5. This is because the value in the fifth position is not affected by the extreme value in this case.

2.1.3 Mean

The most common measure of central tendency is probably the **mean,** or average. Using the income of a poor neighborhood as an example again, if we have n observations, each with an observed value, x_i, then the simple **arithmetic mean** is defined as

$$\bar{x} = \frac{\sum_{i=1}^{n} x_i}{n}. \tag{2.1}$$

A variable with an over bar such as $\bar{x}$ (read as "X bar") usually refers to the mean of the variable. In this simple calculation of an arithmetic mean, the variable x is first summed through all observations i (from 1 to n). The sum is then divided by the number of observations, n, to derive the mean.

In the South America example, to figure out the mean of population densities among the 17 countries, we must first add up all density values:

$$\sum_{i=1}^{n} x_i = 4.02 + 7.62 + 9.26 + \ldots + 949.68 = 3420.91.$$

The values are summed to 3420.91. When this sum is divided by the total number of observations, 17, the mean is 3429.91/17 = 201.23. This mean

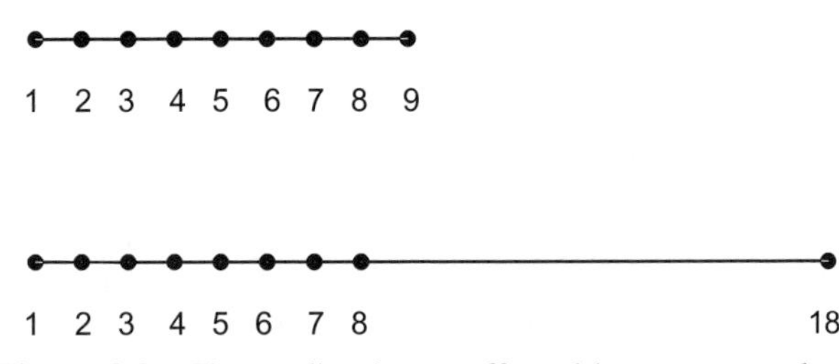

Figure 2.2 The median is not affected by extreme values.

suggests that, on average, we can expect a population density of 201.23 people per square mile among these South American countries.

If the 17 countries are the only ones that we are concerned about, then they constitute the entire set of observations that we want to know about population densities. Using statistical jargon, the 17 countries constitute the *population* of our study. The mean of 17 population densities provides an overall representative view of how densely these South American countries are populated.

Alternatively, if we regard the 17 countries as just representatives of countries all over the world, then the 17 countries constitute a sample of all countries in the world. In that case, the population densities of all countries will be the *population* data and the 17 population densities are the *sample* data.

The **sample mean,** or the mean of the sample, is often a good representative of the **population mean**—provided that the sampling process is unbiased and the sample size is sufficiently large. If countries in South America are used to represent the entire world, we may assume that the range and variation among population densities in the world are similar to those observed in the South American countries. We know this may not be true, but we may need to work under this assumption if we have budgetary or data constraints.

If we compute the mean of this sample, we naturally want to use the sample mean (Equation 2.1) to represent the population, that is, these and those other countries that are not counted. This is known as the sample mean, denoted $\bar{x}$. If we can enumerate the entire population, assuming that the population size is N, then we can compute the population mean, usually denoted as μ (pronounced as "mu"), in the following manner:

$$\mu = \frac{\sum_{i=1}^{N} x_i}{N}. \tag{2.2}$$

In studying geographic phenomena, deciding when to use sample means and when to use population means depends on the case at hand. This is a difficult issue that often causes great debate, as discussed in Costanzo (1983), Summerfield (1983), and other works. At this point, however, we will use the mean as a descriptive tool for central tendency and leave the concepts of sample mean and population mean for further discussion in later chapters on inferential statistics.

The arithmetic mean is a reasonably simple measure of central tendency in most situations. It indicates an overall or average magnitude of the data values in a variable along the continuous value line. It possesses several noteworthy properties that should be discussed.

Based on Equation 2.1, the sum of the differences from the mean should be zero, or formally, $\Sigma_{i=1}^{n} (x_i - \bar{x}) = 0$. This means that some data values are above the mean (positive mean difference) and some are below it (negative mean difference), but the sum of all negative differences from the mean should be the same as the sum of all positive differences from the mean. The magnitudes of the differences should offset each other and result in 0 for summing all differences from the mean. In statistics $\Sigma_{i=1}^{n} (x_i - \bar{x})$ is known as the **first moment.** Consequently, the arithmetic mean is called the **first moment statistic.**

Another interesting property of the arithmetic mean is that the sum of the squared differences between the mean and all data values is smallest when compared to the sum of the squared differences between any given value and all data values. Formally,

$$\min\left[\sum_{i=1}^{n} (x_i - p)^2\right] = \sum_{i=1}^{n} (x_i - \bar{x})^2. \tag{2.3}$$

That is, if one chooses any number p other than $\bar{x}$ on the left hand side in Equation 2.3, the sum will be larger than the one obtained by using $\bar{x}$ in Equation 2.3. Similarly $\Sigma_{i=1}^{n} (x_i - \bar{x})^2$ is called the **second moment;** we will discuss some **second moment statistics** in later sections.

Even though an arithmetic mean is a good representative of a set of data values, it has some shortcomings. When there are outliers in a dataset, the arithmetic mean is often not a good measure of central tendency. This is because the arithmetic mean is sensitive to these extreme values. In Figure 2.2, one of the two value lines has its mean appropriately representing the central tendency, while the mean of the other line is pulled to the right by only one much larger value.

Just as an arithmetic mean is derived by dividing the sum of data values by the number of data values, a **geometric mean** can be calculated by taking the nth root of the product of n data values. Formally, the geometric mean, $\dot{x}$, of a set of data values, x_i, $i = 1, 2, \ldots, n$, can be derived by

$$\dot{x} = \sqrt[n]{\prod_{i=1}^{n} x_i} \tag{2.4}$$

or

$$\dot{x} = \left(\prod_{i=1}^{n} x_i\right)^{1/n} = \sqrt[n]{\prod_{i=1}^{n} x_i}\,. \tag{2.5}$$

For example, if we take the first set of nine values in Figure 2.2, the product of them is 362,880. Taking the ninth root of this product gives us 4.147. The geometric mean of the nine values will be 4.147. The geometric mean for the set of nine values with an outlier is 4.479, just slightly larger than that of the set without any outlier. This example shows that the geometric mean may be more robust, or less sensitive, to handle outliers.

2.1.4 Grouped or Weighted Mean

The above calculations of the mean for interval or ratio data are appropriate if all values are counted individually. But if data values are grouped into classes (or categories), then all data within each group are represented by one value as the overall value in that class. In this case, the calculation of the mean will be slightly different from the calculation of simple arithmetic means. A mean derived from the grouped data is called a **grouped mean** or a **weighted mean.**

In a set of data tabulated into classes or groups, the value midway between the lower and upper bounds of each class is the representative value, or the **midpoint** value for that class. If x_i is the midpoint of the ith class with f_i as the number of data values in that class (frequency), the weighted mean, $\bar{x}_w$, can be calculated as

$$\bar{x}_w = \frac{\sum_{i=1}^{k} f_i x_i}{\sum_{i=1}^{k} f_i}, \tag{2.6}$$

where k is the number of classes or groups.

In other words, the representative value in each class must first be multiplied by the frequency of that group and then summed with the products of other groups to make up the numerator in Equation 2.6. This sum is then divided by the total number of observations, which is the sum of all class frequencies in the denominator in Equation 2.6.

A question may be: why is this step of grouping data values even needed when all data values can be easily keyed in to a computer and have the computer calculate the arithmetic mean—even if we have hundreds of thousands of observations? Before computers were widely available, the grouped mean provided a feasible way of estimating the overall mean of very large datasets. In this procedure, data are divided into groups according to their values. A value from each group, typically the midpoint value between the lower and upper bounds, is used to represent the value of all data in the group. When calculating the group mean, the number of observations in each group is used as the weight, that is, f_i in Equation 2.6. That is why the grouped mean is often referred to as the weighted mean.

Using the South American countries example again, the countries are now tabulated in Table 2.3 by the ordinal variable "Category" based upon their population density levels. The frequencies of the three categories of low, medium, and high are reported (f_i). The ranges of the classes are also shown

TABLE 2.3 Frequencies of 17 South America Countries in Three Density Categories

Category	Frequency (f_i)	Range	Midpoint (x_i)	$f_i x_i$
Low density	9	4.01–50	27	243
Medium density	3	50.01–200	125	375
High density	5	200.01–1000	600	3000
Total (Σ)	17			3618

in the table. The midpoint value of each class (x_i) was computed. The products of the midpoint values and the corresponding frequencies ($f_i x_i$) are recorded in the last column.

After midpoints are multiplied by their corresponding frequencies, the sum of the products is 3618. This sum is divided by 17, the sum of the total frequency (or the n in the ungrouped data), to obtain the weighted mean of 212.824. Note that this weighted mean is much higher than the arithmetic mean ($\bar{x}$) of 201.23.

It is expected that the simple arithmetic mean and the weighted mean will be different, and the arithmetic mean should be a more accurate value to represent the central tendency of the data because the weighted mean is based on less precise weak-ordered data. The discrepancy between the arithmetic and weighted means is mainly a function of how the midpoint value in each class represents the overall values in each class. In the example of population densities in the South American countries, the medium category is represented by the value of 125, which is higher than all the population density values of the countries in that category (Table 2.2). As a result, the population densities in that group based on the midpoint value are overestimated.

2.2 MEASURES OF DISPERSION

While the mean is a good measure of the central tendency of a set of values in terms of where they are generally located along the value line, it captures no information about how the values are concentrated or scattered around the mean and along the value line. Specifically, we know the overall or average value by referring to the mean, but we do not know if most of the data are very far from or close to the average. In other words, we do not know if most data values are very similar to the mean, or quite different from the mean, or if they are just scattered around. To explain the need for more information beyond what we can obtain from the mean, let us examine two data series, x_a and x_b, by comparing their means while observing how the distributions of the two data series differ:

x_a: 1, 3, 5, 7, 9, 11, 13

$$\bar{x}_a = \frac{1 + 3 + 5 + 7 + 9 + 11 + 13}{7} = 7.$$

x_b: −11, −5, 1, 7, 13, 19, 25

$$\bar{x}_b = \frac{-11 - 5 + 1 + 7 + 13 + 19 + 25}{7} = 7.$$

Both data series give the mean of 7, but apparently they have very different distributions. The x_b series even has negative values. If we only rely on the two means and have no further information on the two series, we might speculate that the two data sets are very similar to each other simply because the means are identical. However, by comparing the actual data values across the two series, we see that the first series has a relatively narrow range, while the second series has a wider range of values. The mean tells us only the central location of the data; it provides no information on the distribution of the data. These two very different data series have the same mean, but we obviously need more than their means to tell their full stories.

To better understand how values in a data series distribute, a number of descriptive statistics have been developed. In the above two data series, the range (maximum − minimum) was used to distinguish the series in terms of their magnitude of dispersion. Other sets of statistics commonly used with a range are minimum, maximum, and percentiles. Beyond these, we can use **mean** deviations, standard deviations, skewness, and kurtosis. These statistics provide additional information on the degree of dispersion among values in a data series, often in reference to the mean. Some of them also indicate in which direction the values cluster. Using some or all of these dispersion statistics will help to provide a comprehensive description of the distribution of the data. They will also allow us to compare different data series.

2.2.1 Range, Minimum, Maximum, and Percentiles

Given a data series, the smallest value is the **minimum** and the largest value is the **maximum**. Using these values, the **range** of a set of data value is basically the difference between the maximum and the minimum or the maximum minuses the minimum. If a data series has a minimum value of 1 and a maximum value of 13, the range in this data series will be 12, because 13 − 1 = 12. Taking another data series as an example, if the minimum is −11 and the maximum is 25, then the range is 25 − (−11) = 36.

Intuitively, the greater the range in a data series, the more dispersed the data are. If the two previous data series, x_a and x_b, have the same number of data values, we would feel confident in stating that the distribution of the data values in the second series is more dispersed than that at the data values in the first data series. However, remember that the range only indicates how far the values are scattered; it offers no information on how they are scattered. In general, data values rarely distribute evenly over a range. They may cluster around a certain value, skew toward larger values, or distribute randomly between the maximum and minimum values.

To provide additional information on how data values are distributed, we often identify some "milestones" within the range of data and compare them between data series. A common practice in statistics is to identify the data values at the 25th, 50th, and 75th percentiles. In other words, the data are arranged in a sorted (ascending) sequence first. Then, counting ¼, ½, and ¾ of the total observations from the minimum, the values of the corresponding observations are, respectively, the 25th, 50th, and 75th percentiles.

What are **percentiles?** Let's revisit the concept of median in the previous section. The median is the data value halfway between the minimum and maximum values in a sorted sequence of data values. In other words, the median is basically the 50th percentile. That is, half (50%) of the data values are smaller than the median and the other half are larger. In the same way, 25% of the data values are smaller than the 25th percentile and 75% of the data values are larger. Of course, the 75th percentile is the value that is greater than 75% of the values in the dataset.

Percentiles are the corresponding data values that have certain percentages of the data smaller than these values. In practice, we must first rank the data values in ascending order. Then we find the value that has approximately 25%, for instance, of the data smaller than it. We say "approximately" because a data series may not always have a large number, say 100 or more values in it. We often need to find the value in the ordered series that is closest to the desired percentile.

Using the previous example again, data series x_a and data series x_b both have 7 as the median, but each has different values for the 25th (3 for series x_a and -5 for series x_b) and 75th percentiles (11 for series x_a and 19 for series x_b). Of course, 3 and 11 in series x_a are closer than -5 and 19 in series x_b, supporting our previous claim that series x_b is more dispersed than series x_a.

x_a: 1, 3, 5, 7, 9, 11, 13
x_b: -11, -5, 1, 7, 13, 19, 25

2.2.2 Mean Deviation

Unlike the dispersion measures discussed so far using one or a few data values in the x_a series, the **mean deviation** takes into account all data values. It is calculated by summing all the differences that individual data values have from the mean and then dividing this sum by the number of observations. To avoid positive differences offsetting negative differences, all differences are calculated as absolute values, converting all negative differences to positive differences.

Conceptually, the mean deviation is the average difference from all values to their mean, hence the name mean deviation. Formally, the mean deviation, $\overline{D}$, is defined as

$$\overline{D} = \frac{\sum_{i=1}^{n} |x_i - \overline{x}|}{n}. \tag{2.7}$$

For data series x_a, the mean deviation is

$$\begin{aligned}\overline{D}_a &= \frac{|1-7| + |3-7| + |5-7| + |7-7| + |9-7| + |11-7| + |13-7|}{7} \\ &= \frac{6+4+2+0+2+4+6}{7} \\ &= 3.4286.\end{aligned}$$

For data series x_b, the mean deviation is

$$\begin{aligned}\overline{D}_b &= \frac{|-11-7| + |-5-7| + |1-7| + |7-7| + |13-7| + |19-7| + |25-7|}{7} \\ &= \frac{18+12+6+0+6+12+18}{7} \\ &= 10.285.\end{aligned}$$

The symbol $|x_i - \overline{x}|$ in Equation 2.7 denotes the absolute difference between the ith value of x and the mean, $\overline{x}$. So the procedure is to first take the absolute difference between each x value and the mean of x. Next, add up all the absolute differences to obtain the value for the numerator in Equation 2.7. Then divide the sum of the absolute differences by the number of values or observations to obtain the average absolute deviation from the mean, or the mean deviation. For data series x_b, the mean deviation is 10.285, which is much greater than the 3.4286 of series x_a, indicating a more dispersed distribution in series x_b.

The mean deviation is easy to calculate and understand. It provides a convenient and concise summary of the dispersion of a set of data values using the information from all values. In this manner, each value influences the mean deviation. A value that is closer to the mean contributes less to the mean deviation. A value farther away from the mean contributes more. However, this measure concerns only the distance of the values from the mean, not the direction, since negative differences are also converted to positive values. Values such as 3 and 11 in series x_a contribute equally to the mean deviation even though they are on opposite sides of the mean. With this measure, the presence of extreme values (very large or very small) will have an important bearing on the calculation and eventually will affect the value of this statistic.

2.2.3 Variance and Standard Deviation

In calculating the mean deviation, we use the absolute values of the mean differences as the deviations. If we do not use the absolute values, we will end up having a mean deviation of zero or a value very close to zero (due to rounding error). This is because, according to the definition of an arithmetic mean, positive deviations ($x_i - \bar{x} > 0$) will be offset by negative deviations ($x_i - \bar{x} < 0$). This is true even if data values have a wide range. In that case, the calculated mean deviation will not reflect the extent to which data values deviate from the mean. Consequently, absolute values are used.

Another way to avoid the offsets caused by adding positive and negative deviations from the mean together is to square all deviations from the mean before summing them. Since all squared values are positive, we will be able to accumulate the squared differences without any offsets. A statistic calculated in this way is called **variance,** σ^2, defined formally as

$$\sigma^2 = \frac{\sum_{i=1}^{n} (x_i - \mu)^2}{n}, \tag{2.8}$$

where σ^2 is the variance, μ is the population mean, and all other notations are defined previously. The equation for calculating the variance actually computes the average of the squared deviations from the population mean.

The notation σ^2 is for the variance when the entire population is enumerated or it represents the variance of the population. When a sample is enumerated, the sample variance is defined as

$$S^2 = \frac{\sum_{i=1}^{n} (x_i - \bar{x})^2}{n - 1}. \tag{2.9}$$

The major differences between the population variance in Equation 2.8 and the sample variance in Equation 2.9 are the use of population mean μ and the denominator. Instead of dividing the sum of squared differences with n for the population variance, the sample variance divides the sum of squared differences with $n - 1$ to adjust for sampling error, as the sample mean $\bar{x}$ is used in estimating this sample variance.

The above formulas are easy to understand and help explain the concept of variance quite effectively. However, they are not efficient in computation. A more computationally efficient formula for the variance calculation is

$$\sigma^2 = \frac{\sum_{i=1}^{n} x_i^2}{n} - \mu^2 \tag{2.10}$$

for the population variance and

$$S^2 = \frac{\sum_{i=1}^{n} x_i^2}{n - 1} - \frac{\left(\sum_{i=1}^{n} x_i\right)^2}{n(n - 1)} \tag{2.11}$$

for the sample variance.

At first sight, Equations 2.10 and 2.11 seem to be much more complicated than Equations 2.8 and 2.9. However, by going through the procedures carefully, it becomes clear that Equations 2.10 and 2.11 operate more quickly. For instance, instead of calculating the differences, squaring the differences, and then summing the squared differences in Equation 2.9, the formula for S^2 in Equation 2.11 just needs to sum all squared x_i and square the sum of all x_i. This process will reduce rounding errors, thus yielding a more accurate and faster calculation of the variance.

Although variance measures the magnitude of dispersion in a dataset, it is not commonly used because of its large numeric value. The process of squaring the deviations often leads to large numbers that cannot be compared directly to the original data values. Therefore, the square root of the variance is often used to describe the dispersion of a data set. This measure is known as the root mean square deviation or simply the **standard deviation.** It is calculated by taking the square root of the variance,

$$\sigma = \sqrt{\frac{\sum_{i=1}^{n} (x_i - \mu)^2}{n}}, \tag{2.12}$$

for the population standard deviation or

$$S = \sqrt{\frac{\sum_{i=1}^{n} (x_i - \bar{x})^2}{n - 1}} \tag{2.13}$$

for the sample standard deviation.

By taking the square root of the variance, which is the averaged squared deviations, the statistic returns to the magnitude or scale of the original dataset and thus is used more often than the variance.

As an example, Table 2.4 shows the calculation of the variance and standard deviation of the population density values of the 17 South American countries. As reported in the previous section, the mean density is 201.23. Table 2.4 shows the calculation using the original formula and the computational formula. The variance is 88,432.30, which has a scale much larger than the population density values of the data. By taking the square root of this variance, the standard deviation is 297.38, which is at a scale comparable to that of the original values.

The standard deviation has other useful properties to help describe the distribution of the data values. If the distribution of data values closely resembles a normal distribution (which will be discussed in detail in Chapter 4), then we can establish several relationships between the distribution of the data and the value ranges bounded by certain standard deviations from the mean:

1. About 68% of the data values are within one standard deviation from the mean. That is, 68% of the values fall within an interval bounded by $\bar{x} - \sigma$ and $\bar{x} + \sigma$.

TABLE 2.4 Calculation of the Variance and Standard Deviation of the Population Density of the 17 South America Countries

Country	Population Density (x)	$x_i - \bar{x}$	$(x_i - \bar{x})^2$	x^2
Argentina	31.48	−169.75	28,816.56	990.71
Aruba	949.68	748.45	560,177.42	901,892.13
Bolivia	18.17	−183.06	33,511.82	330.06
Brazil	46.13	−155.10	24,055.38	2,128.17
Chile	48.06	−153.17	23,462.49	2,309.31
Colombia	78.05	−123.18	15,172.50	6,092.31
Ecuador	106.27	−94.96	9,018.01	11,292.63
French Guiana	4.02	−197.21	38,890.15	16.19
Grenada	675.00	473.77	224,455.67	455,621.66
Guyana	9.26	−191.97	36,853.97	85.68
Netherlands Antilles	616.84	415.61	172,730.02	380,489.14
Paraguay	30.90	−170.33	29,011.86	954.89
Peru	48.92	−152.31	23,198.15	2,393.23
Suriname	7.62	−193.61	37,485.11	58.05
Trinidad and Tobago	649.56	448.33	200,995.90	421,922.57
Uruguay	44.85	−156.38	24,455.38	2,011.33
Venezuela	56.11	−145.12	21058.65	3148.78
Σ	3,420.91	0	1,503,349.07	2,191,736.85
				$\sigma^2 = 88{,}432.3$
				$\sigma = 297.38$

2. About 95% of the data values are within two standard deviations from the mean. That is, 95% of the values fall within an interval bounded by $\bar{x} - 2\sigma$ and $\bar{x} + 2\sigma$.
3. About 99% of the data values are within three standard deviations from the mean. That is, 99% of the values fall within an interval bounded by $\bar{x} - 3\sigma$ and $\bar{x} + 3\sigma$.

2.2.4 Weighted Variance and Weighted Standard Deviation

Similar to the calculation of the weighted mean, a **weighted variance** and the associated **weighted standard deviation** can be derived from data representing values (or observations) grouped into classes. Adopting the same notations used earlier for calculating the weighted mean;

f_i is the frequency for the ith group or class,
x_i is the midpoint value in the ith group,
$\bar{x}_w$ is the weighted mean, and
k is the number of groups.

Then the weighted variance is defined as

$$\sigma_w^2 = \frac{\sum_{i=1}^{k} f_i(x_i - \bar{x}_w)^2}{\sum_{i=1}^{n} f_i}. \tag{2.14}$$

Note that in the numerator, the deviation from the weighted mean is squared before it is multiplied by its frequency or weight for each class or category. This intuitive, meaningful formula also has its computational counterpart for efficient computation. For more efficient computation of the grouped variance, the following formula should be used:

$$\sigma_w^2 = \frac{\sum_{i=1}^{n} f_i x_i^2 - \bar{x}_w^2}{\sum_{i=1}^{n} f_i}. \tag{2.15}$$

The above formulas are for the population variance. For the weighted sample variance, a slightly modified formula accounting for the sampling error is required:

$$S_w^2 = \frac{\sum_{i=1}^{n} f_i x_i^2 - \bar{x}_w^2}{\sum_{i=1}^{n} f_i - 1}. \tag{2.16}$$

Then the standard deviation for the grouped data is simply the square root of the weighted variance. Here, we do not distinguish the weighted mean between the sample and the population.

To further demonstrate the procedures for calculating the variance and standard deviation from grouped data, Table 2.5 shows the computation steps using the data of population densities of the 17 South America countries. Recall from Table 2.3 that the weighted mean is $\bar{x}_w = 212.824$. Using Equation 2.15, we only need to calculate a new column in the table, $\Sigma_{i=1}^{n} f_i x_i$, which is the sum of the products of each frequency and its corresponding midpoint value. In this manner,

$$\sigma_w^2 = \frac{\sum_{i=1}^{n} f_i x_i^2 - \bar{x}_w^2}{\sum_{i=1}^{n} f_i} = \frac{71436 - 45294.054}{17} = \frac{26141.946}{17} = 1537.7615.$$

Therefore,

$$\sigma_w = \sqrt{1537.7615} = 39.21.$$

As a result, the standard deviation of this set of group data is 39.21 and the variance is 1537.76.

Note that these weighted variance and standard deviation are very different from their unweighted counterparts. Overall, they are dramatically smaller than the unweighted statistics. The primary reason is that when observations are grouped into classes (low, medium, and high), the variations in the original data are removed, as the observations are now represented by the midpoint values instead. When one evaluates the midpoint values (27, 125, and 600), one finds that there is not much variability among them. Therefore, one can

TABLE 2.5 Categorized Density Data for the 17 South America Countries

Density Category	Frequency (f_i)	Midpoint (x_i)	x_i^2	$f_i x_i^2$
Low	9	27	729	6,561
Medium	3	125	15,625	46,875
High	5	600	3600	18,000
Total	17			71,436
Mean		$\bar{x}_w = \frac{\sum_{i=1}^{n} f_i x_i}{\sum_{i=1}^{n} f_i} = \frac{3618}{17} = 212.824$		
		$\bar{x}_w^2 = (212.824)^2 = 45294.054$		

expect the weighted variance and standard deviation to be smaller than their unweighted versions in general.

2.2.5 Coefficient of Variation

A major problem with the variance and standard deviation is that they are sensitive to the scale of the data. If we take the data series in x_a: 1, 3, 5, 7, 9, 11, 13, with a mean of 7, and then scale up the data in the series by 10, that is, multiplying each value in x_a by 10, we will obtain another data series y_a: 10, 30, 50, 70, 90, 110, 130, with a mean of 70. The standard deviation σ for x_a is 4, and for y_a it is 40, 10 times larger than that of x_a. But the difference in the two standard deviations is attributable to the difference in the scales or bases of the two data series. Therefore, it is misleading to conclude that the data in y_a are more dispersed than the data in x_a by relying only on the measure of standard deviation. A more appropriate statistic is the **coefficient of variation** (CV), a modification of the standard deviation, which is defined as

$$CV = \frac{\sigma}{\bar{x}} \tag{2.17}$$

for the population or

$$CV = \frac{S}{\bar{x}} \tag{2.18}$$

for the sample.

Referring to the two data series, x_a and y_a, both have the same coefficient of variation of 0.5714 (4/7 or 40/70). In other words, the two data series have the same magnitude of variation, as measured by CV. Sometimes CV is multiplied by 100 to make it resemble a percentage for ease of comparison. For example, multiplying 0.5714 by 100 will give us 57.14 instead of expressing it in decimal form.

The mean describes where the center of the data is located, and the standard deviation indicates how much dispersion the data have. Together, they provide a preliminary summary of certain statistical properties of the data and offer a basis for understanding a dataset and for comparing multiple datasets.

2.3 ARCVIEW EXAMPLES

ArcView GIS allows users to calculate descriptive statistics, including the sum, count, mean, maximum, minimum, range, variance, and standard deviation. In general, these statistics provide a quick glimpse of the distribution of data in an attribute field. The example below outlines the steps used to calculate such statistics.

ArcView Example 2.1: Descriptive Statistics

Step 1 Start ArcView GIS

- Start ArcView GIS to create a new project with a new view.

We suggest that users start ArcView GIS with an empty view. Next, add data themes. Finally, load the extension needed for analysis.

Step 2 Add data theme

- Click the **Add Theme** button to start the process of adding data sets to ArcView view documents.
- Navigate to the data folder and add the `C:\Temp\Data\Ch2_data\ccdb2000.shp` shapefile. This shapefile contains a collection of census data for all counties in the United States.

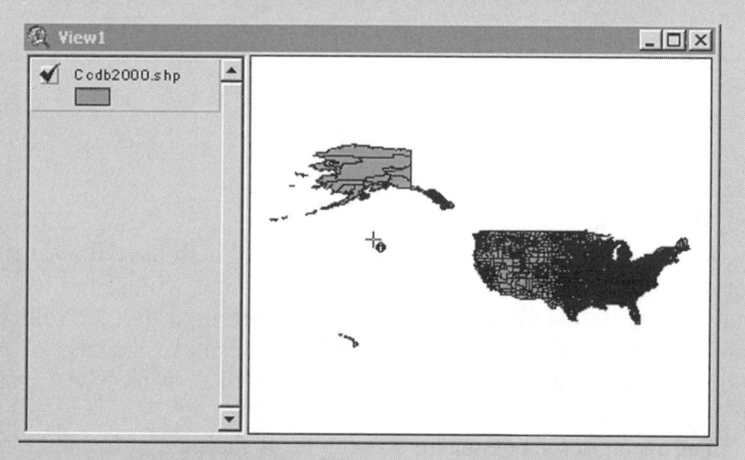

Step 3 Attribute fields

- To determine the content of the attribute field, add the `variables.dbf` as a table.
- First, click the **Tables** icon, Tables, in the **Project** window and then click the **Add** button.
- In the **Add Table** dialog box, navigate to the `C:\Temp\Data\Ch2_data` folder, select `variables.dbf`, and click **OK** to add it to the ArcView project.

- Scroll up and down in the added table to examine the attribute fields in the ccdb2000.shp data theme, as

variables.dbf

Col01-1	Land_area
COF01-2	POPULATION 1992
COF01-3	POPULATION RANK 1992
COF01-4	1990 POPULATION PER 1990 SQUARE MILES
COF01-5	POPULATION 1990
COF01-6	POPULATION 1980
COF01-7	POPULATION CHANGE 1980-1992
COF01-8	POPULATION, PERCENT CHANGE 1980-1992
COF02-9	POPULATION BY RACE, WHITE 1990 (1C)
COF02-10	POPULATION BY RACE, BLACK 1990 (1C)
COF02-11	POPULATION BY RACE, AMERICAN INDICAN, ESKIMO, OR ALEUT 1990 (1C)
COF02-12	POPULATION BY RACE, ASIAN OR PACIFIC ISLANDER 1990 (1C)
COF02-13	HISPANIC ORIGIN POPULATION (MAY BE OF ANY RACE) 1990 (1C)
COF02-14	HISPANIC ORIGIN, POPULATION PERCENT OF TOTAL POPULATION 1990 (1C)
COF02-15	POPULATION BY AGE, PERCENT UNDER 5 YEARS 1990 (1C)

- Close variables.dbf after examining its content.

Note that an additional listing of variables is available if variables.xls is opened with Excel, a spreadsheet program.

Step 4 Calculate descriptive statistics

- Click the title bar of the View document to make the View active.
- Use the **Open Theme Table** button, [button], to open the attribute table of ccdb2000.shp data theme.
- Click the title button of the field, `B1_pop01`, to select it. This field contains the 2001 (July) population counts for counties.

Attributes of Ccdb2000.shp

Shape	Fips	B1_Ind01	B1_pop01	B1_pop03	B1_pop05	B1_pop06
Polygon	27077	1297	4443	4522	4	4076
Polygon	53019	2204	7296	7260	3	6295
Polygon	53065	2478	40641	40066	16	30948
Polygon	53047	5268	39543	39564	8	33350
Polygon	53051	1400	11965	11732	8	8915
Polygon	16021	1269	9926	9871	8	8332
Polygon	30053	3613	18664	18837	5	17481
Polygon	30029	5098	76269	74471	15	59218
Polygon	30035	2995	13125	13247	4	12121
Polygon	30101	1911	5151	5267	3	5046
Polygon	30051	1430	2096	2158	2	2295

- From the **Field** menu, select **Statistics** . . . . The descriptive statistics are calculated.

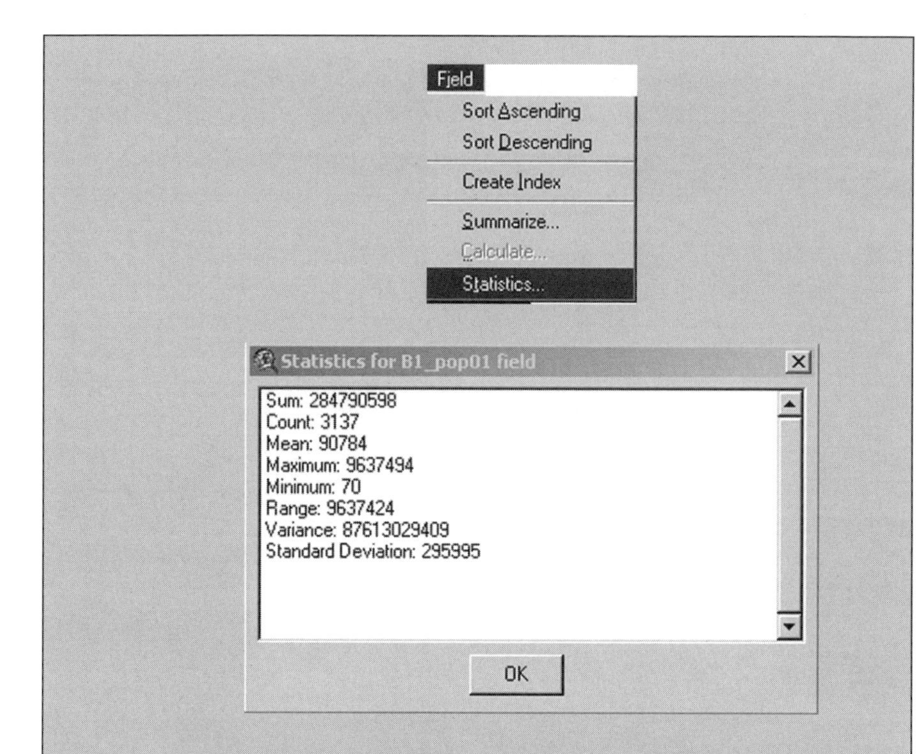

Step 5 Additional variables

- In the attributes of ccdb2000.shp table, scroll to the end of the list to find the last variable, `B13_gvt07`, local government finance for 1996–1997, direct general expenditure.
- Click the title button of the field `B13_gvt07` and then click **Field/Statistics.** The descriptive statistics are calculated:

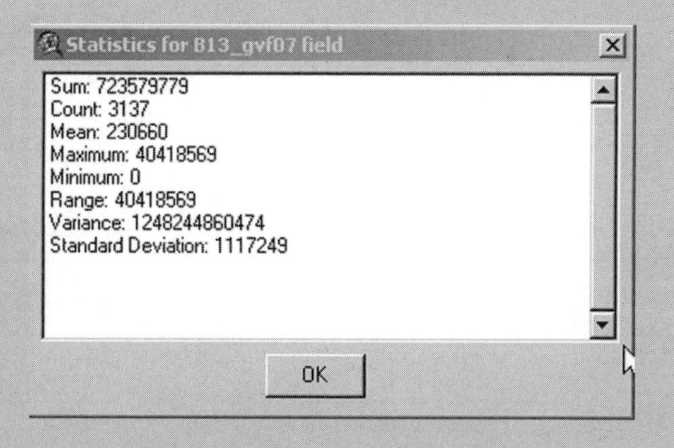

Step 6 End

2.4 HIGHER MOMENT STATISTICS

For a given set of values, the mean indicates its central tendency and the standard deviation shows the spread of data over the numeric range around the mean. While the mean is regarded as the *first moment statistic* in statistical jargon, the variance and standard deviation are regarded as *second moment statistics* because they are based upon the squared deviation from the mean, $(x_i - \bar{x})^2$. Different moment statistics capture different statistical characteristics of the data. The first moment statistic reflects the overall location or the average of all values, while the second moment statistics summarize the distribution characteristics of the data.

2.4.1 Skewness and Kurtosis

In this section, we will discuss two higher moment statistics. The first statistic is a *third moment statistic,* **skewness,** which measures the directional bias of the distribution of the data in reference to the mean. The other is a *fourth moment statistic,* **kurtosis,** which measures the sharpness of the distribution compared to a rather standard distribution, the *normal distribution.* By combining the information provided by the mean, standard deviation, skewness, and kurtosis, we can obtain a detailed, comprehensive description of the distribution characteristics of the set of values.

To understand how the skewness and kurtosis of a distribution are calculated, it is necessary to discuss the concept of **frequency distribution.** The frequency distribution of a given set of data values is often shown in a histogram in which one axis (often the horizontal or x-axis) shows the numeric range of the data and the other axis (often the vertical or y-axis) shows the frequency, or the number of values that falls into each interval.

Figure 2.3 includes five examples of frequency distribution with different levels of skewness and kurtosis. At the top of the figure is a symmetrical distribution with low skewness and medium kurtosis. The two distributions in the middle row have opposite directional biases but similar (moderate) kurtosis levels. The two bottom distributions are symmetric but have different kurtosis levels.

Figure 2.4 shows the frequency distribution of population densities of the 17 selected South American countries. Note that the population densities are highly positively skewed because most countries have relatively low population density levels.

Before we explain the calculation of skewness and kurtosis statistics, let us discuss further the issues related to the histogram. The histograms shown in Figures 2.3 and 2.4 are sometimes confused with bar graphs. Both bar graphs and histograms can be used to show frequency distributions, and both types of graph can have vertical or horizontal bars. The major difference between them is the variable used to tabulate the frequency distribution. For

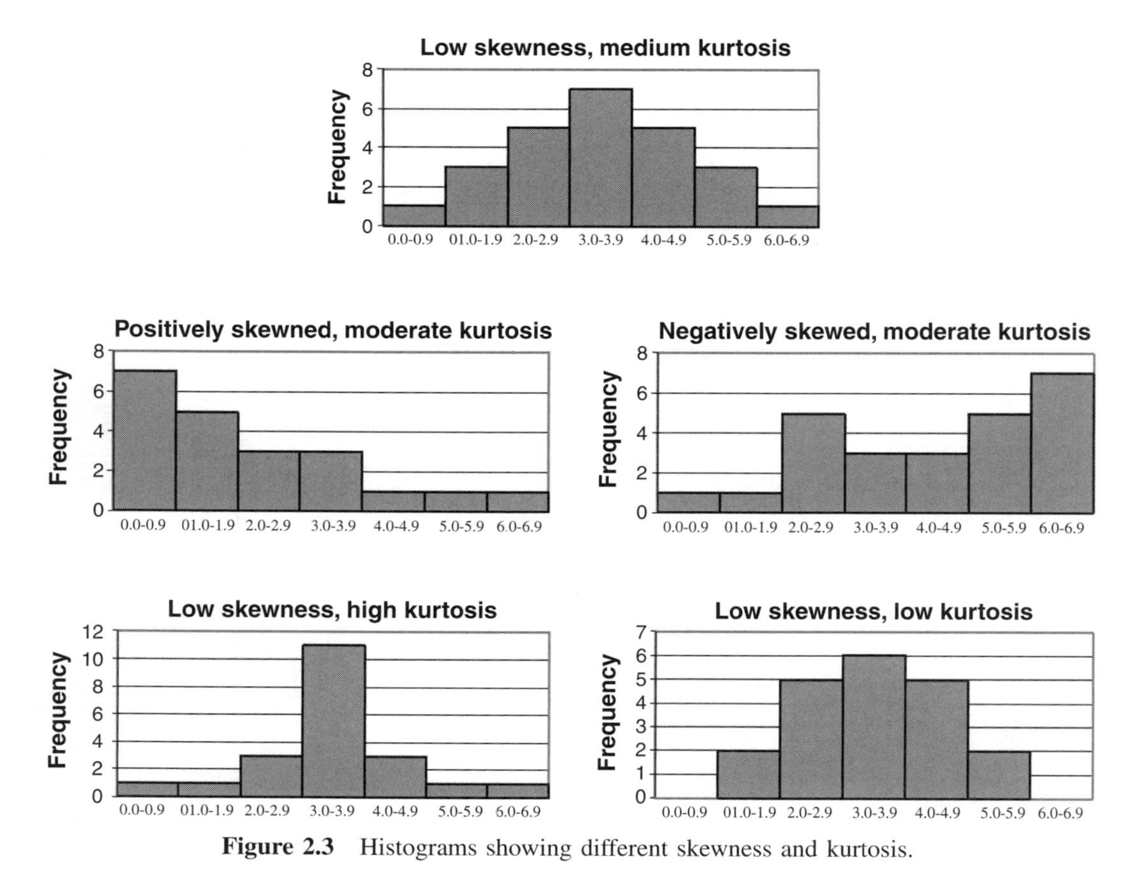

Figure 2.3 Histograms showing different skewness and kurtosis.

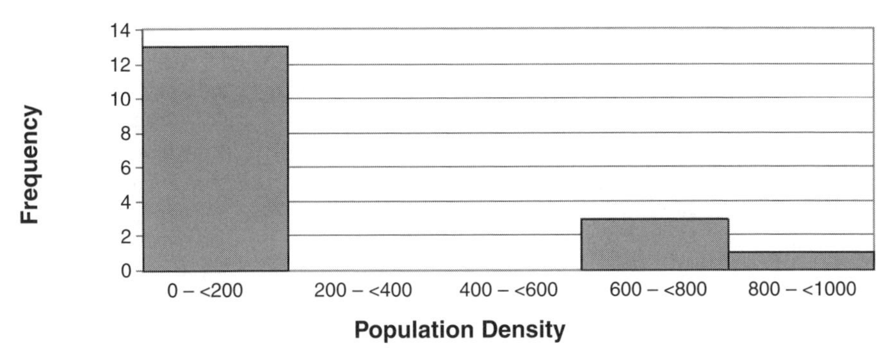

Figure 2.4 Histogram of the population density of 17 South America countries.

histograms, the variable should be in interval or ratio scale or continuous in nature. For bar graphs, the variable can be individual observations or classes of observations, most likely in nominal scale.

In our South American countries example, population density is the variable used to tabulate the frequency distribution. Consequently, using a histogram to show how population densities distribute is appropriate. If the variable is in nominal or ordinal scale, such as different modes of transportation or different land use types, then a bar graph should be used. In that case, the axis showing the variable should have discrete classes. For bar graphs, there is no definite ordering of the classes. The bars can be rearranged in different order without altering the information content.

Two other issues related to the construction of a histogram are the number of classes and the class definition. In principle, we should construct as many classes as possible if our goal is to display the distribution of data at a level that is as detailed as possible. In fact, using more classes will depict the distribution more accurately. But using too many classes, or, to the extreme, using individual observations may not show the overall distribution, but pick up noises.

The histograms in Figures 2.3 and 2.4 possess another essential characteristic. Even though the variables are continuous in scale, we create discrete classes to simplify the representation of the distribution. These classes should be *mutually exclusive* but *collectively exhaustive* (refer to the discussion on this topic in Chapter 1). The range in each class of the histogram should be the same or similar. To ensure that no observation will be counted twice, the upper bound of one class cannot be the same as the lower bound of the next class. For example, in the South American countries example, we cannot use 200 as the upper bound of the first class because it will overlap with the lower bound of the second class. On the other hand, to ensure that no observation will be left out between any two classes, we need to eliminate any gap. One method is to use the scheme in Figure 2.4. Another method is to use more decimal values in the categories (such as 299.9999 instead of 299.9) to ensure that no observation will be left out.

Skewness measures the extent to which the bulk of data values in a distribution cluster to one side or the other of the mean. When most values are less than the mean, the distribution is said to be *positively skewed.* Alternatively, a *negatively skewed* distribution has the bulk of the values greater than the mean. Specifically, skewness, *Sk*, is calculated by

$$Sk = \frac{\sum_{i=1}^{n} (x_i - \bar{x})^3}{n\sigma^3}, \qquad (2.19)$$

where all terms are defined as before. Notice that the measure of skewness is based on the cubic value of the standard deviation, or σ (in the denominator). It is also based on the cubic value of the deviations from the mean or mean deviation (in the numerator). Therefore, skewness is considered a third moment statistic.

Because σ is always positive, the denominator of the skewness formula is always positive. The numerator, however, can be positive or negative. If most of the values are smaller than the mean, the numerator will be positive (this is somewhat counterintuitive) and thus the distribution will be positively skewed with the tail pointing to the right. If most values are larger than the mean, the numerator will be negative and the skewness measure will be negative with the tail pointing to the left. The skewness of a symmetrical distribution is zero.

Using the dataset of population density in South American countries as an example, Table 2.6 show that

$$\sum_{n=1}^{17} (x - \bar{x})^3 = 629{,}070{,}579.33.$$

Because $\sigma = 297.38$, as derived earlier, the skewness is calculated as follows:

$$Sk = \frac{\sum_{i=1}^{n} (x_i - \bar{x})^3}{n\sigma^3} = \frac{629{,}070{,}579.33}{17 * 297.38^3} = \frac{629{,}070{,}579.33}{447{,}078{,}919.60} = 1.41.$$

The distribution is thus highly skewed positively, as most countries have population density levels smaller than the mean density.

Skewness is most useful when it is used to compare distributions. For example, two distributions can have similar standard deviations, but their skewness levels can be quite different if they have different directional biases.

TABLE 2.6 Calculations of Skewness and Kurtosis Using the Population Density of 17 South American Countries

Country	Population Density (x)	$(x - \bar{x})$	$(x - \bar{x})^2$	$(x - \bar{x})^3$	$(x - \bar{x})^4$
Argentina	31.48	−169.75	28,816.56	−4,891,739.16	830,394,355.90
Aruba	949.68	748.45	560,177.42	419,264,799.74	313,798,745,066.81
Bolivia	18.17	−183.06	33,511.82	−6,134,752.97	1,123,042,277.23
Brazil	46.13	−155.10	24,055.38	−3,730,939.73	578,661,130.47
Chile	48.06	−153.17	23,462.49	−3,593,860.17	550,488,478.69
Colombia	78.05	−123.18	15,172.50	−1,868,899.19	230,204,867.31
Ecuador	106.27	−94.96	9,018.01	−856,379.29	81,324,526.06
French Guiana	4.02	−197.21	38,890.15	−7,669,366.00	1,512,443,935.59
Grenada	675.00	473.77	224,455.67	106,339,808.86	50,380,348,476.26
Guyana	9.26	−191.97	36,853.97	−7,075,000.64	1,358,215,383.06
Netherlands Antilles	616.84	415.61	172,730.02	71,787,983.04	29,835,661,325.64
Paraguay	30.90	−170.33	29,011.86	−4,941,552.79	841,688,226.34
Peru	48.92	−152.31	23,198.15	−3,533,295.17	538,153,983.09
Suriname	7.62	−193.61	37,485.11	−7,257,517.65	1,405,133,109.49
Trinidad & Tobago	649.56	448.33	200,995.90	90,111,623.02	40,399,353,561.93
Uruguay	44.85	−156.38	24,455.38	−3,824,385.60	598,065,705.26
Venezuela	56.11	−145.12	21,058.65	−3,055,946.98	443,466,774.02
Σ	3,420.91	0.00	1,503,349.07	629,070,579.33	444,505,391,183.14

With kurtosis, we can measure the extent to which values in a distribution are concentrated in one part of the frequency distribution. If the bulk of the values in a distribution are highly concentrated over a range of values, then the distribution is said to be *very sharp* or to have a *high peak.* Alternatively, a flat distribution is one without a significant concentration of values within a narrow range of the distribution. Kurtosis, K, is usually computed as follows:

$$K = \frac{\sum_{i=1}^{n} (x_i - \bar{x})^4}{n\sigma^4} - 3. \tag{2.20}$$

Kurtosis is based on the fourth power of the mean deviation. Therefore, it is considered a fourth moment statistic. Empirically, we know that normal distributions tend to have a K value of about 3. By subtracting 3 from the first part of the equation, a symmetrical, bell-shaped distribution will have a kurtosis equal to 0. This way, a relatively sharp distribution will have a positive kurtosis value (*leptokurtic*) and a relatively flat distribution will have a negative kurtosis value (*platykurtic*).

Using the selected South American countries data again, Table 2.5 provides $\Sigma_{i=1}^{17} (x_i - \bar{x})^4 = 444{,}505{,}391{,}183$. Given that $\sigma^2 = 88{,}432.30$,

$$
\begin{aligned}
K &= \frac{\sum_{i=1}^{n} (x_i - \bar{x})^4}{n\sigma^4} - 3 \\
&= \frac{444{,}505{,}391{,}183}{17 \times 88{,}432.3^2} - 3 \\
&= 3.34 - 3 \\
&= 0.34.
\end{aligned}
$$

This gives the distribution a relatively low to moderate level of sharpness.

2.5 ARCVIEW EXAMPLES

Using the charting functions in ArcView GIS, bar graphs can be generated easily. Note that ArcView does not provide the function to generate histograms in a simple step. To generate bar graphs, we suggest following the steps outlined below.

ArcView Example 2.2: Statistical Charts

Step 1 Data preparation

- Also using ccdb2000.shp as in ArcView Example 2.1, navigate to the data folder, `C:\Temp\Data\Ch2_data`, and add the shapefile, `ccdb2000.shp`, to a View document with the **Add Theme** button.

Step 2 Calculate descriptive statistics

- Open the attribute table of ccdb2000.shp by using the **Open Theme Table** button, .
- In the Attributes of Ccdb2000.shp, scroll to find the variable, `B5_inc02`, Median Household Income, 1989. Click its title button and calculate its descriptive statistics by selecting **Statistics** from the **Field** menu. The result should be similar to this:

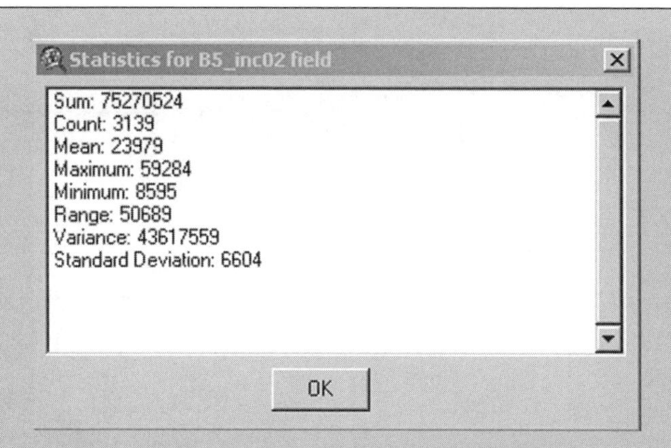

- Using the mean and standard deviation, we can structure these income intervals:
  8595 ~ 17375,
  17376 ~ 23979,
  23980 ~ 30583, and
  30584 ~ 59284.
  These intervals are
  (min, $\bar{x}$ − *st.dev.*),
  ($\bar{x}$ − *st.dev.*, $\bar{x}$),
  ($\bar{x}$, $\bar{x}$ + *st.dev.*), and
  ($\bar{x}$ + *st.dev.*).

Step 3 Reclassify income data

- From the **Table** menu, select **Start Editing** to make the table editable.
- From the **Edit** menu, select **Add Field** to add a new field to the table. Name the new field `Inc_class` with number type, width of `8` with `0` decimal places. Click **OK** to finish.
- Click the **Query Builder** button, , to bring up the Attribute of Ccdb2000.shp dialog box. In the **Fields** list, double click `B5_inc02`. Then click [<] and, finally, type `17375` to complete this condition:

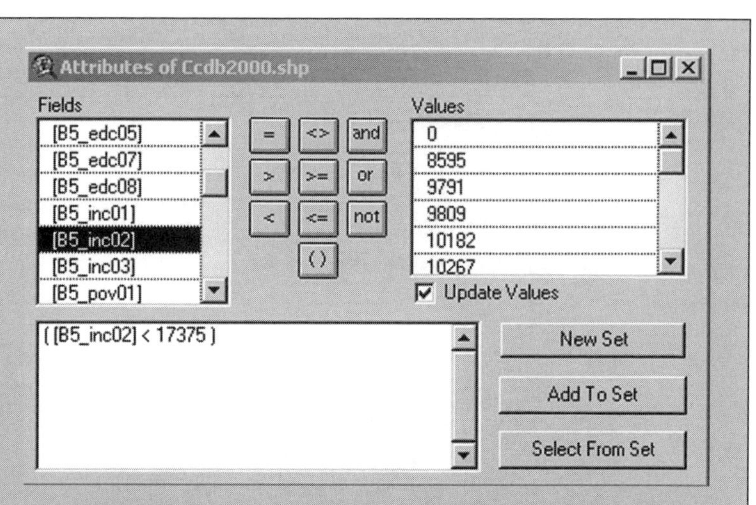

- Click the **New Set** button to make the selection.
- In the attribute table of Ccdb2000.shp, click the **Promote** button, , to bring all selected records to the top of the table.
- Make sure that the title button of Inc_class is clicked.
- Click the **Calculate** button, , to bring up the **Field Calculator** dialog box. Notice that the left size of the calculation equation is defined as `[Inc_class] =`.
- In the text box below `[Inc_class] =`, type `1` and then click the **OK** button.

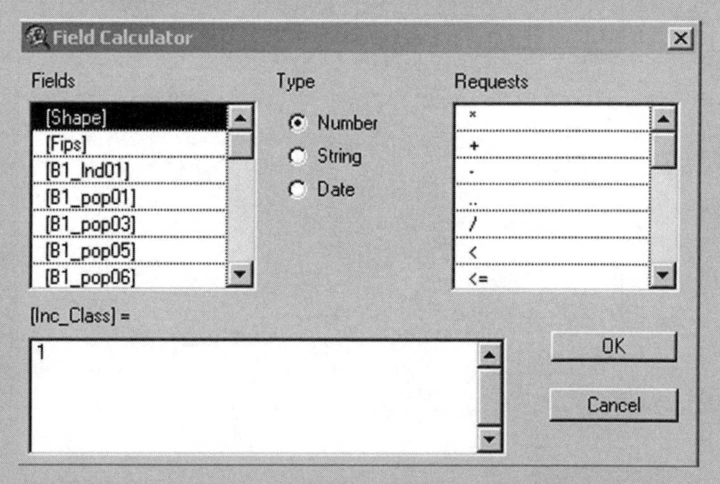

- The selected records will have the value of 1 entered in the field of Inc_class.
- Click the **Select None** button, , to clear the selection.

- Use the **Query Builder** to select records whose values of B5_inc02 fall between 17375 and 23979. Next, use the **Field Calculator** to assign a value of 2 to Inc_class of these selected records. Finally, use the **Select None** to clear the selection before making the next selection.

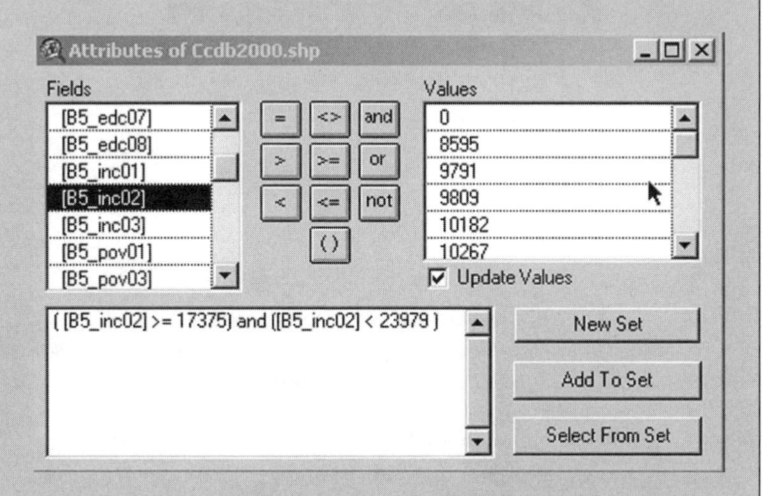

- Repeat this process to assign the value of 3 to Inc_class of those records whose values of B5_inc02 fall between 23979 and 30583.

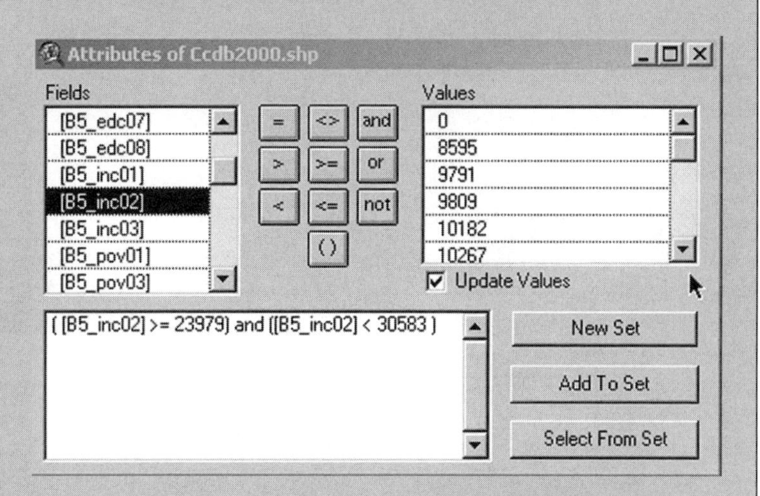

- Similarly, assign 4 to Inc_class for those with values of B5_inc02 falling between 30584 and 59284.

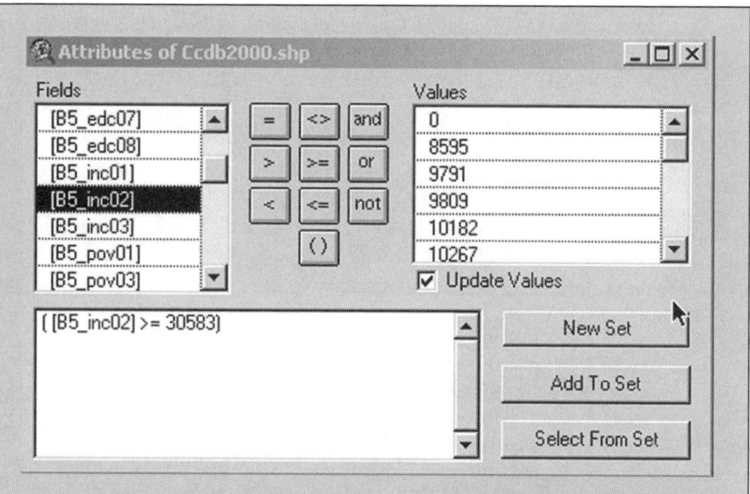

- Click the **Select None** button to clear the selection.
- Once you have completed the calculation, choose **Stop Editing** from the **Stop** menu. Click **Yes** to save the edits.

Step 4 Generate frequencies

- With the attribute table opened and the Inc_class field clicked, click the **Summarize** button, Σ, to invoke the **Summary Table Definition** dialog box.
- In the Field list, select `B5_inc02`.
- In the **Summarized** by list, select `Average`.
- Click the **Add button** to add `Ave_B5_inc02` to the list on the right. In the resulting table, sum1.dbf, this field will hold the averaged household income 1989 for each income class as defined in Step 3.

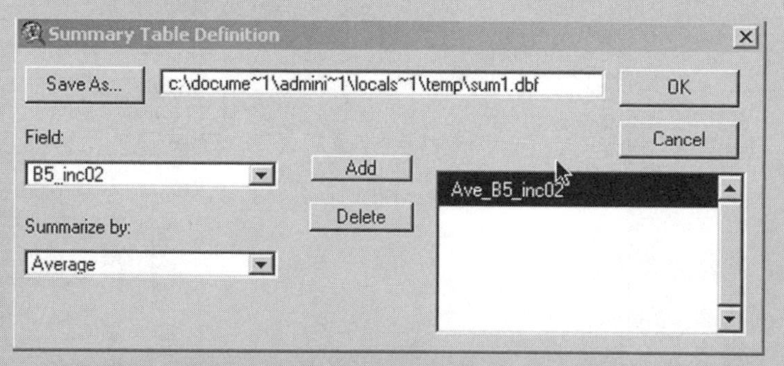

- Click **OK** to create sum1.dbf.
- Use **Query Builder** again to select records whose Inc_class attributes have nonzero values, or [Inc_class] > 0. After the selection, the content of sum1.dbf should be similar to this:

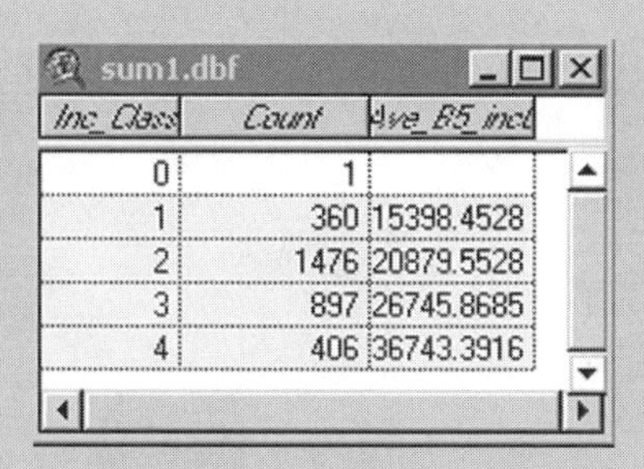
sum1.dbf

Inc_Class	Count	Ave_B5_inc
0	1	
1	360	15398.4528
2	1476	20879.5528
3	897	26745.8685
4	406	36743.3916

Step 5 Charting frequencies

- With sum1.dbf opened, click the **Create Chart** button, [icon].
- In the **Chart Properties** dialog box, type `County Household Income Distribution, 1989` in the **Name** textbox.
- Next, select `Count` from the **Fields** list and then click the **Add** button to add it to the **Groups** list.
- From the **Label series using:** list, select `Ave_B5_inc02` to be the labeling field.

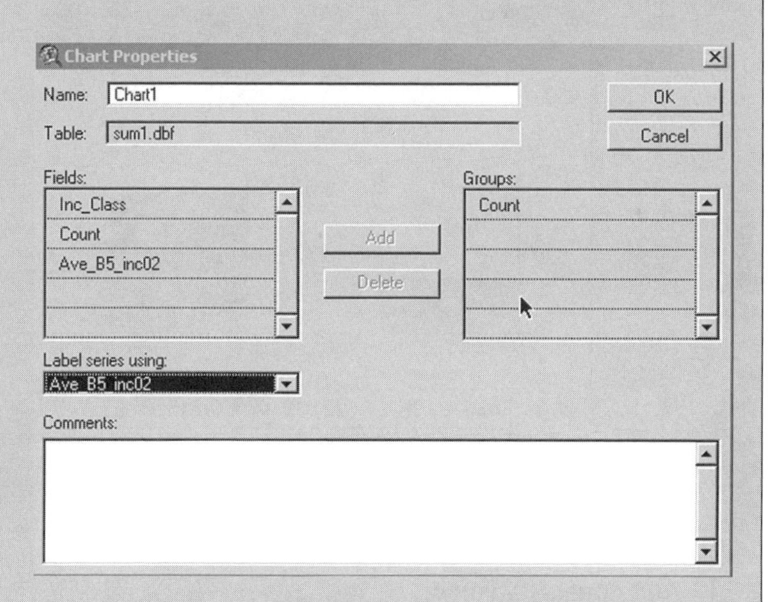

- Click **OK** to generate the chart.

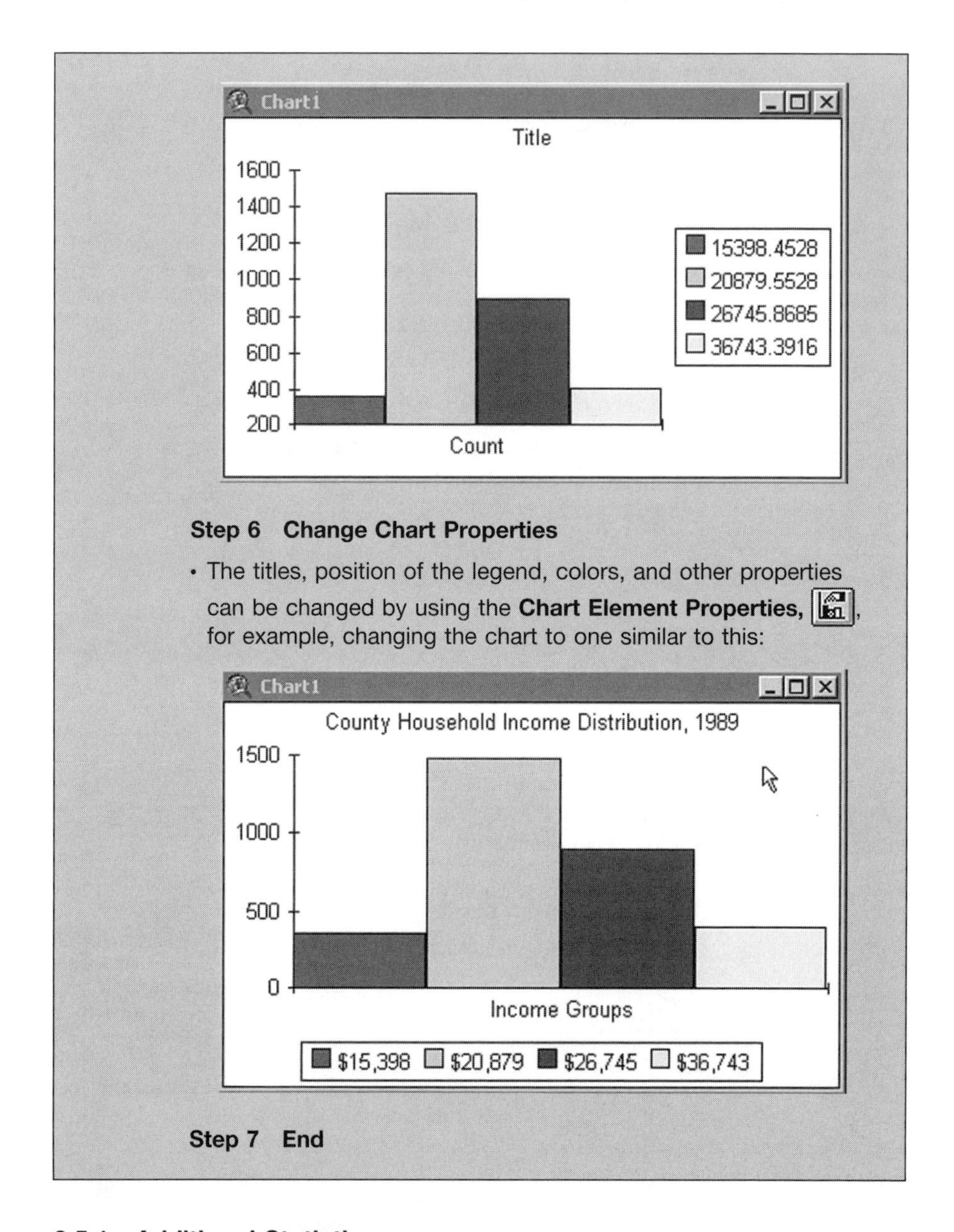

Step 6 Change Chart Properties

- The titles, position of the legend, colors, and other properties can be changed by using the **Chart Element Properties,** , for example, changing the chart to one similar to this:

Step 7 End

2.5.1 Additional Statistics

In cases where higher moment statistics are needed, the Ch2 extension can be loaded for calculating additional statistics. Similar to using the Ch1 extension, the Ch2.avx needs to be pasted into the EXT32 folder within the ArcView installation folder. Once the extension is loaded, additional statistics can be calculated easily, as in the example below.

ArcView Example 2.3: Additional Statistics

Step 1 Start ArcView GIS and add a data theme

- Start ArcView GIS and create a new project with an empty view.
- Use the Add Theme button to add `C:\Temp\Data\Ch2_data\ccdb2000.shp.`

Step 2 Load the Ch2 Extension

- From the **File** menu, select **Extensions.**
- Click the check box beside `Ch2 Extension` and then click **OK** to load this extension.

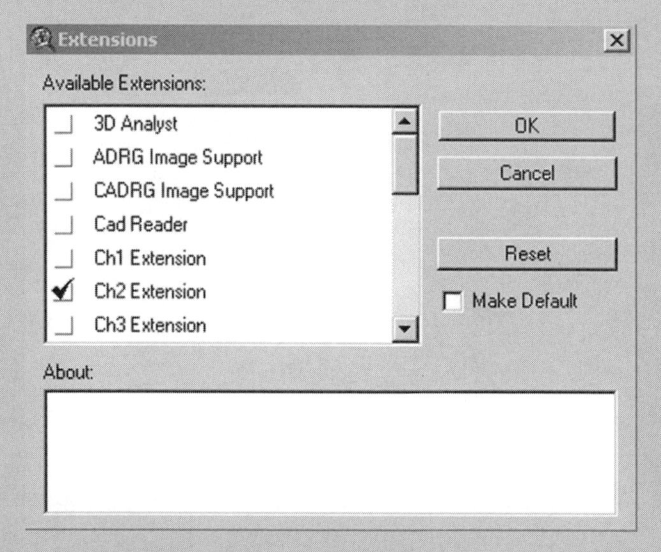

- Once it is loaded, a new menu item, **Ch.2,** appears.
- Note that the **Ch.2** menu appears in both user interfaces of View documents and Tables documents.

Step 3 Calculate univariate statistics

- In this data theme, not all counties have data for all attribute fields. Therefore, it is necessary to select those records with valid data values in the attribute that is to be used as the variable for calculation.
- Click the **Query Builder** button, , to bring up its dialog box. Select records with the condition ([B1_pop01] > 0).
- Use the **New Set** button to create a new set of selected records.

- From the **Ch.2** menu, choose **Statistics-univariate** to invoke the **Univariate Statistics** dialog box.
- In the **Select a Theme** list, click `ccdb2000.shp` to select it.
- Click the **Show Variables** button to refresh the list of variables in the **Select a Variable** list.
- In the **Select a Variable** list, click to highlight `B1_pop01`.
- As an example, click all check boxes to request output for all statistics. At this point, leave the check box of **Weighted by a variable** unchecked.

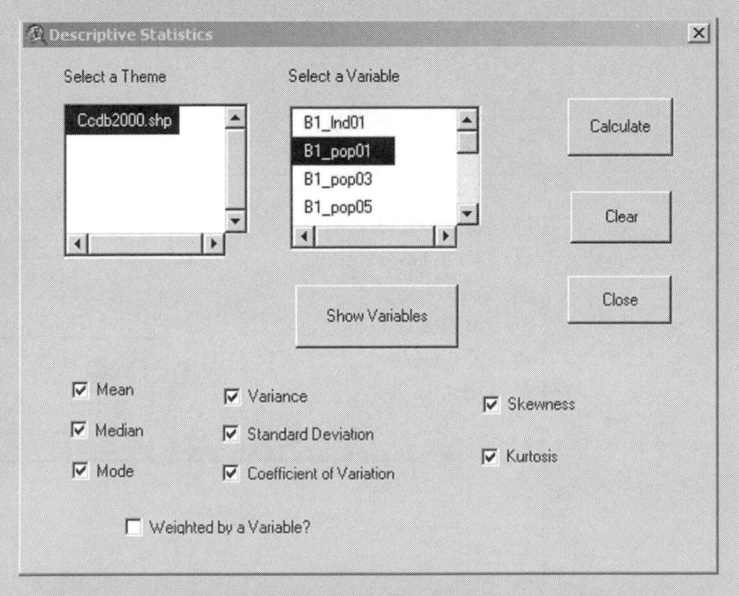

- Click the **Calculate** button to perform the calculation. Since the calculation involves over 3000 records for many statistics, it will take some time to complete. When it is completed, the statistics may be similar to this:

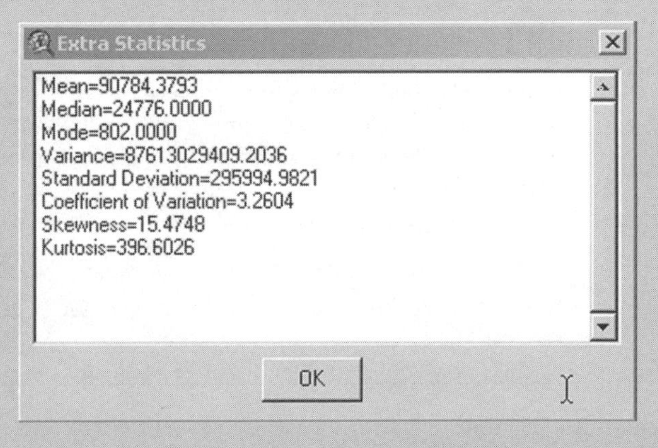

- When prompted to save the results in a text file, click **No** to skip that step.

Step 4 Weighting a variable

- Using **Query Builder,** select counties with the condition B1_pop01 > 0.
- Now we can calculate extra statistics using another variable as a weighting variable.
- From the **Ch.2** menu, select **Statistics-univariate** to invoke the **Univariate Statistics** dialog box.
- Select `Ccdb2000.shp` as the data theme and then click the **Show Variables** button to refresh the list attributes in this data theme. From the list of attributes, select `B1_pop01` as the variable.
- Check all statistics and the check box of **Weighted by a Variable.**

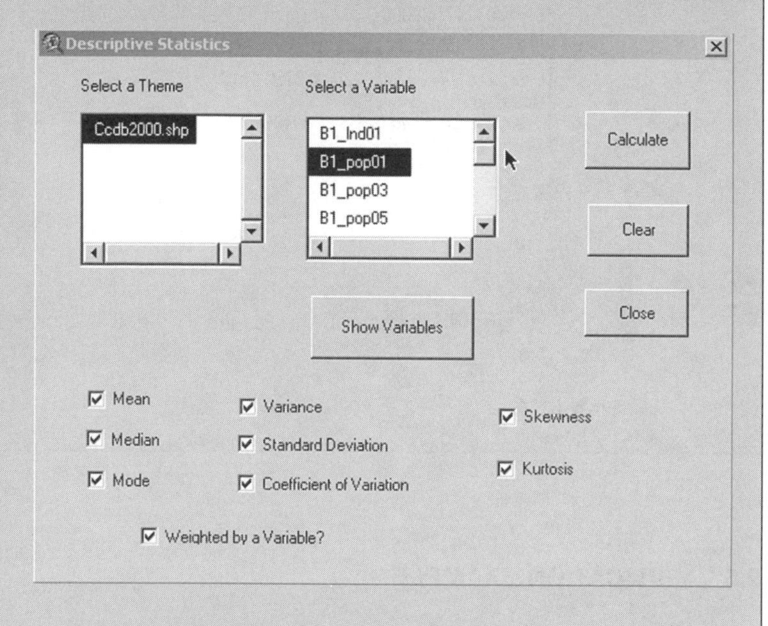

- Click the **Calculate** button to proceed.
- In the **Choose weight** dialog box, select `B5_inc02` as the weighting variable. Again, the calculation will take some time, much more than without a weighting variable.

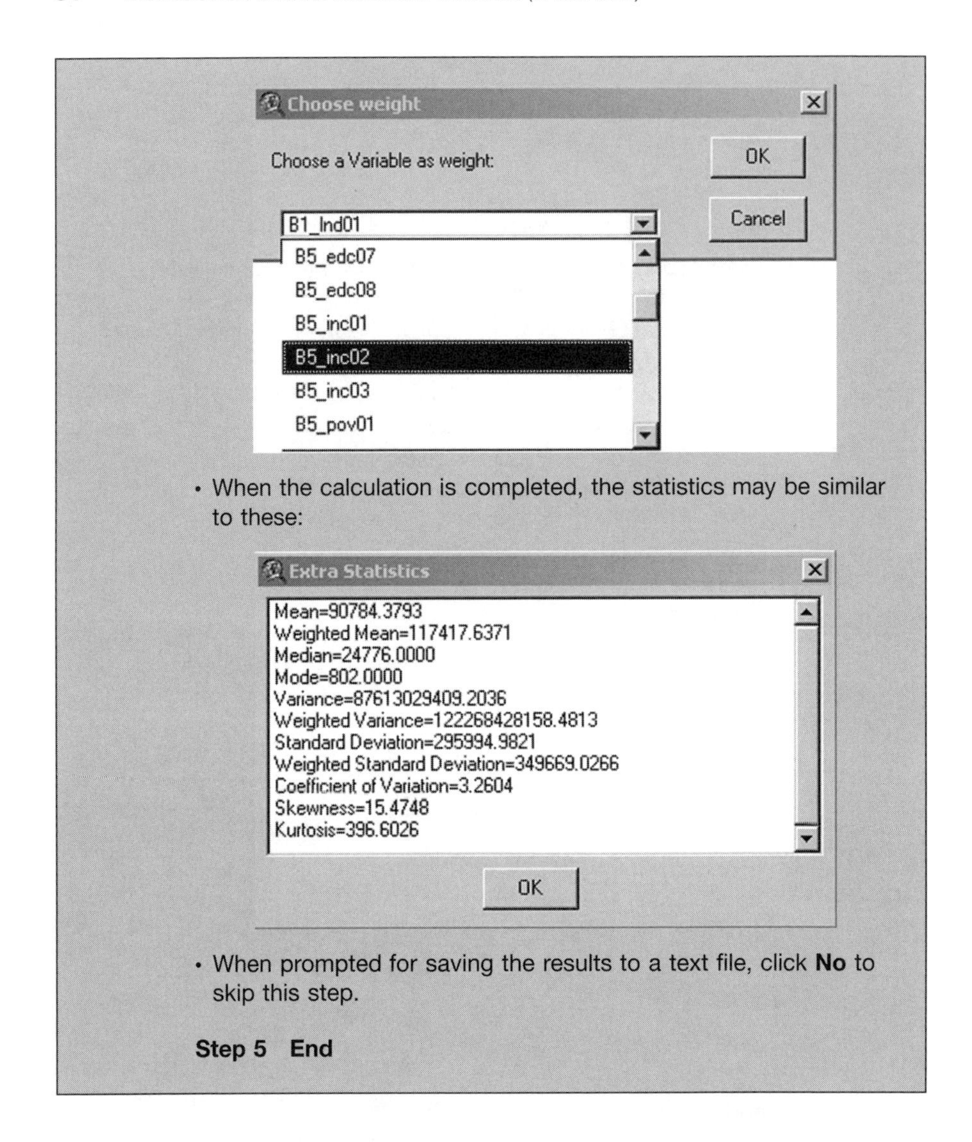

- When the calculation is completed, the statistics may be similar to these:

- When prompted for saving the results to a text file, click **No** to skip this step.

Step 5 End

2.6 APPLICATION EXAMPLE

Below is an application in which the impacts of MAUP, as discussed in Chapter 1, will be demonstrated. In general, MAUP refers to the inconsistencies of analytical results due to the use of data tabulated at different geographic scale levels (scale effect) or according to different zoning configurations (zoning effect). The inconsistencies among analytical results apply to many statistics.

In the following application, we will use income data from the 2000 Census for Washington, D.C., at the census tract and block group levels to demonstrate the MAUP impacts on descriptive statistics covered in this chapter. In both data themes, three income variables are extracted from Census 2000 Summary File 3: Median Household Income (Medhouinc), Median Family Income (Medfaminc), and Per Capita Income (Percapinc). Though all three variables are income-related, they reflect different aspects of the income statistics. In addition, the variable Pop2000, which represents the population count in the 2000 Census, is included for each areal unit (tract or block group).

ArcView Example 2.4: The MAUP effects on Descriptive Statistics

Step 1 Load Ch1.avx and Ch2.avx extensions

- Load Ch1 and Ch2 extensions into an ArcView Project as the discussed in the previous sections.

Step 2 Add data themes

- In the project window (Untitled), click the **View** icon, , and then the **New** button to create a new View document.
- While the View document is opened, use the **Add Theme** button, , to bring in

  `C:\Temp\Data\Ch2_Data\dc_trt_inc.shp` and
  `C:\Temp\Data\Ch2_Data\dc_bg_inc.shp`.
- The first theme is the census tract level data, and the second theme is the block group level data. Both themes include the population counts and the three income variables.

Step 3 Visualizing the MAUP effects

- Double-click the legend of the theme `dc_trt_inc.shp` to invoke the **Legend Editor.** Choose `Graduated Color` for **Legend Type.** In the **Classification Field,** choose `Percapinc` (for Per Capita Income). Click the **Apply** button to implement the legend classification.
- Double-click the legend of the theme `dc_bg_inc.shp` to invoke the **Legend Editor** again. Choose `Graduated Color` for **Legend Type** and choose `Percapinc` as the **Classification Field.** In the **Value** column, change the value ranges to match those used in the `dc_trt_inc.shp` theme. Click the **Apply** button to implement the legend classification.
- Two maps with the same classification values will be developed, as

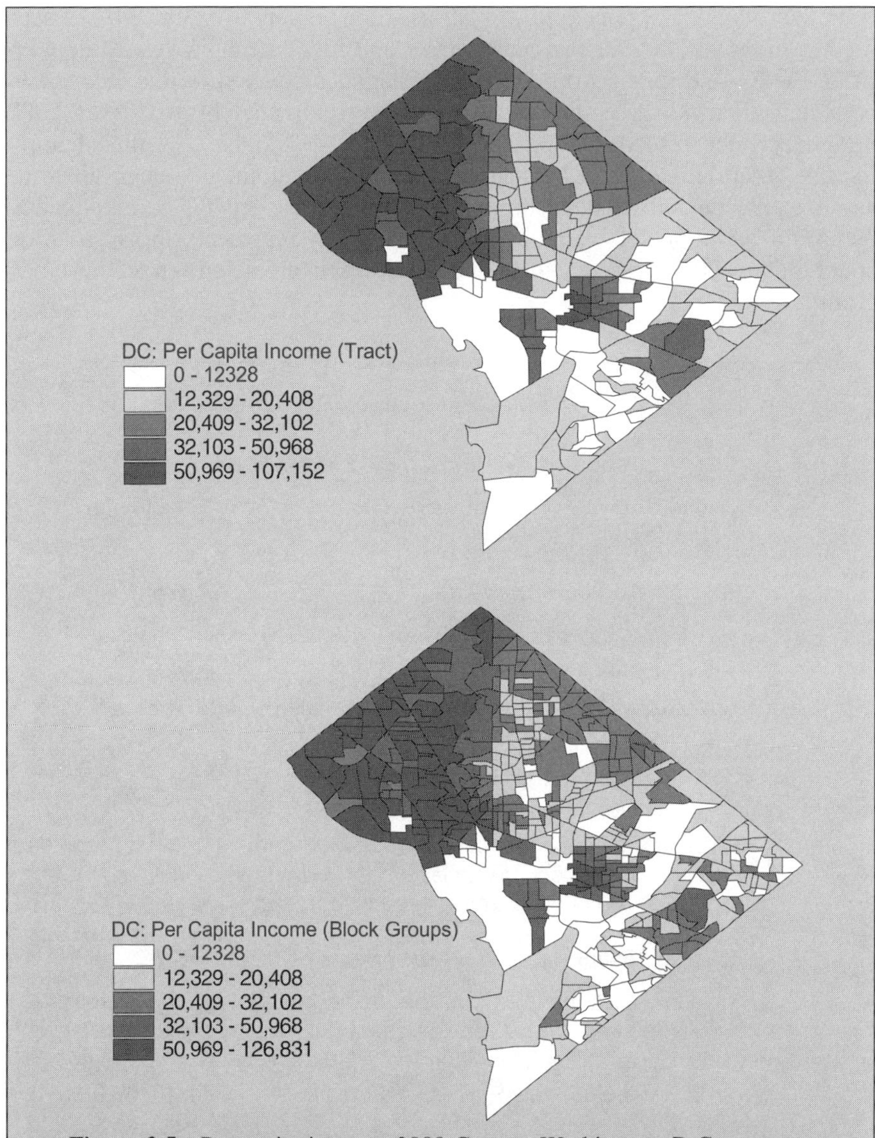

Figure 2.5 Per capita income, 2000 Census, Washington, D.C, census tract and block group levels.

- Though the overall spatial patterns of per capita income based upon census tract data and block group data do not differ dramatically, we do recognize differences at the local scale. The census tract data tend to show more areas with higher income (in the Northwest), while the block group data reveal more local variations, especially in identifying local areas with lower income.

- Note that the classification intervals used in the two maps are identical. Therefore, the differences between the two maps are not due to cartographic tricks, but rather to real differences in the data at the two census levels.

Step 4 Descriptive statistics: block group-level analysis

- From the **Ch2** menu, select **Statistics-univariate,** as
- In the **Descriptive Statistics** dialog box, click `Dc_bg_inc.shp` to use this theme for the analysis. Also, click the **Show Variable** button to display all variables in the attribute table. From the **Select a Variable** window, select `Percapinc` as the variable from which descriptive statistics will be computed. Also, check the boxes for all statistics *except* the one for **Weighted by a Variable,** as

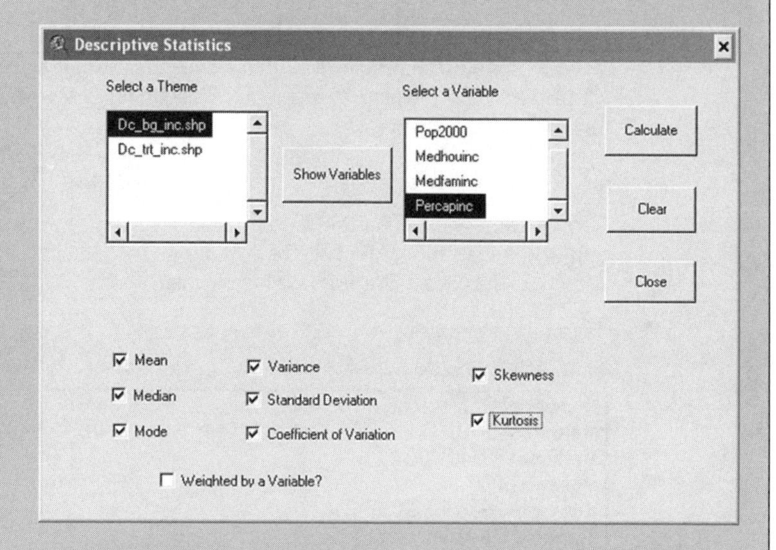

- Click the **Execute** button to proceed.
- When the **Message Box for Selection** appears, asking for **Selecting All Record,** click **Yes** if all records should be included in the computation.
- The results will be displayed by the **Extra Statistics** window, as

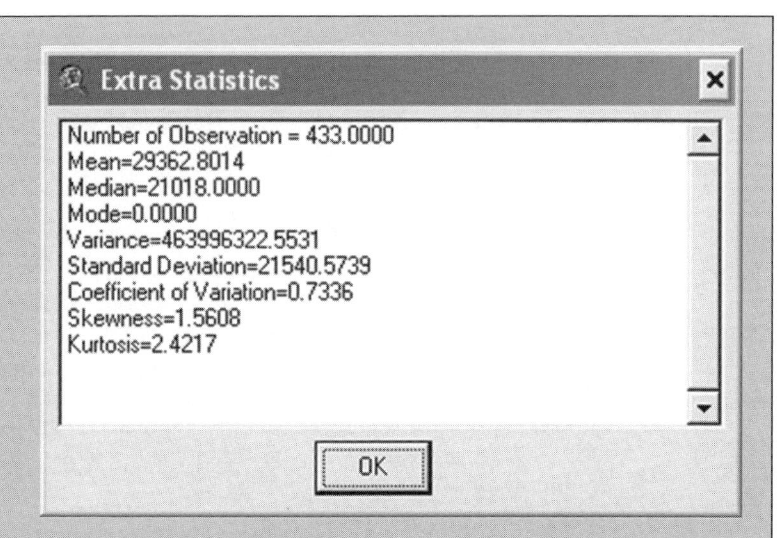

- Click the **OK** button to proceed
- If readers want the results to be written into a text file, click **Yes** when the **Write txt** file box appears. Results will be written to a text file named by the user.

Step 5 Descriptive statistics: tract-level analysis

- Repeat Step 4, but use the theme `Dc_trt_inc.shp` to compute the same set of descriptive statistics. The results will be shown in the **Extra Statistics** window, as

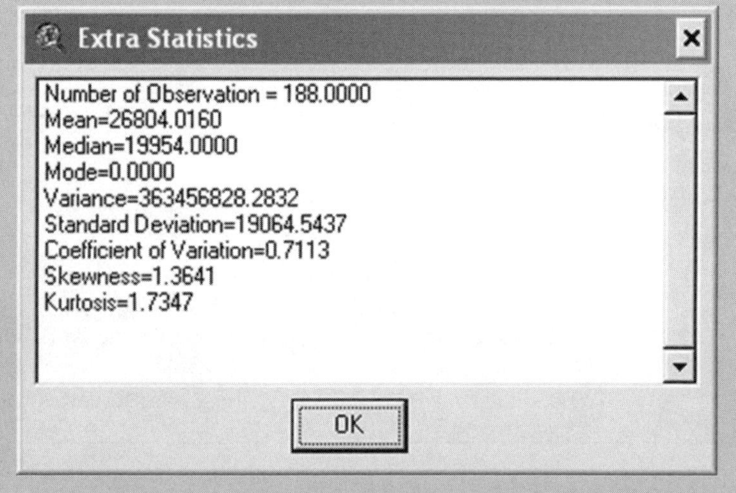

- It is clear that the results of most statistical analyses are different when tract level and block group level data are used, except for the mode, which is zero in both cases. The central tendency measures, mean and median, are both higher from the block group level data than from the tract level data. In measuring dispersion, block group data exhibit a higher level of variability than tract level data in all relevant statistics—variance, standard deviation, and coefficient of variation. In general, data gathered from lower spatial resolution (larger areal units) likely have lower variability (less deviation) than data from higher resolution (smaller areal units). For higher moment statistics, block group level data are more positively skewed than tract level data. Also, block group level data are more concentrated around the mean (kurtosis) than are tract level data.

- In summary, due to the scale effect of MAUP, data gathered at different spatial resolutions or from units of different sizes will yield different results with respect to descriptive statistics.

Step 6 Weighted mean and variance calculations: census tract level

- From the **Ch2** menu, select **Statistics—univariate** again. In the **Descriptive Statistics** dialog box, click `Dc_trt_inc.shp` to select the tract theme for the analysis. After clicking the **Show Variable** button to display all variables, select `Percapinc` as the variable.
- For descriptive statistics, choose only the **mean, variance,** and **standard deviation,** but also check the box for **Weighted by a Variable,** as

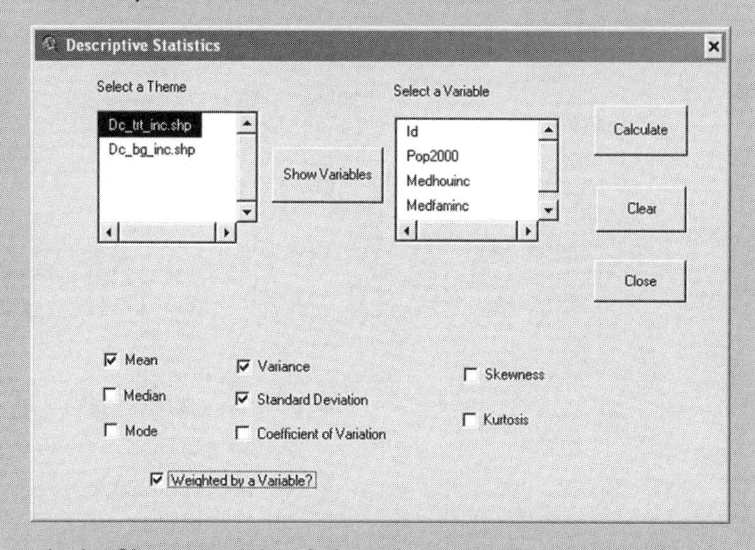

- In the **Choose weight** window, choose the variable `Pop2000` as the weight, as

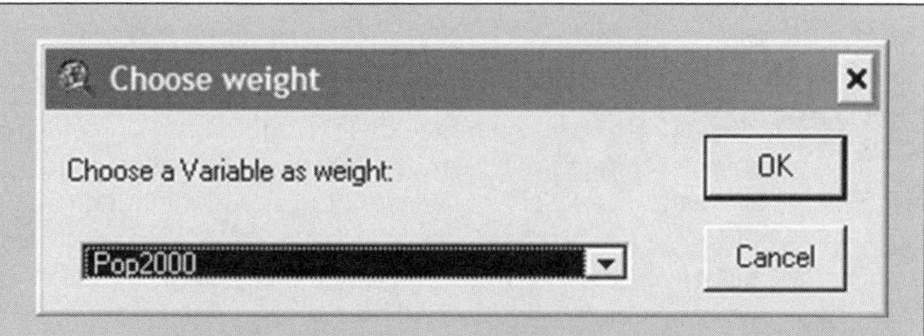

- The results will be shown in a window, as

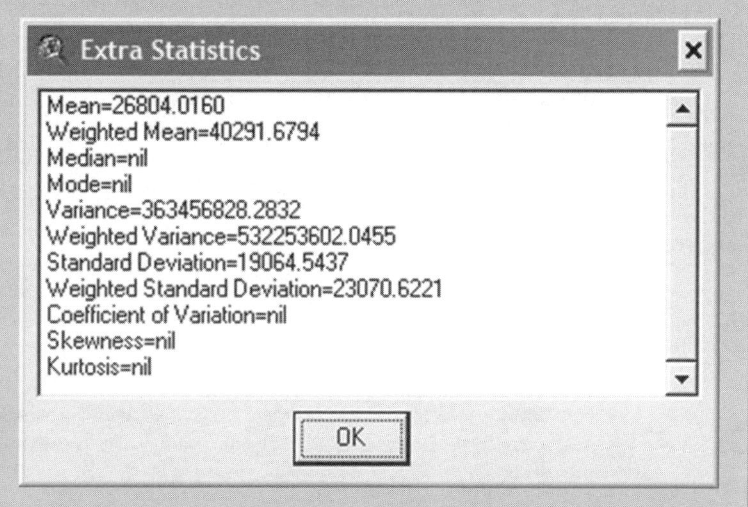

- The **Extra Statistics** window includes results from both the unweighted and weighted statistics. If we calculate the average per capita income by census tract, the average (mean) is only $26,804.016 across census tracts. If the per capita income of each census tract is weighted by its population size, the weighted mean is $40,291.6794, much higher than the unweighted mean.
- Interpretation: the unweighted mean is the average income per capita income obtained in each census tract. The weighted mean is the average per capita income by person. The units of observation are different in the two cases.
- The weighted variance and standard deviation are also higher than their unweighted counterparts, and these results are expected.

Step 7 Weighted mean and variance calculations: census block group level

- One may perform the same process on the block group level data to assess how the scale effect influences the calculation of the weighted statistics.

2.7 SUMMARY

The discussion in this chapter introduces several simple but important statistics: mean, median, mode, minimum, maximum, range, percentile standard deviation, variance, skewness, and kurtosis. Each of these statistics describes a certain distribution aspect of the data values. Collectively, they provide a detailed illustration of how data values in a series distribute. They also make it possible to compare the distributions of two or more data series. Together, the statistics discussed in this chapter are known as descriptive statistics.

REFERENCES

Costanzo, C. M. 1983. Statistical inference in geography: modern approaches spell better times ahead. *The Professional Geographer* 35(2): 158–164.

Summerfield, M. A. 1983. Populations, samples, and statistical inference in geography. *The Professional Geographer* 35(2): 143–148.

EXERCISES

1. Identify four numbers between 1 and 6 such that their mode, range, and median are all equal to 3. There is more than one answer to this question.
2. Given a list of values, 7, 11, 2, 5, 21, and 10, what is the median?

To complete the following set of questions, use the Washington, D.C., census tract data in the application example in this chapter. First, bring the data into ArcView. Also, load the Ch1 extension into the ArcView project. Using the function "Sampling Features: Random" under the Ch.1 menu, select 20 census tracts. Bring up the attribute table of the census tract theme, go to File, and export the selected records to a DBF file, which can then be opened by any spreadsheet program, such as Excel.

3. Choose the variable Per Capita Income (Percapinc); compute the mean, variance, standard deviation, coefficient of variation, skewness, and kurtosis using a calculator. You can also use a spreadsheet program like Excel to perform the algebraic calculation. If you use a spreadsheet, be sure not to use the statistical functions.
4. Using the subset of records, find the median and mode. You can use the sorting functions in ArcView (for the Table document) or the spreadsheet program.
5. Load the Ch2 extension into the ArcView project file. Use the "Statistics-univariate" extension functions to compute the same set of statistics as in Questions 3 and 4. Be sure that the same set of tracts used in Question 3 is still selected. In case the selection of records for Question 3 has been

clear, bring the exported DBF table into the ArcView project. Functions for the Ch.2 extension are also accessible from Table documents. Determine if the results of Questions 3 and 4 are the same as those obtained from the Ch.2 extension.

6. Clear the selected features in ArcView if census tracts are still being selected. Go to the attribute table. Sort the attribute variable Percapinc. Find the median and mode, if any.

7. Run the "Statistics-univariate" function in the Ch.2 extension. Check the mean, median, mode, standard deviation, skewness, and kurtosis statistics. Be sure that all observations are used. In other words, examine the entire population. Determine if the results of the median and mode calculations are identical to your results in Question 4. If they are different, why?

8. Do the mean and median have values similar to those in Question 7? If not, why not? Can other statistics, such as standard deviation and skewness, offer some insights? You may want to examine the distribution of the variable. Which statistic, mean or median, is a better measure of central tendency?

9. In Questions 3 and 4, you computed a sample mean based upon 20 observations. In Question 7, you computed the population parameters. Are they different? If so, why?

10. Let $x = \{2, 7, 9, 11\}$ and let $y = \{4, 14, 18, 22\}$. Do these two sets of values have the same magnitude of dispersion?

11. There are three income variables in the Washington, D.C., data set for the application example: Median Household Income (Medhouinc), Median Family Income (Medfaminc), and Per Capita Income (Percapinc). Similar to Question 10, do these three income variables have the same magnitude of dispersion? If not, why not?

CHAPTER 3

RELATIONSHIP DESCRIPTORS: TWO VARIABLES (BIVARIATE)

Chapter 2 discusses the statistics that describe the distributional characteristics of a set of values or a sample. The mean and its variations address the issue of *location,* that is, where the observations distribute along the continuous value line. Several other measures, such as median and mode, also consider this *central tendency* issue. Variance, standard deviation, and percentiles address the issue of *dispersion,* that is, how much the values of observations spread around the mean. Skewness deals with *directional clustering,* that is, the extent to which the values distribute symmetrically around the mean or not. Kurtosis addresses the issue of *concentration,* that is, how sharply or flatly the values are clustered around the mean. All these measures focus on the distribution of the values using one variable at a time.

These descriptive statistics are useful for understanding and comparing how values are distributed within one dataset or across datasets. Although the mean, standard deviation, skewness, and kurtosis provide information enabling the comparison of different distributions, they cannot measure the relationships between different distributions quantitatively. To do so, we must apply the statistics described in this chapter. One of these statistics is based on the concept of *correlation,* which measures statistically the *direction* and *strength* of the relationship between two sets of data or two variables for a number of observations. We also discuss the concept of *regression,* which measures the *dependence* of one variable on another. Even though we cannot determine causal relationships among the variables with these statistics, they are excellent tools with which to examine any associations between data sets before performing more advanced statistical analysis.

3.1 CORRELATION ANALYSIS

Education is traditionally regarded as an asset. It enriches a person's life in many ways. Among other benefits, the more years of schooling a person has, the more likely this person is to hold a better job or to earn a higher income. Therefore, we believe that education and income are somewhat related and change in the same direction. Of course, there are obvious exceptions to this trend—many millionaires were high school dropouts, and there are unemployed persons with Ph.D. degrees. If we recognize the value of education in eventually achieving a higher income, it would be nice to know how strong this relationship is, that is, how these two aspects of life are related or correlated.

We found that in Kent, Ohio, and Fairfax, Virginia, a house with greater square footage tends to be more expensive than a house with less square footage. It seems reasonable, then, to assume that the larger the house is, the more expensive it is. Of course, one can easily find exceptions to this observation: a big house with acres of land in a rural area, such as rural West Virginia, may cost less than a small upscale townhouse in a densely populated city such as New York City. Again, if housing prices are proven to be related to square footage, we would need to know how strong this relationship is and whether these two variables change in the same direction.

After a new product appears on the market, the number of manufacturers and the product's sale prices seem to move in opposite directions: the more suppliers participate in the manufacture of the product, the lower the price will be. This is a well-known law of supply and demand. Between these two measures an obvious relationship exists, but the measures move in different directions.

In these three examples, we can see that each relationship has two important aspects: the *direction* and *strength* of the relationship. Between two related variables, the relationship is typically measured as **correlation**—a statistical measure indicating how values in one variable are related to values in the other variable. If high values in one variable are associated with high values in the other variable, we say that a *positive* or *direct* correlation exists. Alternatively, a *negative* or *inverse* correlation indicates that high values in one variable are associated with low values in the other variable. In essence, the stronger the correlation is, the more predictable the association will be. A conceptual way to illustrate the concept of correlation is the three-variable example in Figure 3.1.

In this example we have three variables: x, y, and z all have nine observations. Each dot in the figure represents an observation. When we focus on variables x and y, low values in x are associated with low values in y, and high values in x are associated with high values in y. Therefore, the relationship between these two variables is positive. In contrast, the relationship between x and z is just the opposite. When values of x are high, the corresponding values in z are low, and vice versa. Consequently, the relationship between x and z is an inverse one, similar to that between y and z.

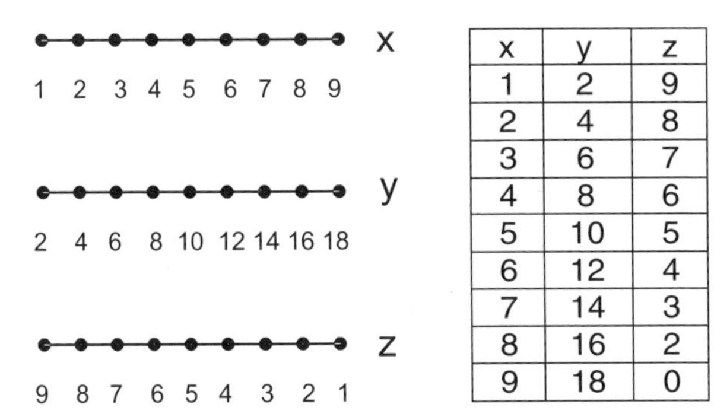

x	y	z
1	2	9
2	4	8
3	6	7
4	8	6
5	10	5
6	12	4
7	14	3
8	16	2
9	18	0

Figure 3.1 A hypothetical three-variable situation illustrating the concept of correlation.

Another way to represent a relationship between two variables is to use a **scatter diagram** or **scatterplot.** In Figure 3.2 we have three scatterplots showing relationships among four variables—two in each diagram. These variables are median household income, median house value, percentage of population in poverty, and percentage of elderly population (i.e., age 65 and older). These variables are tabulated by the 51 states in the United States. based on the Summary File 3A data of Census 2000. The three scatterplots in Figure 3.2 show

1. The relationship between median household income and median house value,
2. The relationship between percentage of population in poverty and median household income, and
3. The relationship between median house value and percentage of elderly population.

It is apparent that median household income and median house value are positively correlated, as shown in the top scatterplot. This indicates that a state with a relatively high median household income level will also have a relatively high median house value. On the other hand, the middle plot shows that percentage of population in poverty and median household income are inversely related, that is, high income is associated with a lower percentage of population in poverty. The third scatterplot shows that the two variables do not have a specific relationship, which is indicated by the scattering of points almost horizontal to the x-axis. That is, regardless of how median house value (x) increases, the percentage of the elderly population (y) does not vary accordingly or systematically.

Beyond the direction of the relationship between the values of two variables, the strength of the relationship can be estimated both visually from the scatterplots and quantitatively. The first two scatterplots in Figure 3.2 are different in the magnitude of their relationship because the points are spread

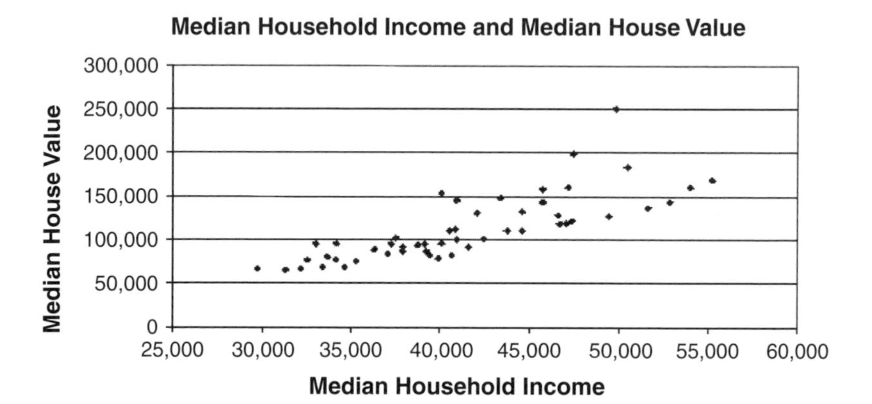

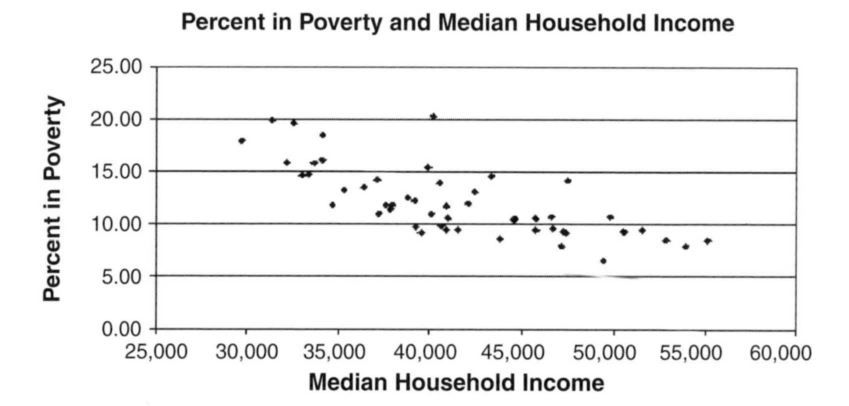

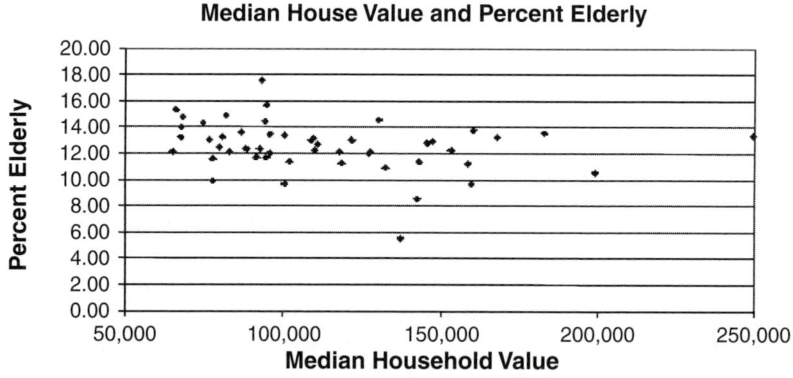

Figure 3.2 Scatterplots of median household income, median house value, percent elderly (i.e., age 65 or older), and percent of population in poverty by states. Source: 2000 Census.

around to different extents. In the first scatterplot, points fall tightly together from the lower left to the upper right in a somewhat linear fashion. The closeness to the linear pattern indicates a strong correlation. In the second scatterplot, points are slightly more scattered than those in the first plot, indicating a slightly weaker correlation.

To quantitatively measure the strength and direction of the relationship between two variables, we can use correlation coefficients. Different coefficients are needed partly because of the different scales of measurement of the data (review the related discussion of scales of measurement in Chapter 1).

For nominal data, the phi (or ϕ) coefficient is used when each variable has only two possible values (binary, or dichotomous). The chi-square (or χ^2) coefficient is used when one or both variables in categorical scale are with more than two possible values (polychotomous). If the two variables are at ordinal scale with a rank assigned to each observation, then Spearman's rank coefficient, r_s, should be applied. Finally, if the variables are at interval/ratio scale, then the product moment correlation coefficient (or Pearson's correlation coefficient), r, should be used. Table 3.1 summarizes how different correlation coefficients are used for variables at different measurement scales.

3.2 CORRELATION: NOMINAL SCALE

For categorical variables, such as modes of transportation (car, subway, bus, etc.) or standings of students in a college (freshman, sophomore, junior, senor, graduate), it seems difficult to understand what is meant by correlation between two variables in nominal scale. To help illustrate this concept, we use data from the 2000 Census in Figure 3.2 to create two categorical variables. First, a state is assigned to be "above average" if its median household income is higher than the average of all 50 states plus the District of Columbia (D.C.). Alternatively, those states with a median household income lower than the average of the 50 states plus D.C. are categorized as "below average." Similarly, each of the 51 median house values can be classified as either above average or below average.

3.2.1 Nominal Scale and Binary: Phi Coefficient

The relationship between nominal (or categorical) variables is often represented by a frequency matrix or a contingency table similar to the one in

TABLE 3.1 Different Correlation Coefficients and Measurement Scales

Scale of Measurement	Correlation Coefficient	Variable Characteristics
Nominal	ϕ (phi)	Binary/dichotomous variables
	χ^2 (chi-square)	Polychotomous
Ordinal (strong ordering)	Spearman's rank (r_s)	Ranked variable
Interval/ratio	Pearson's coefficient (r)	

Table 3.2. A frequency matrix reports the frequencies of data values that are cross-tabulated by the two nominal variables. In our example, both variables are binary because each variable has only two outcomes: above average or below average. Consequently, the **phi (ϕ) coefficient** should be used to evaluate the level of correlation between these two variables as captured by the 2 by 2 matrix.

Each cell in Table 3.2 reports the frequency, or the number of states (including D.C.), in the corresponding category defined by the two nominal scale variables. For instance, there are 27 states with both median household income and median house value below the national averages.

In the context of correlation analysis, the frequencies captured by a table such as those in Table 3.2 give clues to how two variables correlate. Apparently, states with below-average median household income are also likely to have below-average median housing value—as is evident by the existence of 27 out of 50 states plus D.C. falling into that category. States with above-average median household income are also likely to have above-average median housing value. Indicating a strong correlation, most of the states are found in the major diagonal cells (27 and 18) in the matrix, with off-diagonal cells including just a few states (2 and 4). By examining this cross-tabulation of frequencies, one can evaluate the correlation qualitatively simply by using the direction and strength of the correlation between the two categorical variables. We now need to evaluate the correlation quantitatively in order to provide a more objective assessment of the strength and direction of the correlation.

To facilitate the calculation and explanation of the phi coefficient, let us assume that we have two binary variables, x and y, both with only 0 and 1 as possible values. Let's also denote the frequencies of the cross-tabulated categories as a, b, c, and d, as shown in Table 3.3.

The phi (ϕ) coefficient is defined as

$$\phi = \frac{(a \times d - b \times c)}{\sqrt{(a + b)(c + d)(a + c)(b + d)}}. \tag{3.1}$$

The formula seems tedious, but mechanically it is straightforward. In the numerator, it is essentially the product of the major diagonal elements, $a \times$

TABLE 3.2 Frequencies for the Cross-Tabulated Binary Variables

		Housing Value	
		Below Average	Above Average
Household Income	Below Average	27	2
	Above Average	4	18

TABLE 3.3 Notations for the Frequency Tabulation

		y	
		0	1
x	0	a	b
	1	c	d

d, minus the product of the minor diagonal elements, $b \times c$. In the denominator, it is the square root of the product of the row sums, $(a + b)$ and $(c + d)$, and the column sums, $(a + c)$ and $(b + d)$. Numerically, the coefficient ranges between -1 and $+1$. When $\phi = 1$, every 1 in x is matched with a 1 in y and every 0 in x is matched with a 0 in y, if the matrix lays out in the same way as in Table 3.3. On the other hand, $\phi = -1$ indicates that every 1 in x is matched with a 0 in y, and vice versa. When no or little correlation exists between the variables, the coefficient should be 0 or a very small absolute value.

When all the observations fall into the major diagonal elements of the matrix and no observations are in the minor diagonal elements, the numerator will be reduced to $a \times d$, as will the denominator. Thus, the coefficient will equal 1.

$$\phi = \frac{a \times d - 0 \times 0}{\sqrt{(a + 0)\,(0 + d)\,(a + 0)\,(0 + d)}} = \frac{a \times d}{\sqrt{a \times d \times a \times d}} = \frac{a \times d}{a \times d} = 1.$$

If all observations fall into the off-diagonal elements, then the numerator will become $-(b \times c)$ and the denominator will be $(b \times c)$. Then the coefficient will be -1.

$$\phi = \frac{0 \times 0 - b \times c}{\sqrt{(0 + b)\,(c + 0)\,(0 + c)\,(b + 0)}} = \frac{-b \times c}{\sqrt{b \times c \times b \times c}}$$
$$= \frac{-b \times c}{b \times c} = -1.$$

Note that we do not label these extremes as perfect positive or perfect negative correlations. The interpretations of these coefficients depend on the labels of the columns and rows in the matrix. We can easily swap the two columns or rows in Table 3.2, and the resultant coefficient will have the opposite sign. Though the coefficient shifts from positive to negative or from negative to positive, we cannot say that the correlation of the two variables has changed in direction—they are still the same. The relationship of the two variables will not change if we interpret the coefficient according to the labels of the rows and columns in the matrix. So, putting aside the issue of labeling and interpretation, using the frequencies in Table 3.2, the coefficient will be

$$\phi = \frac{(27 \times 18 - 2 \times 4)}{\sqrt{(29)\,(22)\,(31)\,(20)}} = \frac{478}{\sqrt{395{,}560}} = 0.76.$$

Given that the phi coefficient ranges from -1 to 1, a value of 0.76 indicates a moderate correlation. In this chapter, we will not discuss how to test the significance of the phi coefficient. This topic will be covered in Chapter 4.

3.2.2 Nominal Scale and Polychotomous: Chi-Square Statistic

Nominal scale variables with binary outcomes do exist, but they are not very common in real-world analyses. Often nominal scale variables will have more than two outcomes. Consequently, **chi-square,** or χ^2, is used more often. For variables with multiple categories, the most appropriate coefficient is the χ^2 statistic. Using the data for the 50 states plus D.C. in Table 3.1, we can divide the observations into three categories according to the level of median household income. Similarly, we can classify these observations into three categories according to median housing values. Each of the 50 states plus D.C. will be assigned to one of the categories. The resulting frequencies are reported in Table 3.4.

According to the convention used in the χ^2 statistic, the frequencies in Table 3.4 are denoted O_{ij}, or the observed frequency found in cell (i,j), where i refers to a category in variable x and j refers to a category in variable y. For instance, O_{23} is equal to 3, which means that there are three states with medium income value (x) and high housing value (y). In Table 3.4, the row sums and column sums of the frequencies are also included. These sums can be denoted R_i for a row sum and C_j for a column sum. For instance, R_1 is 17, the sum of the first row, and C_2 is 18, the sum of the second column. The total of all frequencies, that is, either the sum of all row sums or the sum of

TABLE 3.4 Frequencies of States plus D.C. in the Categories Cross-Classified by the Two Three-Outcome Categorical Variables

Income	Housing Value (y)			
Value (x)	Low	Medium	High	Row Sum
Low	13	4	0	17
Medium	6	8	3	17
High	0	6	11	17
Column Sum	19	18	14	51
				Grand Total

all column sums, is the grand total, denoted N for the total number of observations.

Formally, the χ^2 statistic is calculated in the following manner:

$$\chi^2 = \sum_i \sum_j \frac{(E_{ij} - O_{ij})^2}{E_{ij}}, \tag{3.2}$$

where O_{ij} is the observed frequency for the cell (i,j) and E_{ij} is the expected frequency for cell (i,j), and it is computed by

$$E_{ij} = \frac{R_i \times C_j}{N}. \tag{3.3}$$

The expected frequency of a cell in the ith category in x and the jth category in y is derived by first multiplying the total number of observations in the ith category in x, R_i by the total number of observations in the jth category in y, C_j and then dividing this product by the total number of observations, N. This process can be repeated for all cells in the table to construct a table of expected frequencies.

E_{ij} indicates the frequency we can expect for the table cell at (i,j) when the two categorical variables are independent. The meaning of independence, using our example in Table 3.4, is that a state having a low income level will not affect its chance of having a high housing value. The χ^2 value in Equation 3.2 is the sum of all standardized differences between the actual and expected frequencies in the tables. This sum of differences indicates how far the observed frequencies are from the expected frequencies when two variables are independent.

In other words, the expected frequency is computed by taking the product of the corresponding row sum and column sum and then standardizing or normalizing this product by the total number of observations. The expected frequencies generated by using Equation 3.3 are the frequencies expected when two variables are independent of each other—no association exists between them.

If the observed frequencies are very similar to the expected frequencies, χ^2 will be relatively small or close to zero. In this case, we can claim that the two variables likely do not have a relationship or a correlation. If the observed and expected frequencies are very different, then χ^2 tends to have a large value, indicating a possible relationship between the variables. In terms of the range of χ^2, this coefficient has a lower bound of zero but no upper bound.

Using the 50 states plus D.C. data again, the frequencies reported in Table 3.4 can be regarded as O_{ij}. Table 3.5 shows the expected frequencies, E_{ij}. For instance, E_{11} equals to $(19 \times 17)/51 = 6.333$. Repeating this process for all cells, we have Table 3.5 showing the expected frequencies.

TABLE 3.5 Expected and Observed (in Parentheses) Frequencies of the 51 States Data

Income Value (x)	Housing Value (y) Low	Medium	High	Row Sum
Low	6.333 (13)	6 (4)	4.667 (0)	17
Medium	6.333 (6)	6 (8)	4.667 (3)	17
High	6.333 (0)	6 (6)	4.667 (11)	17
Column Sum	19	18	14	51
				Grand Total

With E_{ij} and O_{ij}, we can compute the χ^2 statistic by comparing the observed frequencies with the expected frequencies. These comparisons are reported in Table 3.6. Each cell value is $(E_{ij} - O_{ij})^2/E_{ij}$. Therefore, by summing all these differences, we can obtain the χ^2 statistic of 28.56, which is clearly much larger than zero.

$$\chi^2 = \frac{(6.333 - 13)^2}{6.333} + \frac{(6 - 4)^2}{6} + \frac{(4.667 - 0)^2}{4.667}$$
$$+ \frac{(6.333 - 6)^2}{6.333} + \frac{(6 - 8)^2}{6} + \frac{(4.667 - 3)^2}{4.667}$$
$$+ \frac{(6.333 - 0)^2}{6.333} + \frac{(6 - 6)^2}{6} + \frac{(4.667 - 11)^2}{4.667}$$
$$= 7.019 + 0.667 + 4.667 + 0.018 + 0.667 + 0.595$$
$$+ 6.333 + 0 + 8.594$$
$$= 28.56.$$

As a result, the two variables are clearly associated in some way. This is because the value of χ^2 would have been very small if the two variables were

TABLE 3.6 Comparing the Observed and Expected Frequencies Using χ^2 (Equation 3.2)

Income Value (x)	Housing Value (y) Low	Medium	High	Row Sum
Low	7.019	0.667	4.667	12.353
Medium	0.018	0.667	0.595	1.280
High	6.333	0.000	8.594	14.927
Column Sum	13.370	1.334	13.856	28.560
				Grand Total

not related. In this chapter, we will not discuss if a χ^2 value of 28.56 is big enough for the relationship to be significant. We will describe how to test the significance of this statistic in Chapter 4.

3.3 CORRELATION: ORDINAL SCALE

Very often we work with data that have assigned ranks for observations according to certain subjective criteria instead of objective quantitative measurements. For example, we may want to make up a list of cities according to our preference to live there. In that case, the preference may be based on aspects of life that are not directly measurable, such as scenic views, reputation of the cities, historical importance, and so on. Yet, we could still rank the cities according to our perceptions or reports and then use the assigned ranks for our preferences.

In fact, even with quantifiable attributes, we sometimes prefer to rank the quantitative measurements before using these data. For instance, we can use crime rates as an indicator of how safe each city is. The question, however, is, at what rate do we begin to feel safe? Could the cutoff be 2 murders per 100,000 people or 5 murders per 100,000 people per year? How does the difference of 3 murders per 100,000 per year translate into a feeling of safety? One way to deal with this issue is to rank the cities by their crime rates so that a ranking structure is available to help us determine our perception of safety. However, note that we will lose some details of the data in the process as we convert data in ratio (or interval) scale to ordinal scale.

If we have two ranked ordinal variables for a set of observations, we can determine if these variables are correlated or not. In this case, we can use **Spearman's rank correlation coefficient,** or r_s. Another appropriate statistic for ordinal-scale variables is the Kendall's tau statistic. But because Spearman's coefficient is more intuitive, we will discuss only it here.

Let's assume that there are two ordinal variables, x_i and y_i and each observation (i) has an assigned rank according to each variable. For each observation, let d_i be the difference between the ranks from the two variables for observation i, or $d_i = x_i - y_i$. Then Spearman's rank correlation coefficient, r_s, can be computed for a set of n observations:

$$r_s = 1 - \frac{6\left(\sum_i d_i^2\right)}{n(n^2 - 1)}, \tag{3.4}$$

where n is the number of observations. The d_i value is the difference between two ranks of observation i, x_i and y_i (one rank in each variable). We square d_i so that it does not matter which rank is higher or lower. We are only concerned with how much they differ. If the rank pairs of all observations are identical (i.e., highly correlated), the numerator of Equation 3.4 will become zero, or will become very small if the two sets of ranks are similar but not identical. If they are identical, r_s will be 1, indicating a perfectly positive correlation. If the rank pairs of all observations have opposite rankings (e.g., one observation has a rank of 1st in one variable and the rank of nth in the other variable), then $\Sigma_i d_i^2$ will equal to $[n(n-1)(n+1)]/3$. In this case, the numerator of Equation 3.4 will be equal to two times the denominator and thus r_s will be -1, indicating a perfectly negative correlation in their rankings.

To see how Spearman's correlation coefficient can be calculated, let's use the following data as an example. Based on a survey of students commuting from Cleveland to Kent, Ohio, Table 3.7 lists their rankings of six neighborhoods in Cleveland in terms of safety and amenity.

Using Equation 3.4, we have a correlation coefficient of -0.086, an almost negligible correlation.

$$r_s = 1 - \frac{6\left(\sum_i d_i^2\right)}{n(n^2-1)} = 1 - \frac{6(38)}{6(36-1)} = 1 - \frac{228}{210} = 1 - 1.086 = -0.086.$$

Based on this low correlation, we may conclude that the rankings of the two attributes of the six neighborhoods are not related to each other.

Using two variables (median household income and median house value) in the example of 50 states plus D.C. data again, data have been ranked as in Table 3.8. These ranks are recorded in the columns labeled "Rank (Income)" and "Rank (Value)." The last column reports the squared of the rank difference.

At the bottom of the table, the squared of the rank differences are summed to give a total of 2956. Using Equation 3.4, Spearman's rank correlation coefficient will be

TABLE 3.7 Rankings of Six Neighborhoods in Cleveland By a Random Survey

Neighborhood	Rank (Safety)	Rank (Amenity)	d	d^2
Cleveland Heights	6	3	3	9
Bedford Heights	5	6	−1	1
Mayfield Heights	2	5	−3	9
South Euclid	3	2	1	1
Shaker Heights	1	4	−3	9
University Circle	4	1	3	9
				$\Sigma d^2 = 38$

TABLE 3.8 Steps for Calculating Spearman's Rank Correlation Coefficient

States	Households: Median Household Income in 1999	Owner-Occupied Housing Units: Median Value	Rank (Income)	Rank (Value)	Squared Difference (d_i^2)
Alabama	34,135	76,700	9	7	4
Alaska	51,571	137,400	48	39	81
Arizona	40,558	109,400	25	28	9
Arkansas	32,182	67,400	3	3	0
California	47,493	198,900	44	50	36
Colorado	47,203	160,100	42	46	16
Connecticut	53,935	160,600	50	47	9
Delaware	47,381	122,000	43	34	81
Dist. of Columbia	40,127	153,500	24	44	400
Florida	38,819	93,200	18	19	1
.	.	.	.	.	.
.	.	.	.	.	.
.	.	.	.	.	.
Wisconsin	43,791	109,900	34	29	25
Wyoming	37,892	91,500	16	17	1
				Total	2,956

$$r_s = 1 - \frac{6(2{,}965)}{51(51^2 - 1)} = 1 - \left(\frac{17{,}736}{132{,}600}\right) = 1 - 0.13376 = 0.866.$$

Given that the value of Spearman's coefficient ranges from −1 for a perfectly negative correlation to +1 for a perfectly positive correlation, with 0 indicating no correlation, we can say that the rankings of the two variables are strongly and positively correlated. Similar to the other statistics discussed in this chapter, we will reserve the discussion of testing the significance of this statistic for Chapter 4.

3.4 CORRELATION: INTERVAL/RATIO SCALE

In statistical analysis, variables measured at interval or ratio scale probably are the most common. If we want to assess the magnitude and direction of correlation between two interval or ratio scale variables, the most common and appropriate statistic is **Pearson's product moment correlation coefficient,** or r. This coefficient can be calculated between two variables, x_i and y_i, $i = 1, 2, 3, \ldots, n$, as

$$r = \frac{\sum_{i=1}^{n} (x_i - \bar{x})(y_i - \bar{y})}{(n - 1)S_x S_y}, \tag{3.5}$$

where S_x and S_y are the standard deviations of x and y, respectively.

The numerator of Equation 3.5 is essentially a covariance, indicating how the two variables, x and y, vary together. Each x_i and y_i is compared with its corresponding mean. If both x_i and y_i are greater than their means, then the product will be positive. This indicates that x and y of the observation vary in the same direction—a positive correlation. Even if both x_i and y_i are smaller than their means, each of the deviations from the mean will be negative but the product will still be positive, indicating a positive correlation. However, if x_i and y_i are on the opposite sides of their means (i.e., one is greater and the other is smaller), then the covariance will be negative, indicating an inverse relation or a negative correlation. The sum of all these covariance values reflects the overall direction and strength of the relationship between the two variables.

An alternative way to compute the correlation coefficient is

$$r = \frac{\dfrac{\sum_{i=1}^{n} x_i y_i}{n} - \bar{x}\bar{y}}{S_x S_y}. \tag{3.6}$$

Equation 3.6 avoids the needs to compute each of the deviations from the means, but it still requires the computation of $\bar{x}$, $\bar{y}$, S_x and S_y. This is cumbersome and complicated unless the means and standard deviations of the two variables are available. If those statistics are not available, then the preferred formula is Equation 3.7:

$$r = \frac{\sum_{i=1}^{n} x_i y_i - \dfrac{\sum_{i=1}^{n} x_i \sum_{i=1}^{n} y_i}{n}}{\sqrt{\sum_{i=1}^{n} x_i^2 - \dfrac{\left(\sum_{i=1}^{n} x_i\right)^2}{n}} \sqrt{\sum_{i=1}^{n} y_i^2 - \dfrac{\left(\sum_{i=1}^{n} y_i\right)^2}{n}}}. \tag{3.7}$$

This seemingly complicated formula is in fact the most efficient way to calculate r. The numerator requires the sum of the products of x_i and y_i (the last column in Table 3.9) for all i's, the sum of all x_i's, and the sum of all y_i's. These results and those in related columns are shown partially in Table

TABLE 3.9 The Major Steps in Computing Pearson's Correlation Coefficient

State	Median Household Income in 1999 (x)	Median House Value (y)	x^2	y^2	$x_i y_i$
Alabama	34,135	76,700	1,165,198,225	5,882,890,000	2,618,154,500
Alaska	51,571	137,400	2,659,568,041	18,878,760,000	7,085,855,400
Arizona	40,558	109,400	1,644,951,364	11,968,360,000	4,437,045,200
Arkansas	32,182	67,400	1,035,681,124	4,542,760,000	2,169,066,800
California	47,493	198,900	2,255,585,049	39,561,210,000	9,446,357,700
Colorado	47,203	160,100	2,228,123,209	25,632,010,000	7,557,200,300
Connecticut	53,935	160,600	2,908,984,225	25,792,360,000	8,661,961,000
Delaware	47,381	122,000	2,244,959,161	14,884,000,000	5,780,482,000
DC	40,127	153,500	1,610,176,129	23,562,250,000	6,159,494,500
Florida	38,819	93,200	1,506,914,761	8,686,240,000	3,617,930,800
.	.	.	.	.	.
.	.	.	.	.	.
.	.	.	.	.	.
Wyoming	37,892	91,500	1,435,803,664	8,372,250,000	3,467,118,000
Σ	2,108,684	5,750,600	89,163,059,156	721,694,880,000	247,300,334,300

3.9. The sum of all x_i's and the sum of all y_i's are then squared, and they are used in the denominator. As for the denominator in Equation 3.7, it requires each of the x_i's and y_i's to be squared (the third and fourth columns in Table 3.9). The squares of the x_i's and y_i's are then summed separately for each variable.

We suggest that, whenever possible, a table should be used to calculate the values in columns. This avoids many potential mistakes and helps to keep the calculation organized. Specifically, the various sums can be easily derived from Table 3.9 for the calculation of r:

$$r = \frac{247{,}300{,}334{,}300 - (2{,}108{,}684 \times 5{,}750{,}600)/51}{\sqrt{89{,}163{,}059{,}156 - \dfrac{2{,}108{,}684^2}{51}}\sqrt{721{,}694{,}880{,}000 - \dfrac{5{,}750{,}600^2}{51}}}$$

$$= \frac{9{,}531{,}741{,}939}{\sqrt{1{,}975{,}839{,}316}\sqrt{73{,}275{,}265{,}098}}$$

$$= 0.79.$$

With $r = 0.79$, a relatively strong relationship seems to exist between the two variables. Another suggestion is to implement this procedure in a spreadsheet program for even more efficient calculation.

This correlation coefficient is structured so that the sign of r indicates the direction of the relationship:

$r > 0$ when the relationship between the two variables is direct (or positive),
$r < 0$ when the relationship between the two variables is inverse (or negative), and
$r \approx 0$ when there is no clear relationship between the two variables.

The absolute value of r indicates the strength of the relationship with the numerical range of

$r = -1$ for the strongest or perfectly inverse relationship and
$r = 1$ for the strongest or perfectly direct relationship.

In the previous examples of three scatterplots in Figure 3.2, the corresponding correlation coefficient values are 0.7922 (top), −0.7273 (middle), and 0.2590 (bottom), indicating a strong positive correlation between median household income and median house value, a very strong but negative correlation between median household income and percentage of people in poverty, and a weak but still positive correlation between median house value and percentage of elderly population.

Again, we will reserve the discussion of testing how significant a calculated correlation coefficient is for Chapter 4.

3.5 TREND ANALYSIS

The previous section focuses on the statistics that measure the direction and strength of the relationship between two variables. In this section, we discuss the technique for measuring the trend, as shown by the relationship between the two variables. The technique addresses the *dependence* of one variable on another. Going beyond the strength and direction of the relationship, the technique to be discussed in this section allows us to model the relationship and to estimate the likely value of one variable based on the value of another variable. Models that are constructed with this technique are known as **regression models.**

3.5.1 Simple Linear Regression Models

It should be noted that regression models do not imply causal relationships between the variables being examined. Therefore, they should not be used for prediction. However, regression models do provide the information necessary to estimate the values of one variable when the values of another variable are given (Mark and Church, 1977) if there is a relationship between them. In addition, if the relationship between two variables is not linear, the regression model discussed in this section will not be appropriate because this model assumes a linear relationship between the variables.

Regardless of how strong or weak a relationship is between two variables, we can always construct a regression model to describe it. However, regression models may not make much sense when the relationship is weak. They work well when the relationship between variables has a clear linear trend, such as the relationships in the top and middle scatterplots of Figure 3.2. If a regression model is constructed for the bottom scatterplot of Figure 3.2, it cannot be used to estimate the values of the dependent variables effectively.

To illustrate what this means in measuring the trend of the relationship between two variables, we use one of the scatterplots in the previous section for median household incomes and median house values of the 50 states plus D.C. as the example. This scatterplot with modifications is shown in Figure 3.3. As discussed before, when median household income increases, we expect the median house value to rise also. As demonstrated previously, this relationship is quite strong and positive. In other words, states with lower household income levels will likely have lower median house values. Figure 3.3 has a trend line added to the scatterplot. By visually inspecting how data points scatter around this trend line, we can determine the strength and direction of this relationship. This trend line—the regression line—can be mathematically estimated.

Because we are using only a straight line to model the relationship between two variables, this type of regression model is called a **simple linear regression model.** It is also known as a **bivariate regression model** because it involves only two variables. In other words, we can construct a straight line to describe the relationship between the two variables.

A simple linear regression model is the simplest form of regression model. It is generally presented as

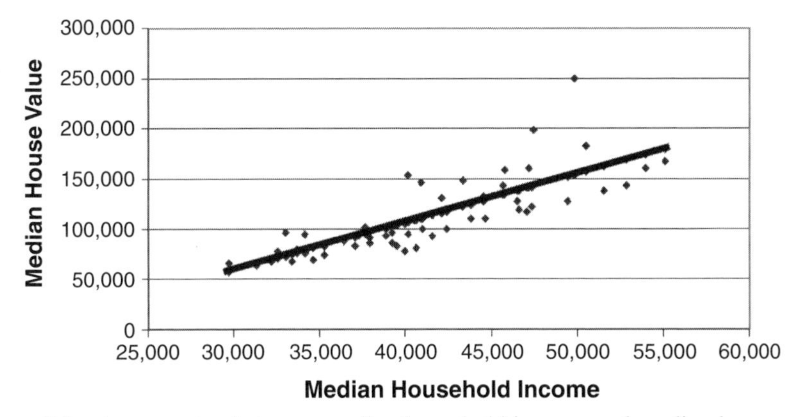

Figure 3.3 A regression between median household income and median house value for 51 states.

$$y = a + bx, \tag{3.8}$$

where

y is the dependent variable,
x is the independent variable,
a is the intercept, or the value of y when the regression line intersects the y-axis, or the value of y in the regression when $x = 0$, and
b is the slope of the trend line, or the tangent of the angle between the x-axis and the trend line.

As in Equation 3.8, the variable that is used to estimate the value of another variable is the **independent variable** (x). The variable that is estimated by the independent variable is the **dependent variable** (y). The **intercept** (a) is the value of the dependent variable when the independent variable (x) has a value of zero. The intercept is also the value of y when the regression line intersects the y-axis. Finally, the **slope** (b) is the rate of change of the dependent variable per unit change of the independent variable—or the amount of changes in y when x changes by 1 unit.

If all parameters of the regression line are available (a and b), then we can estimate the value of y given a value of x. This estimated value is often denoted $\hat{y}$ (known as "y-hat" or "predicted y"). The value of $\hat{y}$ will probably not equal the actual or observed y given the x value. The difference between the observed y and the predicted $\hat{y}$ is sometimes known as the **residual,** or $y - \hat{y}$.

Given a relationship, the regression line that best fits the trend is the line that minimizes the total differences between y's and $\hat{y}$'s or the sum of the residuals. To derive this line, we need to compute the intercept (a) and the slope parameter (b) so that we can determine the line, $y = a + bx$. Specifically,

$$b = \frac{\sum_{i=1}^{n} x_i y_i - n\bar{x}\bar{y}}{\sum_{i=1}^{n} x_i^2 - n\bar{x}^2} \tag{3.9}$$

and

$$a = \bar{y} - b\bar{x}, \tag{3.10}$$

where all notations are the same as before.

In Figure 3.3, showing the relationship between median household income and median house value, the parameters of the simple linear regression model, a and b, are computed. All the terms we need to compute b were calculated when we computed the correlation coefficient in Table 3.9. The slope parameter is

$$b = \frac{247{,}300{,}334{,}300 - \left(51 \times \frac{2{,}108{,}684}{51} \times \frac{5{,}750{,}600}{51}\right)}{89{,}163{,}059{,}156 - \left(51 \times \left(\frac{2{,}108{,}684}{51}\right)^2\right)}$$

$$= \frac{9{,}531{,}741{,}939}{1{,}975{,}839{,}316} = 4.82.$$

Using this b value and plugging in $\bar{x}$ and $\bar{y}$, we can compute the intercept, a (−86.705.97). Then the resulting regression model is

$$\hat{y} = -86{,}705.97 + 4.82x,$$

where $\hat{y}$ is the predicted median house value and x is the median household income.

In this regression, which is shown as the trend line drawn on the scatterplot in Figure 3.3, the intercept is −86,705.97 and the slope is 4.82. With this model, we can calculate a set of estimated values ($\hat{y}$'s) for the dependent variable using the values of the independent variable. The resulting $\hat{y}$'s will be different from the observed y's. The deviations between the observed and estimated values are the residuals ($y_i - \hat{y}_i$). Ideally, a perfect regression model should have zero residuals. The greater the sum of the residuals, the less powerful the regression model is. When the sum of the residuals is small, we typically say that the regression line is a *good fit* for the data. As noted earlier, the regression line is derived from minimizing the total differences between y's and $\hat{y}$'s, or the distances between the observed points and the predicted values.

3.5.2 Coefficient of Determination

Some phenomena may be modeled by the regression reasonable well, and others may not. How can we tell if a specific regression model is better or more powerful for studying one phenomenon than another? To answer this question, we need to compute a **coefficient of determination,** usually denoted r^2. This coefficient of determination is the ratio between the variance of the predicted values ($\hat{y}$'s) and the variance of the observed values (y's) of the dependent variable. Specifically, the coefficient of determination is defined as

$$r^2 = \frac{S_{\hat{y}}^2}{S_y^2}, \tag{3.11}$$

where $S_{\hat{y}}^2$ is the variance of the predicted values of y, also known as the **regression variance,** and S_y^2 is the variance of the observed y values, also known as the **total variance.** In other words, the coefficient indicates how much the total variance of the dependent variable (denominator) can be explained by the variance of the predicted values (nominator) of the regression model.

In the example above, $r^2 = 0.6275$. Converting this ratio to a percentage, we can say that 62.75% of the variance of the dependent variable is accounted for or captured by the regression. Consequently, the higher the r^2 value is, the better is the fit of the regression model to the data. Also note that the square root of the coefficient determination is r, which is Pearson's product moment correlation coefficient discussed in the previous section.

If you are using Equation 3.11 to calculate r^2, $S_{\hat{y}}^2$ and S_y^2 must be computed separately. That will involve estimating all $\hat{y}$'s from y's and then computing $S_{\hat{y}}^2$. Then S_y^2 is computed from y's. But because of the relationship between the correlation coefficient and r^2, we can basically derive the correlation coefficient first and square it to obtain the coefficient of determination. In the previous example of the relationship between median household income and house value, r is 0.7922. Squaring it would give us an r^2 value of 0.627.

Such a level of correlation is relatively high in social science fields but may not be regarded as high in certain science disciplines. A low coefficient of determination has several implications. One is that the regression model may not be appropriate to depict the relationship between the two variables being modeled. Remember that this regression model assumes a linear relationship between the variables. If the relationship is not linear or if the two variables have weak or no realtionship, then the simple regression model will perform poorly. Another characteristic of this regression model is that it is bivariate. That is, only two variables are included: one dependent and one independent. If the dependent variable is affected not just by one independent variable but also by another variable not included in the model, then using just one independent variable will not allow you to estimate the dependent variable very well. In that case, the bivariate model is incomplete. Therefore, we have to use a more complicated regression model known as a **multivariate regression model,** which can accommodate multiple independent variables. Under either circumstance (a nonlinear relationship or missing variables), we may have committed a *model specification error.*

The topic of modeling relationships beyond simple linear regression or with two or more independent variables is often discussed in more advanced statistical textbooks. Similarly, ways to correct model specification errors are beyond the scope of this book.

3.5.3 Empirical Examples

In this section, we will use examples to show how the two techniques, correlation analysis and trend analysis, can be applied empirically. In the first example, we use the data partly discussed in the previous sections, but there the detailed calculations were not shown. In this example, we have two variables from the 2000 U.S. Census (Summary File 3A): median household income by states and percent of population in poverty by states.

First, we will illustrate how the correlation or relationship between the two variables can be evaluated. Assuming that the median household income is variable x and the percent in poverty is variable y, Table 3.10 uses columns to perform the necessary calculation. To compute the correlation coefficient, we need the sum of all x's, the sum of all y's, and the sum of the products of the x's and y's.

Table 3.10 lists the partial values of x and y, their sums, the squared values of all x's and y's, and their sums. The last column is the product of x and y, and the bottom is the sum of the products. With all these values, we can compute r:

$$r = \frac{24{,}779{,}295 - (2{,}108{,}684 \times 618)/51}{\sqrt{89{,}163{,}059{,}156 - \dfrac{2{,}108{,}684^2}{51}}\sqrt{8025 - \dfrac{618^2}{51}}}$$

$$= \frac{-755{,}923}{\sqrt{1{,}975{,}839{,}316}\ \sqrt{546.74}}$$

$$= -0.7273.$$

TABLE 3.10 Calculation Using 2000 Census Data on 50 States and Washington, D.C.

States	Median Household Income in 1999 (x)	Percent in Poverty (y)	x^2	y^2	xy
Alabama	34,135	16.10	1,165,198,225	259	549,711
Alaska	51,571	9.40	2,659,568,041	88	484,630
Arizona	40,558	13.91	1,644,951,364	194	564,335
Arkansas	32,182	15.84	1,035,681,124	251	509,662
California	47,493	14.22	2,255,585,049	202	675,251
Colorado	47,203	9.26	2,228,123,209	86	436,913
Connecticut	53,935	7.86	2,908,984,225	62	424,095
Delaware	47,381	9.21	2,244,959,161	85	436,294
DC	40,127	20.22	1,610,176,129	409	811,197
Florida	38,819	12.51	1,506,914,761	157	485,725
.	.	.	.	.	.
.	.	.	.	.	.
.	.	.	.	.	.
Wyoming	37,892	11.42	1,435,803,664	131	432,883
Σ	2,108,684	618	89,163,059,156	8,025	24,779,295

The numerator is −755,923, which already indicates the direction of the relationship. Remember that, in the original formulation of the correlation coefficient, the numerator is the covariance of the two variables, x and y. It is the product of the deviations from the means of the two variables. If the two variables have a direct or positive relationship, then the covariance will be positive. If they are inversely related, then the covariance will be negative, indicating that a large value of one variable is likely associated with a small value of the other variable for the same observation. In this example, the negative numerator implies that, in a given state, a higher level of median income is associated with a lower percentage of population in poverty and vice versa.

This relationship seems to fit our expectation that richer states will have fewer people in poverty. This negative relationship of the two variables also turns out to be a strong one because their value is not too far from −1. A correlation coefficient closer to zero indicates a weak relationship, whether it is positive or negative. But in this case, the correlation coefficient is −0.7273, a moderately strong negative correlation.

We can further model the relationship of the two variables by treating the median household income as the independent variable (x) and the percent in poverty as the dependent variable (y). Using regression analysis, we can derive a linear trend to describe how the two variables are related. The regression model will look like the following equation:

$$\text{Percent in poverty } (y) = \text{intercept } (a) + b \times \text{median household income},$$

where b is the slope. To formulate this regression model, we first need to calculate the value of the slope, b. Once we have this value, we can compute the value of a, the intercept.

Table 3.10 gives us almost all of the values we need to compute the regression parameters. To facilitate the process, we need to compute the means of x and y first. Using the sums of x and y in Table 3.10, the means are

$$\bar{x} = \frac{2{,}108{,}684}{51} = 41{,}346.75$$

and

$$\bar{y} = \frac{618}{51} = 12.11.$$

Given $\bar{x}$ and $\bar{y}$, the slope parameter can be calculated as

$$b = \frac{24{,}779{,}295 - (51 \times 41{,}346.75 \times 12.11)}{89{,}163{,}059{,}156 - (51 \times 41{,}346.75^2)}$$

$$= \frac{-755{,}923.70}{1{,}975{,}839{,}316} = -0.0003826.$$

Given b, we can calculate the value of a, which is

$$a = 12.11 - (-0.0003826) \times 41{,}346.75 = 27.93.$$

Therefore, the regression model is

$$\hat{y} = 27.93 - 0.0003826x.$$

How well does this regression perform? We must compute r^2, the coefficient of determination. We can derive the variance of y and the variance of the predicted $\hat{y}$, but these procedures are rather tedious because we must compute all predicted values of $\hat{y}$ first. Instead, we can use the relationship that the coefficient of determination is the squared value of the correlation coefficient. We computed the correlation coefficient in the previous section, and r is -0.727. Squaring it gives us the coefficient of determination of 0.5290. This level of the coefficient indicates that almost 53% of the variance of y is explained by the regression model. This is not a very high coefficient, but for many disciplines in social science it is a reasonably good model, given that we are using only one variable. Note that the coefficient of determination does not indicate the direction of relationship but only the performance of the model.

3.6 ARCVIEW NOTES

As in the previous two chapters, an extension to ArcView GIS was developed to allow the calculation of correlation coefficients and a simple linear regression model. The example below illustrates how to use this extension.

ArcView Example 3.1: Correlation Coefficient and Regression Model

Step 1 Data preparation
As before, data for ArcView examples in this chapter can be found under `C:\Temp\Data\Ch3_data`.
Under this folder, `Ccdb1994.shp` is the digital data from the 1994 City and County Data Book, and `Ch3Example.shp`

contains the same county boundaries but only a few attributes—some were from Ccdb1994.shp and some were generated from Ccdb1994.shp.

Step 2 Start ArcView GIS

- Start ArcView GIS to create a new project with a new view.

Again, we suggest that readers start ArcView, add the data theme, load the Ch3.avx extension, and then perform the analysis with added statistical functions.

Step 3 Add data theme

- In the **Add** data dialog box, click the **Yes** button to add data to the **View** document.
- Navigate to the data folder and add the C:\Temp\Data\Ch3_data\Ccdb1994.shp shapefile. This shapefile contains a collection of census data for all counties in the United States.

Step 4 Attribute fields

Similar to Ccdb2000.shp used in the previous chapter, there is a large collection of attribute fields in Ccdb1994.shp. Refer to the listing of attribute fields in Variables.xls for details of what each field contains.

In this example, we use the following fields:

COF01-05: Population 1990
COF14-141: Personal income 1990
COF16-165: Manufacturing earnings 1990
COF17-182: Retail trade sales 1990

- Click the **Open Theme Table** button, , to open the attribute table. Use the scroll bar to navigate around the table to verify the existence of the above-listed attributes.
- You may want to repeat the same procedure discussed in previous chapters to select records in the table so that the values of the attributes are not "blanks" or missing.
- Note that these variables are in interval/ratio scale. Therefore, Pearson's correlation coefficient will be appropriate.
- For variables in other measurement scales, the scale has to be determined first. Then choose the appropriate correlation statistic(s) based upon the scale of measurement.

Step 5 Load Ch3.avx extension

- Load Ch3.avx extension by choosing the menu **File** and then **Extensions . . .**
- Check the box beside the Ch3 extension to load it. Once it is loaded, a new menu item, **Ch3,** appears. It has only one item in the drop-down menu, which is **Bivariate.**

- Note that the Ch3 extension is accessible not only from the View document but also from the Table document. Therefore, the user can run all procedures in the Ch3 extension when working on the Table also.

Step 6 Calculate correlation coefficients
To calculate coefficients, choose the **Bivariate** menu item from the **Ch3** menu. In the subsequent dialog box:

- Check the box using **Pearson's correlation coefficient.**
- Next, select a theme by clicking the name in the list box.
- Click the **Show Variables** button to update the list of attributes in the theme selected.

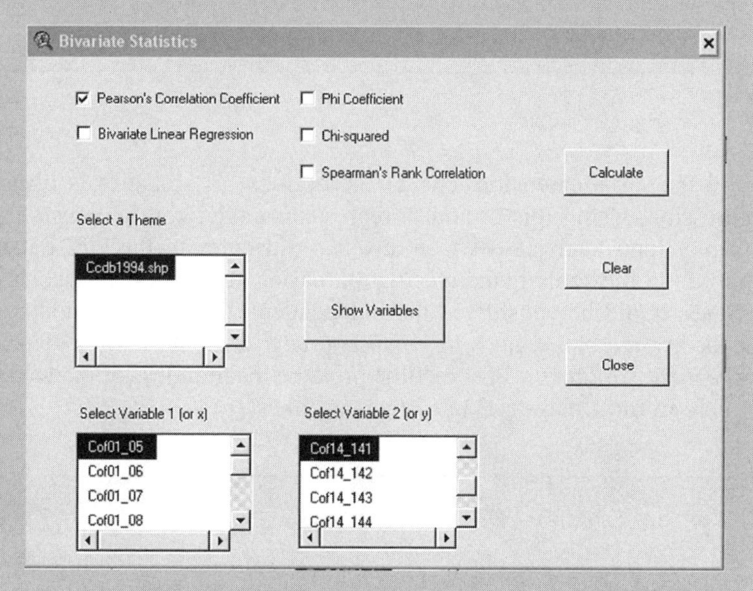

- For **Select Variable 1 (or *x*),** choose `Cof01-05` for Population 1990.
- For **Select Variable 2 (or *y*),** scroll down the list to find `COf14-141`.
- Click the **Calculate** button to compute the value of Pearson's product moment correlation coefficient.

Once the **Calculate** button is clicked, you will be informed that no records are selected and that all records will be used. Of course, if some records are marked as selected (records without missing data), then calculation will only use the selected records.

- Based on the output window, the relationship between Cof01-05 and Cof14-141 has a strong positive correlation, 0.9861.

- Repeat the same process for other variables so that the following table is filled:

	COF01-05	COF14-141	COF16-165	COF17-182
COF01-05	1	0.9861	0.9125	0.9816
COF14-141	0.9861	1	0.9260	0.9884
COF16-165	0.9125	0.9260	1	0.9245
COF17-182	0.9816	0.9884	0.9245	1

From the tabulated correlation coefficient, strong relationships among these variables can be observed.

Step 7 End

In the same extension, one of the options for statistics is bivariate linear regression. Using this option, a regression model can be constructed to perform a trend analysis between any two attributes in the GIS database. Note that while the tools make calculation easier, the user must decide which attribute should be considered the independent variable and which one the dependent variable. Choosing inappropriate attributes for independent and dependent variables will sometime produce meaningless or misleading results, which, in turn, may lead to further mistakes.

ArcView Example 3.2: Trend Analysis

Step 1 Data preparation
Also using Ccdb1994.shp data, let's look at how well the level of manufacturing earnings estimates personal income and how well the level of personal income estimates retail trade sales at the county level.

- Start ArcView GIS if it is not already started.
- Again, from the **File** menu, choose **Extensions** . . . and check the `Ch3 extension.`
- Use the **Add Theme** button, [button icon], to add the `Ccdb1994.shp` to your View document from the **C:\Temp\Data\Ch3_data** folder.
- Use the **Open Theme Table** button, [button icon], to open the attribute table and examine these fields:

`COF14-141`: Personal income 1990
`COF16-165`: Manufacturing earnings 1990
`COF17-182`: Retail trade sales 1990

- You may want to select records with no blanks or missing data before proceeding to the next step.

Step 2 Modeling the first regression trend
First, we will model the trend between manufacturing earnings 1990 (COF16-165, as the independent variable) and personal income 1990 (COF14-141, as the dependent variable).

- From the **Ch3** menu, choose **Bivariate.**
- In the dialog box, check **Bivariate Linear Regression** and make sure that `Ccdb1994.shp` is highlighted in the **Select a Theme** list.
- Click the **Show Variables** button to update the list of variables.
- From the **Select Variable 1 (or *x*)** list, scroll down and click `COF16-165` to make it the independent variable (*x*).
- From the **Select Variable 2 (or *y*)** list, scroll down and click `COF14-141` to make it the dependent variable (*y*).
- Click the **Calculate** button to start the calculation.

When prompted in the **Bivariate Regression** dialog box:

- Click **Yes** to create a scatterplot for the data.

Two windows are created. One is titled **Extra Statistics.** It has the number of observations (*N*), Intercept (*a*), and Slope (*b*) listed.

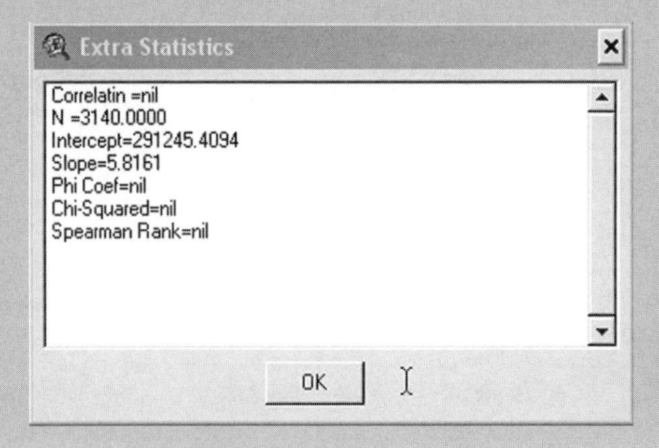

The second window is a **Chart** document that contains the scatterplot requested and the statistics of the regression model, including r^2.

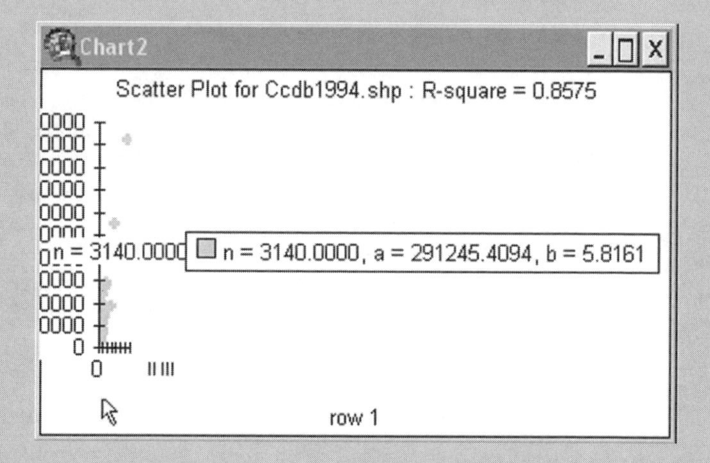

Since r^2 is 0.8575, the regression model is considered a good model because about 86% of the variation in COF14-141 is explained by the variation in COF16-165.

Formally, the regression equation should be

$$\text{COF14-141} = 291245 + 5.8161 \times \text{COF16-165}.$$

Step 3 Modify scatterplot

A closer look at the scatterplot shows that the data points are cramped in the limited space since the chart legend took up most of the space.

To improve the chart, use the **Chart Element Properties** button, [icon], to make changes to how elements are displayed.

- Click the **Chart Element Properties** button to activate it.
- Inside the Chart document, click the chart legend to bring up the **Chart Legend Properties** dialog box.
- In the **Chart Legend Properties** dialog box, click in the space below the chart symbol as the new position for the chart legend. The chart legend should have been moved to the location to reflect the change. Click **OK** to proceed with the change.

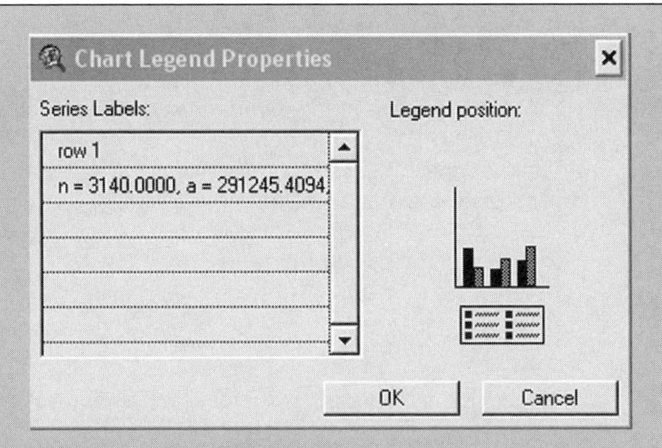

Next, we will change the label for the *y*-axis.

- In the Chart document, click at the *x* label to bring up the **Chart Axis Properties** dialog box.
- In the dialog box, replace the long label for the *y* axis with `COF14-141`. Click **OK** to proceed with the change.

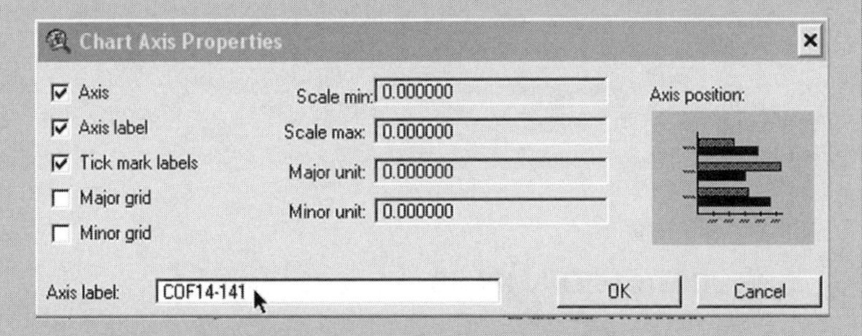

The Chart should show more details of points.
A final way to improve the display is to resize the chart horizontally to make room for points to be displayed.

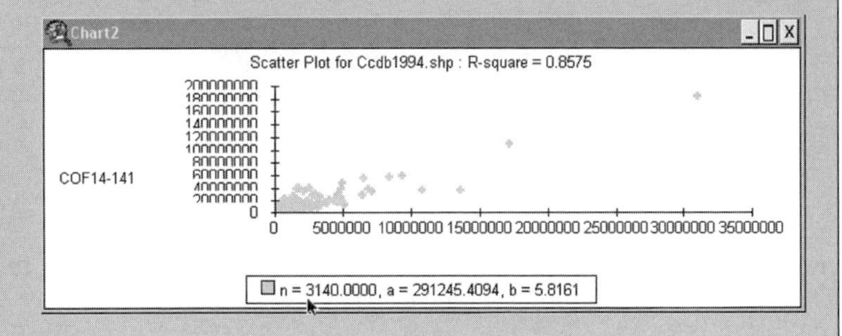

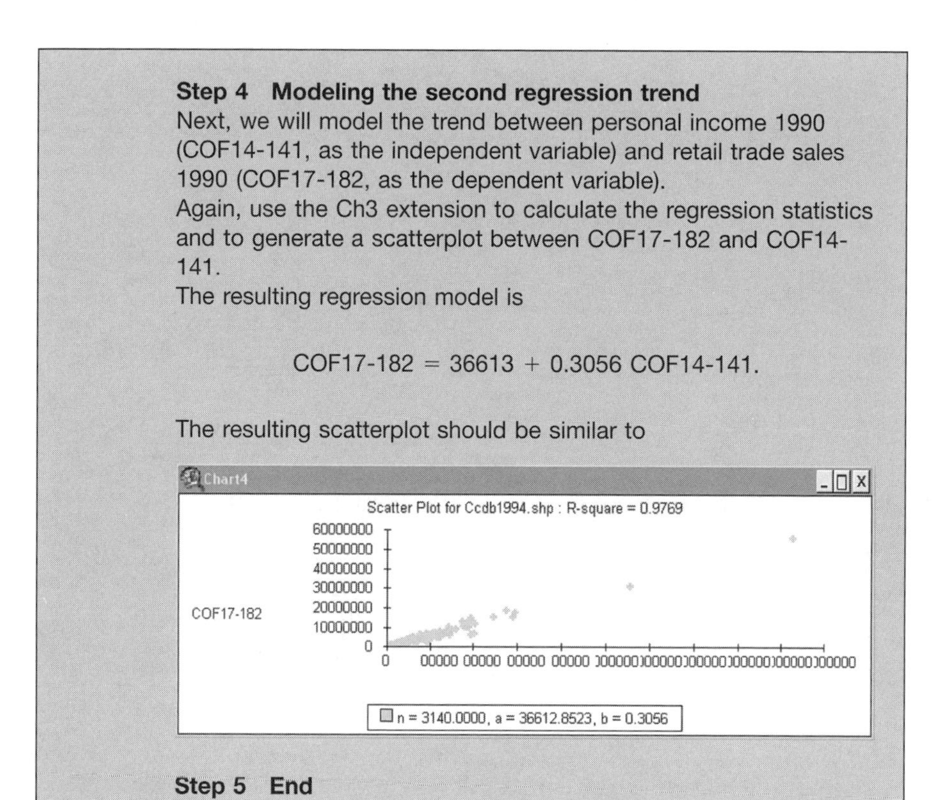

Step 4 Modeling the second regression trend
Next, we will model the trend between personal income 1990 (COF14-141, as the independent variable) and retail trade sales 1990 (COF17-182, as the dependent variable).
Again, use the Ch3 extension to calculate the regression statistics and to generate a scatterplot between COF17-182 and COF14-141.
The resulting regression model is

$$\text{COF17-182} = 36613 + 0.3056\ \text{COF14-141}.$$

The resulting scatterplot should be similar to

Step 5 End

3.7 APPLICATION EXAMPLES

As discussed earlier in this chapter, several correlation coefficients have been developed to measure the strength and direction of the relationship between two variables. Depending on the scales of measurement of the data being analyzed, different correlation coefficients should be used for different types of data. But do all correlation coefficients work the same way, measure the relationships effectively, and give us consistent results? The results are partly dependent on how much information in the original data is retained by the data at other scales of measurement.

To address these issues, this application example uses the results obtained by calculating various correlation coefficients to describe the effectiveness of these correlation coefficients. In this example, we use the data set installed under the folder C:\Temp\Data\Ch3_data. To explore this data set further, add Ch3Example.shp from this folder to ArcView GIS.

First, Table 3.11 lists the attributes included for this application example. Those attributes with names similar to COFxx-xxx are taken from

TABLE 3.11 List of Attributes in Ch2Example Data Set

Areaname	County Name
COF01-5	Population 1990
COF01-6	Population 1980
COF01-7	Population change, 1980–1990
COF01-8	Population percent change, 1980–1992
RPC8092	Ranks of population percent change, 1980–1992
PC8092C	Classes of population percent change, 1980–1992
BPC8092	Binary classes of population percent change, 1980–1992, by mean
COF14-141	Personal income 1990
COF14-142	Personal income percent change 1980–1990
RPIC8090	Ranks of personal income percent change 1980–1990
PIC8090C	Classes of personal income percent change 1980–1990
BPIC8090	Binary classes of personal income percent change, 1980–1990, by mean
COF16-165	Manufacturing earnings 1990
COF16-166	Manufacturing earnings, percent change 1987–1990
RMFGPC8790	Ranks of manufacturing earnings 1987–1990
MFGPC8790C	Classes of manufacturing earnings 1987–1990
BMFGC8790	Binary classes of manufacturing earnings 1987–1990, by mean
COF17-182	Retail trade sales 1987
COF17-183	Retail trade sales, percent change 1982–1987
RRSC8287	Ranks of retail trade sales percent change 1982–1987
RSC8287C	Classes of retail trade sales percent change 1982–1987
BRSC8287	Binary classes of retail trade sales percent change 1982–1987

Ccdb1994.shp. Other attributes are additional information generated using the COFxx-xxx attributes.

First, let's examine the correlation among population change, income change, changes in manufacturing earnings, and changes in retail trade sales. Specifically, we will use attributes that have names starting with B (for binary). These include BPC8092, BPIC8090, BMFGC8790, and BRSC8287. These binary attributes are derived from classifying those values above and below their means into two categories. Table 3.12 shows the results of calculating the phi correlation coefficient in regular font.

Next, using ranked attributes, including RPC8092, RPIC8090, RMFGPC8790, and RRSC8287, Spearman's rank correlation coefficients are calculated and listed in Table 3.12 in *italics.* Finally, with PC8092C, PIC8090C, MFGPC8790C, and RSC8287C, Pearson's product moment correlation coefficients are calculated and listed in Table 3.12 in **boldface.**

As can be seen, different correlation coefficients perform differently. Note that the precise measurements of the data were lost when the data were de-

TABLE 3.12 Example Correlation Coefficients

	Population Changes	Income Changes	Changes in Manufacturing Earnings	Changes in Retail Trade Sales
Population changes		0.5292	0.0422	0.0627
		0.6397	*0.0984*	*0.5439*
		0.5793	**0.0386**	**0.4708**
Income changes	0.5292		0.0269	0.0325
	0.6397		*0.0982*	*0.5243*
	0.5793		**0.0374**	**0.4693**
Changes in manufacturing earnings	0.0422	0.0269		0.0125
	0.0984	*0.0982*		*0.0282*
	0.0386	**0.0374**		**0.0408**
Changes in retail trade sales	0.0627	0.0325	0.0125	
	0.5439	*0.5243*	*0.0282*	
	0.4708	**0.4693**	**0.0408**	

graded. Spearman's rank correlation coefficient tends to dismiss the detailed differences between data values measured at interval/ratio scale. Similarly, converting interval/ratio data to binary data eliminates much detail embedded in the original data. This is not to say that the phi coefficient or Spearman's rank correlation coefficient should never be used. They should be used in studies when the best data available are binary or in rank order or when the nature of the studies requires the use of binary or ranked data.

REFERENCES

Mark, D. M. and M. Church. 1977. On the misuse of regression in earth science. *Mathematical Geology,* 9: 63–75.

EXERCISES

States	Deaths (per 1000)	IMR (per 1000)
Connecticut	868.4	6
Delaware	892.8	10.6
DC	1037.1	10.8
Maine	967	6.2
Maryland	813.9	8.1
Massachusetts	886.6	5
New Hampshire	779.4	3.8
New Jersey	877.8	6.4

States	Deaths (per 1000)	IMR (per 1000)
New York	834.4	6.1
Pennsylvania	1054.4	7.2
Rhode Island	945.7	6.8
Vermont	848.5	5.7
Virginia	782	7.4
West Virginia	1164.2	7.3
Mean	910.87	6.96
Standard deviation	107.54	1.84

In the above table, two health statistics for 2001 for states in the New England and Mid-Atlantic regions have been extracted from the Centers for Disease Control and Prevention (CDC)/National Center for Health Statistics (NCHS) website. The two variables are the general mortality rate expressed as deaths per 100,000 people and the infant mortality rate (IMR) per 1000 live births. The means and standard deviations (std) of both variables are also computed

1. Using the means of the two variables, create two new binary variables, BMR and BIMR, one for each of the original variables. Label the variables' values as high or low and assign the states to the two categories. Tabulate a 2×2 frequency table to show the frequency in each cross-classification cell. Compute the phi coefficient. How do you interpret the relationship of the two binary variables?

2. Create a new nominal variable with three outcomes from the general mortality rate. Use the following boundary values to classify states into the three categories: (a) minimum to (mean−std), and (b) (mean−std) to (mean+std), and (c) (mean+std) to maximum). Use this new nominal variable and the binary IMR (BIMR) created in Question 1 to tabulate a 3×2 frequency table. Compute the χ^2 statistic. Interpret the result.

3. Rank each of the states according to the two original variables (Deaths/1000 and IMR). You will derive two sets of rankings. Use these rankings to compute Spearman's rank correlation coefficient.

4. Use the two original variables to compute Pearson's correlation coefficient.

5. Compare the results from Questions 1 to 4. Do all these correlation statistics provide consistent results. If not, why not?

6. If a bivariate linear regression model is to be constructed for the two variables and you are asked to assess the power of the model, what will be your answer without deriving the model?

7. Using the general mortality rate (Deaths/1000) as the independent variable and the IMR as the dependent variable, compute the parameters for a linear regression model.
8. Bring your data and new variables into ArcView as a Table document, and use the Ch3 Extension from Table to verify if your calculations are correct.

CHAPTER 4

HYPOTHESIS TESTERS

In previous chapters, we suggest using descriptive statistics to summarize and describe the characteristics of sampled observations if the population is too large to be completely enumerated. In an ideal situation in which one is able to examine the entire population and to compute the appropriate *parameters* for the population, such as the mean and standard deviation, it would be possible to cross-check whether or not the calculated sample statistics are different from the parameters calculated for the population. Unfortunately, in most cases, this ideal situation does not exist; otherwise, there would be no need to sample.

Since a sample is not the entire population, we would expect that statistics calculated from the sample are not identical to the parameters calculated from the population. This is because the sample statistics are based on the sample only, and thus are subject to *sampling error.* If we use sample statistics in place of population parameters, then the issue is whether the difference between the sample statistics and population parameters is small enough so that we could claim that this difference is attributable only to the sampling error, not to the real difference between the sample and the population. In this case, we should perform *statistical tests* to decide if the difference is too large to be considered a sampling error.

Another situation in which statistical testing is needed is when we wish to determine if two samples come from the same population. In that case, we calculate same statistics from each sample and test them to see if the difference between them is significant. If so, the difference may support the claim that the two samples are likely to be from different populations. Or perhaps the difference between them is statistically insignificant; in this case, the difference may be attributable to sampling error. Then we may conclude that the two samples are likely to be from the same population. To this end, similar testing can be performed for multiple sets of samples.

In order to perform these statistical tests, we must first discuss basic concepts regarding probabilities and the general procedures for setting up a hypothesis. With these general concepts, we will then be able to discuss several common methods for testing the statistical significance of the differences among tested statistics. Readers may want to refer to other references, including Rogerson (2001). The statistical libraries we developed for ArcView are based on the algorithms from Press (et al. 1992).

4.1 PROBABILITY CONCEPTS

To illustrate several concepts related to probability theory, it is desirable to start with a sampling example. Suppose that we are interested in finding out how widespread a specific type of event is in a neighborhood. The type of event can be theft or the presence of a disease, such as the West Nile virus. In order to assess how common the event is, we could select a number of households (our observations) in the neighborhood and ask them if they have encountered the event. If possible, we should examine every household (i.e., the entire population) or try to select as many households as our resources and time allow. For now, let's assume that we randomly select only four households and ask them if they have experienced any theft in the past six months. The four households are our sample.

We have four **observations** in our **sample,** and each observation can respond with a "yes" (Y) or a "no" (N). What are all of the possible outcomes we can record after questioning four households? Table 4.1 lists all the possible outcomes, which together constitute the **sample space** in this situation. Given that our sample consists of four households and each household can respond with either a Y or an N, there are 16 possible responses. They are arranged by the number of Y's in Table 4.1.

Since our objective is to know how serious theft is in the neighborhood, our concern is with the positive responses—the Y's. Thus, we will focus on these responses. The response is also known as the **random variable.** In this example, the random variable can be the presence of theft or another event such as a vector-borne disease. According to the tabulation in Table 4.1, there is only one possible outcome with no Y's and four Y's. With one Y, there are four possible outcomes. With two Y's, there are six possible outcomes. With three Y's, there are also four possible outcomes. Other possible outcomes are also enumerated in Table 4.1. Together, these 16 possible outcomes define the sample space. The survey result from the four households will be one of these outcomes.

If we have only one household with either a Y or an N, the sample space will have two possible outcomes, or $2^1 = 2$. When we have two households with two possible responses each, we have a sample space of $2 \times 2 = 2^2 =$

TABLE 4.1 All Possible Outcomes of Surveying Four Households

Possible Outcomes	Number of Y's	Number of N's	Frequency
NNNN	0	4	1
NNNY NNYN NYNN YNNN	1	3	4
NNYY NYNY YNNY NYYN YNYN YYNN	2	2	6
NYYY YNYY YYNY YYYN	3	1	4
YYYY	4	0	1

4. Similarly, we can use a quick calculation to find the sample space of four households (or more). Instead of exhaustively listing all possible outcomes, our sample space for four households with two possible responses each is $2 \times 2 \times 2 \times 2 = 2^4 = 16$. As can be easily seen, this is a better way to find the sample space when we have a large number of observations.

So far, we have assumed that the order of responses is important. For example, the sequence NNNY is not treated the same way as NYNN or YNNN—even though all these sequences give us a value of 1 for our random variable. The order of the responses is important when we need to keep the order of households. For example, the order of the four households is the order of their location along a river.

We can generalize the above computation to determine the sample space, or the total number of possible outcomes. Let's use n as the sample size and r as the number of choices for each sample to choose from. Let's also assume that the order of the answers is important. In this case, the number of all possible outcomes will be r^n. In the above example, if five instead of four households are surveyed, then the number of all possible outcomes will be $2^5 = 32$. If four households are surveyed but each household is given three choices—"yes," "no," and "no idea"—then the number of all possible outcomes will be $3^4 = 81$. This computation is also called **permutation.**

In Table 4.1, the possible outcomes are listed to correspond to the given number of Y's and N's. We can derive the number of outcomes for a given number of Y's or the value of the random variable by using combinatorial methods. In general, the **combination** or all possible outcomes of a sample size of n with r Y's is

$$C_r^n = \binom{n}{r} = \frac{n!}{r!(n - r)!}, \tag{4.1}$$

where $n! = n \times (n - 1) \times (n - 2) \times \ldots \times 1$.

If we choose the row with one Y, then we will have $4!/[1!\,(4 - 1)!] = (4 \times 3 \times 2 \times 1)/[1 \times (3 \times 2 \times 1)] = 4$. If we choose the column with two Y's, then we will have $4!/2!\,(4 - 2)! = (4 \times 3 \times 2 \times 1)/(2 \times 1)\,(2 \times 1) = 6$. These enumerations are essentially the same as asking the number of ways we can select r objects out of a total of n objects while the order of the selection is important. That is, YYYN and YNYY are regarded as different outcomes because they do not have the same order.

In the above example, the random variable, the number of Y's, carries *discrete* values, such as 0, 1, 2, 3, or 4, and thus it has a limited number of possible outcomes. Most discrete values imply some degree of categorical distinctions, or are measured at nominal or ordinal scales. A random variable can carry *continuous* values when the values are measured at interval or ratio scales. These continuous values typically generate an infinite number of possible outcomes when they are enumerated in ways similar to that in the above example.

4.2 PROBABILITY FUNCTIONS

As mentioned before, we are not interested in each of the possible outcomes in Table 4.1. Instead, we are interested in how the values of the random variable vary. That is, we would like to know how likely we are to get a low or high number of Y's during the survey. All of the possible outcomes listed in Table 4.1 can be translated into a **frequency distribution,** indicating how many possible outcomes we may have given each number of Y's. The frequency distribution is shown as a bar graph in Figure 4.1. This essentially shows us the likelihood of getting any specific number of Y's given that four households are surveyed.

Based on this distribution, it does not seem likely that we would get 0 Y or all 4 Y's from the survey. This is because either 0 Y or 4 Y's could occur only once out of the 16 possible outcomes. This makes the likelihood of having 0 Y or 4 Y's 1/16 each. Similarly, the likelihood of having 2 Y's is 6 out of 16, or 6/16.

In this manner, the frequency distribution can be translated into a **probability distribution** showing the probability of obtaining a given number of Y's. The probability distribution tells us the likelihood of obtaining a certain value of the random variable out of all possible outcomes. When a mathe-

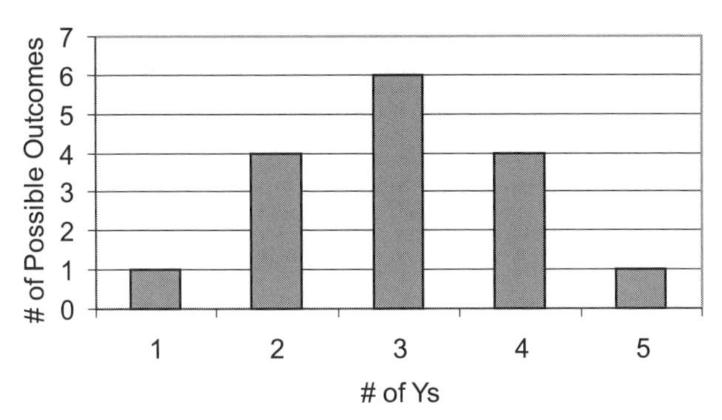

Figure 4.1 Frequency distribution of all possible outcomes.

matical function is used to describe the probability distribution, the function is known as a **probability function.**

Using another familiar example, if we define the random variable as the number on a dice showing up when the dice is thrown, we will have six possible outcomes. This is because the dice has six values, from 1 to 6, on its six faces. If the dice is unbiased, each value has the same probability of showing up when the dice is thrown—that is, one out of six, 1/6.

Let's use $p(x)$ to represent the probability function of the random variable x. The probability function of values of a dice will be $p(x) = 1/6$, $x = 1, 2, 3, \ldots, 6$. Following this, $p(x = 1) = 1/6$, $p(x = 2) = 1/6$, $p(x = 3) = 1/6$, $p(x = 4) = 1/6$, $p(x = 5) = 1/6$, and $p(x = 6) = 1/6$. Together, all possibilities sum to a value of 1, indicating that $\Sigma_{x=1}^{6} p(x) = 1$.

So far, we have focused on the probability distributions and functions of discrete random variables. For continuous random variables, the probability functions will likely be more complicated but more useful for statistical testing. Below, we will discuss three commonly used probability functions, two for discrete random variables and one for continuous random variables.

4.2.1 Binomial Distribution

The first discrete probability function to be discussed is the **binomial distribution.** Several characteristics of a situation warrant the use of this distribution. First, the binomial distribution is for discrete random variables and is appropriate only when an event has only two possible outcomes. In our previous example, if a household is given only two choices, Y or N, it may be appropriate to use a binomial situation. Or, if we toss a coin, normally we will have only two outcomes: head or tail.

Sometimes, we use general terms such as *success* and *failure* to describe the two possible outcomes in a binomial situation. In addition, we often use

p to denote the probability of success and q for the probability of failure. Recalling that probabilities of all outcomes should sum to 1, we should have $q = 1 - p$ or $p + q = 1$. Apparently, in a coin-tossing event, if the coin is normal (unbiased), then $p = 0.5$ and therefore $q = 0.5$, or $(1 - p)$.

The second characteristic of a binomial distribution is that the probability of success or failure for each trial is both known and the same for all trials. This characteristic affects our analysis in two ways. First, it implies that all trials are independent events. What happens in one trial does not affect what happens in the next. This, of course, may not always be the case in real events. Second, the characteristic implies that all trials are subject to the same probability distribution. This, again, may not always be true for real events. Keep these two implications in mind when using binomial distributions in studies. In our earlier example, the chance of contracting the West Nile virus is a function of many factors, including the outdoor activity pattern and the health condition of the household. Therefore, it may not be realistic to assume a constant probability of success (p) across all trials. However, it may be necessary to assume a constant probability when no other data are available or when constraints exist in budget or time.

The final characteristic of a binomial distribution is that the number of trials is fixed or known. Referring to our example, the number of trials is the same as the number of sampled households—four in our example. Obviously, when the number of trials changes, the number of all possible outcomes and their distributions across the values of the random variable will also change. If we survey five instead of four households, we will have 32 possible outcomes, using the permutation of $2^5 = 32$.

Given the number of trials, n, the number of successes, x, which is the binomial variable or random variable in this context, can be computed through the binomial distribution as follows:

$$p(x) = \binom{n}{x} p^x \, (1 - p)^{(n-x)}. \tag{4.2}$$

The first part of the expression is the same as the combinatorial calculation described in Equation 4.1. If we sample four households and we would like to know the number of all outcomes with only one household responding Y or (one Y), then the probability of the binomial variable for $x = 1$ will be

$$p(x = 1) = \binom{4}{1} p^1 \, (1 - p)^{(4-1)} = \binom{4}{1} p^1 (1 - p)^3.$$

The combinatorial of $\binom{4}{1}$ is $4!/1!3! = 4$, consistent with the results shown in Table 4.1. If the chance or probability that any household will be "visited" by a theft is less likely than random, say 0.4, then the probability of having one household visited by a theft $p(x = 1)$ is

$$p(x = 1) = 4 \times 0.4 \times 0.6^3 = 0.3456.$$

The above expression has intuitive meanings. The 0.4 probability indicates the probability that a household was visited. Given this probability, if one household has been visited, the other three will not be and each will have a probability of 0.6. Since this is true for the three remaining households, then the probability will be $0.6 \times 0.6 \times 0.6$. However, the household being visited can be the first household, the second, the third, or the fourth, leading to four possible arrangements of the patterns. Therefore, the product of the 4 probabilities is multiplied by four to give us the probability of one visit, regardless of which specific household among the four has the visit.

Note that we use $x = 1$ as the example above. We can set x equal to 0, 2, 3, or 4 in addition to 1 in order to derive the probabilities for all possible values of the random variable x. Of course, these values of x depend on the probability of success, p.

Once we calculate the probabilities of all combinations, we can construct a histogram of probability distribution for x, given $x = 0, 1, 2, 3, 4$.

$$p(x = 0) = 1 \times 0.4^0 \times 0.6^4 = 0.1296$$

$$p(x = 1) = 4 \times 0.4^1 \times 0.6^3 = 0.3456$$

$$p(x = 2) = 6 \times 0.4^2 \times 0.6^2 = 0.3456$$

$$p(x = 3) = 4 \times 0.4^3 \times 0.6^1 = 0.1536$$

$$p(x = 4) = 1 \times 0.4^4 \times 0.6^0 = 0.0256$$

$$\text{Total} \qquad 1.0000$$

Note that $0.4^0 = 1$ and that the sum of all the probabilities of the binomial random variable for all values of x, or the probabilities of all possible outcomes, is 1.0. This probability distribution describes the situation when $n = 4$ and $p = 0.4$. If the probability of success is different, a different probability distribution will be generated.

4.2.2 Poisson Distribution

The second discrete probability distribution is the **Poisson distribution.** This probability distribution is applicable for count variables. It is especially important and frequently used in testing spatial point patterns. It is different from the binomial distribution in several aspects. The computation of a Poisson distribution relies on an average, which may be derived from the long-term average of occurrences of an event, such as the number of landslides along a mountain slope over a period of 50 years, the number of hurricane landfalls in a region over the past 20 years, or the number of thefts in a neighborhood in the past 10 years.

Given the average number of occurrences of such an event, the Poisson distribution allows us to determine the probability of having a given number of occurrences. For instance, after studying the hurricane statistics over the past 50 years, we can compute an average number of hurricanes having landfalls along the Florida coast per year. With this average applied in a Poisson distribution, we can compute the probability that, for instance, only one hurricane will hit the Florida coast this year or any given year. For another example, if we know the average weekly number of car accidents at a particular street intersection over several years, we can estimate the probability of having no accident, one accident, or n accidents in any given week using the Poisson distribution. When a Poisson distribution is used, the events should be random and independent.

Conventionally, the average number of occurrences in a Poisson distribution is denoted λ. Also, x is conventionally used to denote the random variable, representing the number of occurrences of an event over time or across some geographic extent. With these notations, the probability of observing the event x times in a given condition can be computed using the following Poisson function:

$$p(x) = \frac{e^{-\lambda}\, \lambda^{x}}{x!}, \tag{4.3}$$

where e is the inverse of the natural logarithm (ln) function and e^1 is 2.71828.

From Equation 4.3, we can derive the probabilities of a range of x, theoretically from 0, 1, 2, 3 and up to infinity. For instance, let's assume that the long-term average number of earthquakes for a hypothetical region is three per year. We can compute the probabilities of having different numbers of earthquakes in a year:

$$p(x = 0) = \frac{e^{-3}3^0}{0!} = 0.0498$$

$$p(x = 1) = \frac{e^{-3}3^1}{1!} = 0.1494$$

$$p(x = 2) = \frac{e^{-3}3^2}{2!} = 0.2240$$

$$p(x = 3) = \frac{e^{-3}3^3}{3!} = 0.2240$$

$$p(x = 4) = \frac{e^{-3}3^4}{4!} = 0.1680$$

$$p(x = 5) = \frac{e^{-3}3^5}{5!} = 0.1008$$

$$p(x = 6) = \frac{e^{-3}3^6}{6!} = 0.0505$$

$$p(x = 7) = \frac{e^{-3}3^7}{7!} = 0.0216$$

$$p(x = 8) = \frac{e^{-3}3^8}{8!} = 0.0081$$

We may continue to compute the probability for $x > 8$ to infinity, but it is apparent that the sum of the probabilities for x larger than 8 will be very small—so small that it makes little sense to compute them. This is because probabilities quickly grow smaller as x increases beyond the average. Note that the sum of the probabilities for $x = 0, 1, 2, 3, \ldots \infty$ is 1. Therefore, the total probability for x larger than or equal to a given number k is

$$p(x >= k) = 1 - p(x < k)$$

or

$$p(x >= k) = 1 - [p(0) + p(1) + p(2) + \ldots + p(k - 1)].$$

If one takes a closer look at the procedures for computing $p(x)$, $x = 1, 2, 3, \ldots, 8$, as listed above, one may find that there is a shortcut method for computing the probabilities for a sequence of x's. For instance, the probability of $p(x = 1)$ is essentially

$$\frac{e^{-3}3^0}{0!} \times \frac{3}{1},$$

where $\lambda = 3$ and $x = 1$. For the probability of $p(x = 2)$, it is basically

$$\frac{e^{-3}3^0}{0!} \times \frac{3}{1} \times \frac{3}{2}.$$

In the additional ratio, 3 is λ and the denominator, 2, is x. With this, we can restructure the calculation of Poisson probabilities with this shortcut:

$$p(x + 1) = p(x) \times \frac{\lambda}{x}. \tag{4.4}$$

Thus, in the above example,

$$p(x = 3) = p(x = 2) \times \frac{3}{3}.$$

Consequently,

$$p(x = 3) = p(x = 2).$$

The shortcut formula is especially useful in computing the range of probabilities for x from 0 to k, or $p(x < k)$.

Compared with the binomial distribution, the number of trials, n, is not known in the Poisson distribution. The binomial distribution is used to determine the probability of success (or failure) of a given trial, while the Poisson distribution is used to determine the probability of having a specific number of occurrences of an event. The Poisson distribution is generally used to describe a random process of a low-occurrence event or a low-density pattern. If one is asked to throw darts at a wall without aiming at a specific place (i.e., random throwing), the resulting point pattern can be described very well by a Poisson distribution. This will be the way we use the Poisson distribution in the sections on point pattern analysis.

One other characteristic of the binomial and Poisson distributions deserves mention. In the binomial distribution the mean is $n \times p$, while in the Poisson distribution it is λ. The variance of the binomial is $n \times p \times q$, or $n \times p \times (1 - p)$. For the Poisson distribution the variance is also λ, a unique characteristic that will be explored further later in analyzing spatial point patterns.

4.2.3 Normal Distribution

Both the binomial and Poisson distributions are used for discrete random variables such that the values of the random variable, x, may equal to 0, 1, 2 or any discrete value. But in many situations random variables are continuous, such as income, age, or stream flow. Therefore, it is necessary to use continuous probability distributions for these variables. The most commonly used probability distribution for continuous random variables is the **normal distribution.**

Formally, a normal distribution of a random variable, x, is defined as

$$p(x) = \frac{1}{\sqrt{2\pi\sigma^2}} e^{-(x-\mu)^2/2\sigma^2}, \qquad (4.5)$$

where μ is the mean and σ is the standard deviation of x.

In Equation 4.5, a normal distribution is dependent on the mean, μ, and the standard deviation, σ. When different sets of means and standard deviations are used, different normal distributions will be generated. Suppose that we have two data series:

Series A: 5, 10, 15, 20, 25, 30, 35, 40, 45 ($\bar{x}_A = 25$, $s_A = 12.90994$)
Series B: 41, 42, 43, 44, 45, 46, 47, 48, 49 ($\bar{x}_B = 45$, $s_B = 2.582$)

Figure 4.2 shows the two normal distributions generated by different mean and standard deviation values. Though the distributions look the same, it is obvious that they span differently.

Because the standard deviation that defines how a normal distribution spreads is a relative measure, sometimes it is difficult to compare different normal distributions with similar standard deviations. In other words, we cannot easily compare normal distributions with different numeric ranges but with similar standard deviations. As a result, a set of statistics, called ***z*-scores,** have been used to transform values of different numeric ranges in different distributions so that they can be compared. The z-score statistics can be computed as

$$z_i = \frac{x_i - \bar{x}}{S}, \tag{4.6}$$

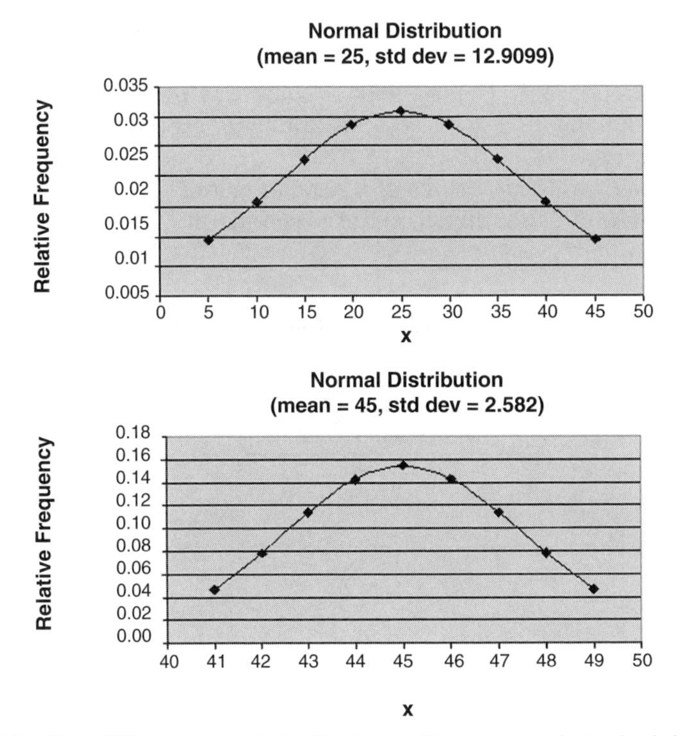

Figure 4.2 Two different normal distributions with means and standard deviations.

where $\bar{x}$ and S are the mean and standard deviation of the original data, x_i, $i = 1, 2, \ldots, n$. Since $\bar{x}$ is used as the reference point and S is used as the scale in Equation 4.6, z_i will follow a standard normal distribution with a mean of 0, and a standard deviation of 1. In this manner, we can convert all numeric values in a normal distribution into values in a standard normal distribution. With this standardization or transformation procedure, it will be convenient to compare different sets of values based on their standardized scores, z_i.

In the above example, when both series (A and B) are transformed and standardized, their z-scores become

Series A: $\bar{x}_A = 25$, $S_A = 12.90994$		Series B: $\bar{x}_B = 45$, $S_B = 2.582$	
5	−1.5492	41	−1.5492
10	−1.1619	42	−1.1619
15	−0.7746	43	−0.7746
20	−0.3873	44	−0.3873
25	0	45	0
30	0.3873	46	0.3873
35	0.7746	47	0.7746
40	1.1619	48	1.1619
45	1.5492	49	1.5492

Apparently, the mean (and the median) of both z-score series is 0 and the standard deviation is 1. This strategy of converting the original values into z-scores is commonly used in hypothesis testing to test the differences between distributions.

Figure 4.3 shows the z-scores of a standard normal distribution. The y-axis can be thought of as the likelihood of observing the value x. For any distri-

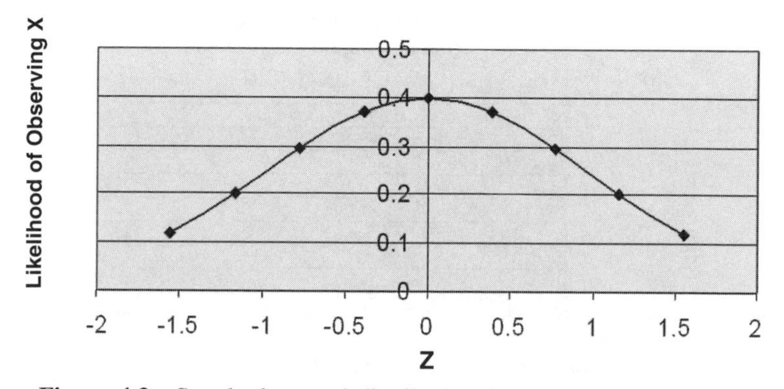

Figure 4.3 Standard normal distribution for data series A and B.

bution of continuous values, there are infinite possible outcomes for x. Therefore, the probability of observing x out of ∞ is very, very close to zero. Thus, we often calculate z-scores only within a given range of z-values, i.e., prob($a < z < b$), where a and b define a numeric range.

The standard normal distribution is symmetrical with respect to the mean. In addition, the sum of the probabilities or the area under the curve of the standard normal distribution should be 1, with the area on either side of 0 being 0.5 (= 1/2). In other words, since half of the x_i's should be equal to or greater than the mean, their z-scores should be equal to or greater than 0. Similarly, the other half of the x_i's should result in z-scores that are equal to or less than 0.

Let's use another, but greatly simplified, data series to illustrate how the standard normal distribution and its associated z-scores can be used to estimate the likelihood of having a certain number of events. Though the normal distribution is for continuous random variables, we use events to illustrate the usage only.

Series C: 10, 15, 20, 25, 30, 35, 40 ($\bar{x} = 25$, $s = 5$)

Assuming that the number of traffic accidents each day in a region follows a normal distribution with a mean of 25 and a standard deviation of 5, the standard normal distribution can be shown by the bell-shaped curve in Figure 4.4.

One can compute the probability of observing fewer than 20 accidents on a particular day. Using Equation 4.6, we can convert 20 into a z-score:

$$z = \frac{(x_i - \bar{x})}{S} = \frac{(20 - 25)}{5} = -1.$$

Then the procedure consists essentially of determining the probability under the standard normal distribution curve that z is less than -1, or $p(z < -1)$. Conceptually, this is the same as accumulating the area under the standard

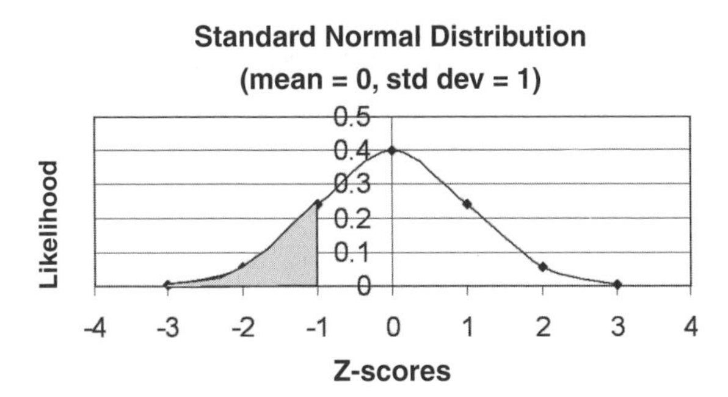

Figure 4.4 Standard normal distribution.

normal curve in Figure 4.4 from $-\infty$ up to the z-score of -1. Graphically, it is the shaded area in Figure 4.4. Note that the shaded area of $p(z < -1)$ is the same as the area of $p(z > 1)$. This is because the standard normal distribution is symmetrical.

To figure out the probability of observing z within a given range, we must perform integration on the standard normal distribution for the range of z values. Often we can refer to the standard normal distribution or z-score table found in most statistics textbooks. Different authors structure these tables in different forms. Some show the cumulative probability from $-\infty$ to z, and others show the cumulative probability from 0 to z. From any of those tables, we can verify that $p(z < -1) = p(z > 1) = 0.1587$. For the probability of z-scores over a range, such as $p(1 < z < 2)$, we need $p(z < 2) - p(z < 1) = 0.997 - 0.841 = 0.156$. Some simple statistical tests essentially use these probabilities to determine how likely two parameters are to be different or if the difference can be attributed to sampling error.

4.2.4 ArcView Notes

In the ArcView extension for Chapter 4, we have included a function to report the standard normal probability of larger than a given z-value. This function will report the probability or the area under the standard normal distribution to the right of the given z-value. This area is on the opposite side of the shaded area in Figure 4.4. Because the probability is only on the right-hand tail, we label it a one-tail probability. For two-tail tests, the significant level, α, must be divided into two halves.

When the Chapter 4 extension is loaded, a new menu, Ch.4, will be added to the ArcView user interface. Under the Ch.4 menu, the menu item "Prob > Z" is available. After clicking on this menu item, the user will be asked to provide a z-score. After the z-score is entered, the probability of a score larger than the z-score under the standard normal distribution will be returned. For instance, if we enter one as the z-score, the function will return a probability of 0.158674, which can be rounded off to 0.1587, the same answer we obtain from a statistical table. If one enters 1.96 as the z-score, the returned probability will be 0.250701, or roughly 0.25. The significance of this specific probability value will be discussed later.

In addition to "Prob > Z," probabilities for values of other statistics, which will be discussed later in this chapter, can be calculated in similar ways by choosing the menu items "Prob > t," "Prob > F," or "Prob > D (KS)."

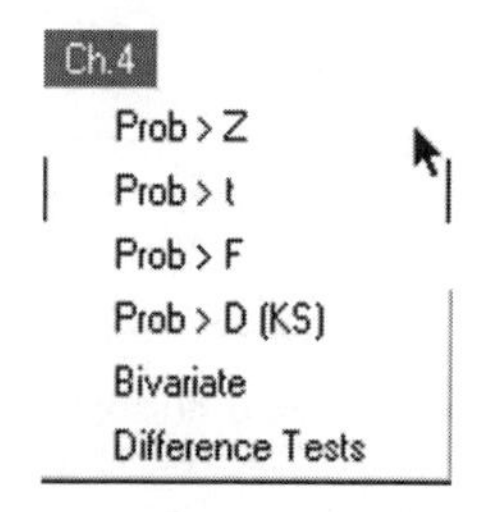

Note that all functions under the Ch.4 menu are accessible from View documents and Table documents.

Finally, ch4.avx and associated files would have been pasted to the correct folders if the procedure to install the CD was followed.

Then, to load `ch4.avx`, choose the **File** menu and then the **Extensions** . . . Click to check the box besides Ch4 Extension and then click OK to finish.

4.3 CENTRAL LIMIT THEOREM AND CONFIDENCE INTERVALS

The standard normal distribution is used in statistical inference in many ways. One possible purpose is to provide some level of confidence about how far a given value is from the true mean of a distribution. Before discussing that, we must describe the main idea of the central limit theorem.

When we compute the mean of a sample, it is based on a set of observations selected as a sample. If we call our sample X, then the associated sample mean is $\overline{X}$. If we select k sets of samples, then we have a set of sample means $\overline{X}_i$, where $i = 1$ to k. In other words, the sample mean is not a constant but a variable. Very likely, we know that sample means are not the same, or $\overline{X}_i \neq \overline{X}_j$, when $i \neq j$. But the question is: how much do the sample means vary? In other words, how much difference between sample means is reasonable when the samples are supposed to come from the same population?

Often we assume that the observations in the population form a distribution in which observations are independent (not dependent on each other) and are in an identical distribution. This is the so-called *iid property* of observations in the population. The independent property implies that there is a sufficiently large number of observations in the population and that the observed value of one individual is not dependent on the observed value of another. Moreover, all observations are drawn from the same frequency of the population, or an identical distribution.

In this case, if we create a large number of subsets or samples from the population in a random manner and compute the sample mean from each set of observations, the **central limit theorem** tells us that the sample means will follow a normal distribution. In addition, the mean of the sample means will approach the true population mean if there are enough sample sets from which the mean of the sample means is computed.

To estimate the variance of sample means, we take the sample variance and divide it by the number of observations in the sample—that is, S^2/n. If we take the square root of the variance of sample means, then we have an estimate for what is known as the **standard error,** or $S/\sqrt{n}$. Note that the standard error is only an estimate. The variability of the sample means is estimated with the variance from only one set of sample observations. The larger this sample size is, the smaller will be the variability of the sample means.

If we know the variability of the sample means, we can determine the likelihood that the true population is within a given range around a sample mean. We can do so because, in a normal distribution, we know the percentage of observations that are found within a given standard deviation around the mean. Using this property, we can estimate the percentage of chances (or probability) that the true population mean falls within a range defined by the estimated standard error. Figure 4.5 presents such information.

Figure 4.5 shows that, within one standard error ($-1 < z < 1$) around the mean of a normal distribution, we can expect to find approximately 68% (rounded from the actual 68.26%) of observations. Within two standard errors around the mean, we can expect to find approximately 95.44% of observations. Based on this property of the normal distribution, we can estimate that if we want to include 95% of the observations, the value range should be approximately 1.96 standard errors around the mean in the z-distribution table. Consequently, we can claim that there is a 95% chance that the true mean, μ, will be within 1.96 times the standard error around the sample mean. In other words,

$$Prob\left[\left(\bar{x} - 1.96\frac{s}{\sqrt{n}}\right) \leq \mu \leq \left(\bar{x} + 1.96\frac{s}{\sqrt{n}}\right)\right] = 0.95.$$

The 95% here is sometime labeled the **confidence interval,** as it indicates how certain we are that the population mean is within the given range. The complement of the confidence interval is 0.05 (= 1 − 0.95), the **significance level,** often denoted α. That is also the total areas or probabilities on both tails of the distribution beyond the range of the confidence interval. Thus, the **confidence level** is $(1 - \alpha)$. If the confidence interval is narrower, then there will be less chance that the population mean will fall within it. In that case, we would have a smaller confident interval. For instance, if we need only a 90% confidence level, then we need only 1.64 times the standard error from the sample mean, or

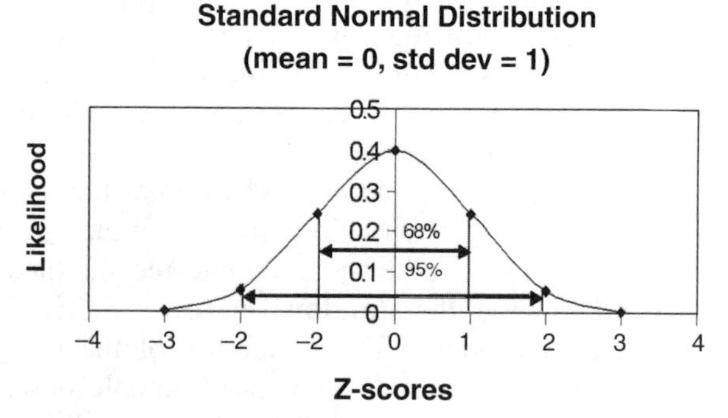

Figure 4.5 Approximated percentage of observations fall within a given standard deviation in a normal distribution.

$$Prob\left[\left(\bar{x} - 1.64\frac{s}{\sqrt{n}}\right) \leq \mu \leq \left(\bar{x} + 1.64\frac{s}{\sqrt{n}}\right)\right] = 0.90.$$

The standard error is basically the z-score of a given sample mean that gives the total probability equal to $(1 - \alpha)$. Therefore, as soon as we can determine the desired confidence level (90%, 95%, or 99%), we need to find out the corresponding z-score (1.64, 1.96, or 2.57, respectively). With the sample standard deviation, s, and the sample size, n, we can determine the range within which the population mean may fall with the given confidence level.

4.4 HYPOTHESIS TESTING

Besides estimating the probability of a given number of events or estimating the range within which we can expect to find the population mean, the probability distributions we have discussed so far are essential for statistical testing—when we want to test, for instance, whether the difference between two sample means is attributable to sampling errors or whether the two samples were selected from different populations. Or we may observe a spatial distribution of points and want to determine if the observed pattern is different from a random pattern or not. In this section, we will describe the standard procedures for testing a hypothesis and some of the associated statistical concepts.

For each hypothesis, we set up an alternative hypothesis. This ensures that we reach a conclusion even when we need to reject the original hypothesis. The original hypothesis to be tested is conventionally referred to as the **null hypothesis (H_0).** The null hypothesis is usually set up so that we can reject it if there is sufficient evidence. In other words, when we test a hypothesis, our goal is to find out if there is enough evidence to support the conclusion that the hypothesis is not true or that the **alternate hypothesis (H_a or H_1)** may be true.

Therefore, the first step in hypothesis testing is to formulate the null hypothesis—the hypothesis that we would like to reject so that we can claim that the alternate hypothesis may be true. That is, the alternate hypothesis represents what we hope to be supported by the evidence. Setting up hypothesis testing this way will give us some pertinent information, such as the presence of a relationship or the evidence of a pattern.

The null hypothesis is usually formulated to state that there is no statistically significant relationship between two variables, no statistically significant difference between two statistics, no specific pattern that is statistically significant, no statistically significant difference from a random pattern, and so on. While our intent is to reject the null hypothesis so that we can support the alternate hypothesis, the alternate hypothesis should state that there is a significant relationship, a difference, or a pattern. Apparently, the alternate hypothesis is what we would expect or what we would like to see in most cases.

After formulating the null and alternate hypotheses, the second step is to choose an appropriate test and compute the suitable test statistic for the specific situations. Different situations require different types of tests and statistics. While the decision is sometimes not straightforward, it is critical to use the right test for the situation. This topic will be addressed in more detail later.

As an example, let's assume that we wish to test if two sample means are statistically significantly different. When the sample size is small, we will use Student's *t*-test and will need to compute the *t*-statistic. If we want to test whether two frequency distributions are significantly different, one possible statistic to use is the chi-square (χ^2) statistic. Each statistic follows a specific probability distribution, such as the normal distribution or the Poisson distribution discussed earlier.

Which statistic or test to use is dependent on the circumstances. Some of the criteria that we can use to determine the appropriate statistic include the nature of the random variable (discrete or continuous), the measurement scale of the variables (nominal or interval/ratio), the nature of the question (looking for a relationship or a difference), the sample size (large or small), the number of variables involved, and so on. In fact, choosing the most appropriate statistic to use in a given situation is probably the most challenging and often the most frequent problem in statistical analysis.

Once the test is determined and the statistic is computed, the third step is to determine the probability that will guide us to reject or fail to reject the null hypothesis. Traditionally, we can resort to statistical tables, such as the standard normal distribution table or the chi-square table, to look up the probability corresponding to the calculated statistic. Today, when statistical packages are widely used, all packages will generate the corresponding probability for the calculated statistic. The probability indicates how likely or unlikely we will be to observe such a value of the statistic. This probability is then compared with our confidence level, α, defined a priori. Sometimes the probability has to be determined together with the **degrees of freedom** (*df*), which is often a function of the sample size and the number of variables involved. Given the calculated statistic and the *df*, we can determine the probability of observing such a value of the calculated statistic. If the probability is relatively low compared to our confidence level (such as $\alpha = 0.05$), then we could argue that the observed situation is not due to chance, but rather to systematic relationships or real differences.

Assuming that the probability of observing a given statistical value is readily available, the third step is to determine how low a probability is low enough to argue that it is unlikely to happen by chance. A customary benchmark is 5% or 0.05. This is also the **significance level,** α. $(1 - \alpha)$ gives us the confidence level discussed earlier.

Conceptually, α reflects the total probability or the area at the tails of the distribution used to define the **critical region(s).** If the calculated statistics fall into the critical region(s), we may argue that the chance that the observed statistic is that large is unlikely and that whatever we are testing is statistically significant. Figure 4.6 graphically shows the critical regions in a standard normal distribution with a total probability or area of 0.05 for the critical regions with the corresponding z-score, ± 1.96. The z-value(s) defining the size of the critical region(s) is/are also known as the **critical value(s).** In this particular example, if the calculated statistic (z-score) is larger than 1.96 or smaller than -1.96, then it falls inside the critical regions (shaded areas in Figure 4.6). It also means that the chance of observing a calculated statistic at this level has a probability of less than 0.05, a very unlikely situation.

Finally, given the calculated statistic, its corresponding probability, and the predefined level of significance, α, we can decide whether to reject the null hypothesis or not. In the case described above where the calculated statistic falls within the critical region, or when the corresponding probability is smaller than the desirable level of significance, α, we can argue that the observed situation does not occur by chance. Some relationships or differences do exist, and therefore we can reject the null hypothesis.

Note that when we reject the null hypothesis, we *cannot* say that we accept the alternate hypothesis completely. The alternate hypothesis is true only to a certain extent with a probability of $(1 - \alpha)$. If the critical value falls outside of the critical regions in the **region of acceptance,** or if the associated probability of the calculated statistic is greater than α, then we will fail to reject the null hypothesis and may conclude that there is evidence to support the conclusion that the null hypothesis may be true.

In summary, the major steps for hypothesis testing are as follows:

1. Set up the null and alternate hypotheses.
2. Choose an appropriate test and compute the associated test statistics.

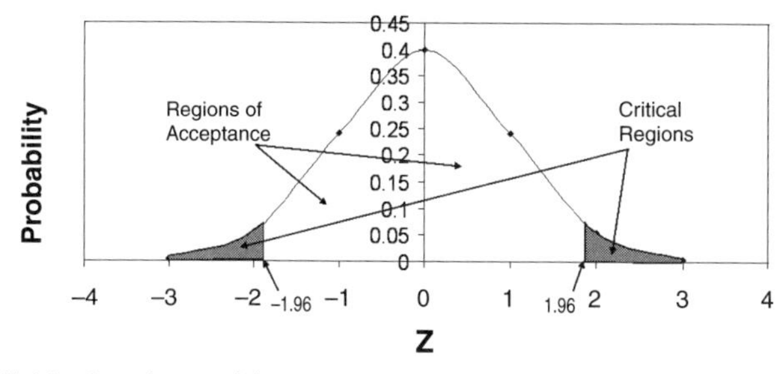

Figure 4.6 Critical regions with $\alpha = 0.05$ and the corresponding Z-score of ± 1.96.

3. Determine the level of significance, α, and compare the probability corresponding to the calculated statistic with α.
4. Draw a conclusion to reject or fail to reject the null hypothesis.

4.4.1 ArcView Notes

The menu item "Prob > Z" will return the probability of the critical region on the right-hand tail for a given z-value. Therefore, after one computes the z-score, one can enter the z-score into the "Prob > Z" function. The function will return a one-tail probability. If the probability is smaller than the traditional markers or α levels (e.g., 0.05 or 0.1), then one may claim that the chance of observing such a z-score is very unlikely, less than 5% or 10%.

4.4.2 Types of Error

In testing a hypothesis, we may commit an error and fail to reach a correct decision. In general, the error can be divided into two types. If the null hypothesis is correct but is rejected, then we have committed a **Type I error.** To minimize the chance of rejecting H_0 when in fact it is true, we may want to lower α, say from 0.05 to 0.01, to make it more difficult to reject H_0 or to be more conservative. But if H_0 is false and we fail to reject it, then we commit a **Type II error.** In general, the Type I error is more detrimental because, if we assume no relationship or no difference in H_0 and we reject it, we are saying that a relationship or difference exists. On the other hand, if we commit a Type II, we just fail to reject H_0, and we will refine or restate it and keep trying to reject it. In other words, we just fail to confirm what we think may be true instead of accidentally affirming what we think was true but in fact was false, committing a Type I error. Therefore, in general, we would like to minimize the chance of making a Type I error by being conservative and choosing a smaller α.

4.5 PARAMETRIC TEST STATISTICS

In the following sections, we will describe several commonly used statistical tests. At this point, we focus on the so-called parametric statistics. These statistics are appropriate to use with interval/ratio data. They are often used with the assumption that the distribution described by the statistics is a normal distribution with known parameters. The tests for nonparametric statistics, such as the chi-square test, which use the same concept, will be addressed later.

4.5.1 Difference in Variances

In geographical analysis, it is often assumed that the spatial process is homogeneous or stationary over a large study region even though we can observe a certain level of variability of a phenomenon or events across subregions. In other words, we often assume that the levels of variability across the subregions are relatively uniform, or **stationary.** Of course, this is essentially the claim that there is only one spatial process across the entire region. For example, we could assume that there is a nationwide trend toward decreasing manufacturing activities and increasing service activities. We naturally expect some levels of variation among subregions. But if the magnitude of variation is not uniform across subregions, we have a non-stationary situation.

Under the stationary assumption, what do we do with any variation we observe among our samples (or subregions)? In this case, we should test to see whether the differences in variances among the subregions are due to sampling errors (H_0) or to nonstationary processes over space (H_a). For this purpose, we can test the difference in variances first.

Formally, the null hypothesis is formulated as a ratio in the following manner:

$$H_0\colon \frac{\sigma_1^2}{\sigma_2^2} = 1$$

and the alternate hypothesis is

$$H_a\colon \frac{\sigma_1^2}{\sigma_2^2} > 1,$$

where σ_1^2 is the larger of the two variances. Empirically, we know that this ratio tends to follow an F-distribution. Setting up the null hypothesis this way allows us to test if the differences between two variances are large enough to reject the hypothesis that

$$\sigma_1^2 = \sigma_2^2 \quad \text{or} \quad \frac{\sigma_1^2}{\sigma_2^2} = 1.$$

Traditionally, the larger variance will be in the numerator, so that the ratio will be at least equal to 1. Thus, we only need to test if the ratio is significantly larger than 1 or not. Since the test focuses on one direction (i.e., $\sigma_1^2/\sigma_2^2 > 1$), this type of test is often known as a **one-tail test,** as opposed to the two-tail test situation in which we test for differences in both directions, that is, larger or smaller than the values defined by the critical statistics.

To compute the calculated F-ratio, we need the variances of the two samples:

$$F = \frac{S_1^2}{S_2^2}, \tag{4.7}$$

where S_1^2 is the larger of the two sample variances. The calculated F-ratio will be compared with a critical F value as defined by the level of significance, α. To determine the critical value, three parameters are needed: the significance level, α; υ_1, which is $n_1 - 1$; and υ_2, which is $n_2 - 1$, where n_1 and n_2 are the sizes of the two samples. Using these three parameters, the critical value for the F-ratio, $F_{(\alpha,\upsilon_1,\upsilon_2)}$, can be determined. The null hypothesis should be rejected if the calculated F-ratio is greater than the critical F value. This means that the difference between the two variances is large enough so that, statistically, we can say that the difference does not exist by chance.

The following is a simple example illustrating the F-test:

$$n_1 = 11 \qquad n_2 = 16$$

$$S_1^2 = 6.2 \qquad S_2^2 = 3.9$$

Because the first sample has a larger variance, its variance will be in the numerator of the F-ratio. The calculated F-ratio is

$$F_{\text{calculated}} = \frac{6.2}{3.9} = 1.590.$$

To determine the critical value, let's use the traditional marker $\alpha = 0.05$ as the level of significance. Together with $\upsilon_1 = 11 - 1$ and $\upsilon_2 = 16 - 1$, we can look up an F-table to identify the critical value. The critical value, $F_{(0.05,\,10,\,15)} = 2.5437$, can be found from any F-table. This critical value is greater than the calculated F-ratio of 1.590, or

$$F_{\text{calculated}} = 1.590 < F_{\text{critical}} = 2.5437.$$

Therefore, we fail to reject the null hypothesis and conclude that there is evidence showing that the null hypothesis is true; the difference in variances is due to sampling error.

4.5.2 ArcView Note

The Chapter 4 extension includes a new menu item, "Difference Tests." This item provides the capability to test the difference in variance. Check the "Difference of Variance Test" box; then choose the theme containing the data and click the "Show Variables" button. Variables of the selected theme will appear in the two variable boxes. Then select two variables and click the

"Calculate" button to start the computation. The program will compute the variances of the two selected interval/ratio variables, use the larger of the two variances as the numerator of the F-ratio, compute the F-ratio, and compute the corresponding probability according to the F-distribution. The function will return the calculated F-ratio and the probability of observing an F-ratio larger than the calculated one. Because the function will automatically use the larger variance in the numerator of the F-ratio, we do not have to consider which variable has the larger variance when choosing the two variables for comparison.

If we already have the variances of two variables computed by other means (e.g., using the functions in the extension of Chapter 3), then we can use the "Prob $> F$" function in the Ch.4 menu to test the difference in variance. After choosing "Prob $> F$," we will be asked to enter the two variances and the corresponding sample sizes. The function will return the F-ratio and the associated probability. Using the example above, we can enter 6.2 as the first (larger) variance and 3.9 as the second (smaller) variance. Then we enter 11 as the sample size for the first group and 16 for the second group. The function will return an F-ratio of 1.5897 and the associated probability of 0.1938, which is larger than the standard markers of $\alpha = 0.05$ or $\alpha = 0.1$. Therefore, we fail to reject the null hypothesis.

ArcView Example 4.1: Test of Variances

Step 1 Data source

In this example, required data can be found in
`C:\Temp\Data\Ch4_data\ch4data.shp.`
These shapefiles contain county boundaries of the 48 states plus Washington, DC, in the continental United States.

Step 2 Explore data

- Start ArcView GIS with an empty view.
- Using the **Add Theme** button, [icon], add the shapefile, `ch4data.shp`, to the opened View document. Next, use the **Open Theme Table,** [icon], to open the attribute table of ch4data.

Within the attribute table, the following attribute fields are available:

`Shape`: Polygon.
`Name`: County name.
`State_name`: Name of the state for each county.
`FIPS`: FIPS code for each county. For this example, all counties in Ohio have FIPS codes starting with 39, and all counties in Virginia have FIPS codes starting with 51.

`Svc_ann_p`: Annual payroll for service industries in each county (000's) in 1997.
`Svc_empl`: Number of employees in service industries in each county in 1997.
`Pcpopch`: Percentage change in population between 1997 and 2001.
`Pctcpinc`: Percentage change in per capita income between 1997 and 2001.
`Mfg_valadd`: Value added in manufacturing industries in each county in 1997.
`Mfg_empl`: Number of employees in manufacturing industries in each county in 1997.
`Mfg_ann_p`: Annual payroll for manufacturing industries in each county (000's) in 1997.

Step 3 Load ArcView extension, ch4.avx

- As with extensions in previous chapters, choose the Ch4 extension from the menu item of **Files,** then **Extensions** . . .

After the extension is loaded, a new menu item, **Ch.4,** is available from View documents and from Table documents.

Step 4 Formulate hypothesis

In this example, we test if the annual payrolls in service industries among 3111 counties vary significantly differently from those in manufacturing industries. If we use σ_1^2 to represent the variance of annual payrolls in service industries and σ_2^2 to represent the variance of annual payrolls in manufacturing industries, assuming that the variance of the service industries is the larger of the two variances, we can hypothesize that

$$H_0\colon \frac{\sigma_1^2}{\sigma_2^2} = 1.$$

The alternative hypothesis would be

$$H_a\colon \frac{\sigma_1^2}{\sigma_2^2} > 1.$$

Step 5 Test hypothesis

First, we need to select counties where we have data on annual payrolls for both manufacturing and service industries. To do this,we will use the **Query Builder.** Using the **Query Builder** button, choose counties with `[Svc_ann_p] > 0` and `[Mfg_ann_p] > 0`:

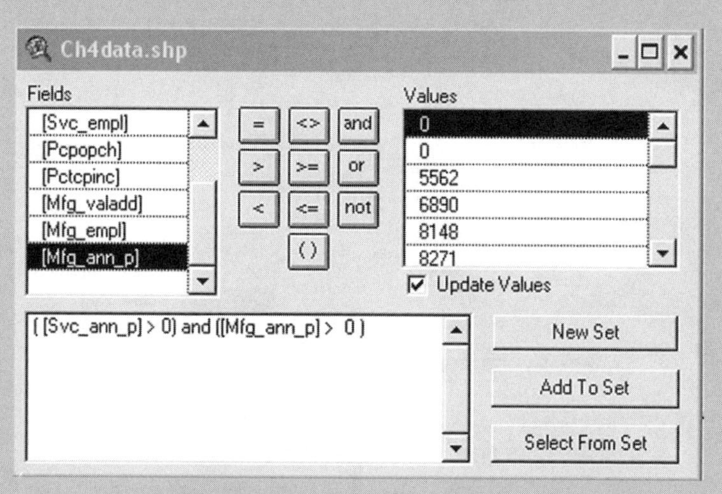

Next, we use the **Ch.4** menu to do the test.

- From the **Ch.4** menu, choose **Difference Tests.**

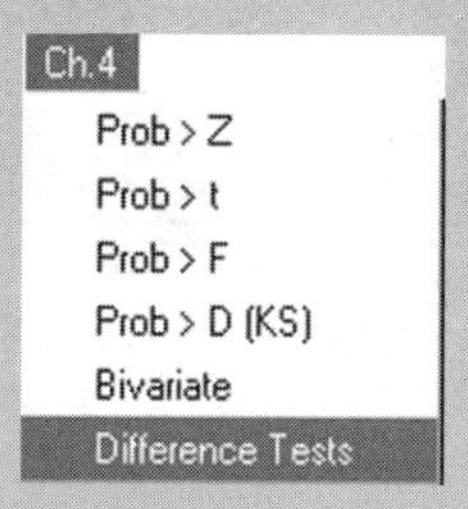

The Difference Tests dialog box appears.

- Check the box beside **Different of Variance Test.**
- Click the `Ch4Data` theme.
- Click the **Show Variables** button to list all attribute fields.
- For Variable 1, select `Svc_ann_p`.
- For Variable 2, select `Mfg_ann_p`.

Your dialog box should look similar to the one here:

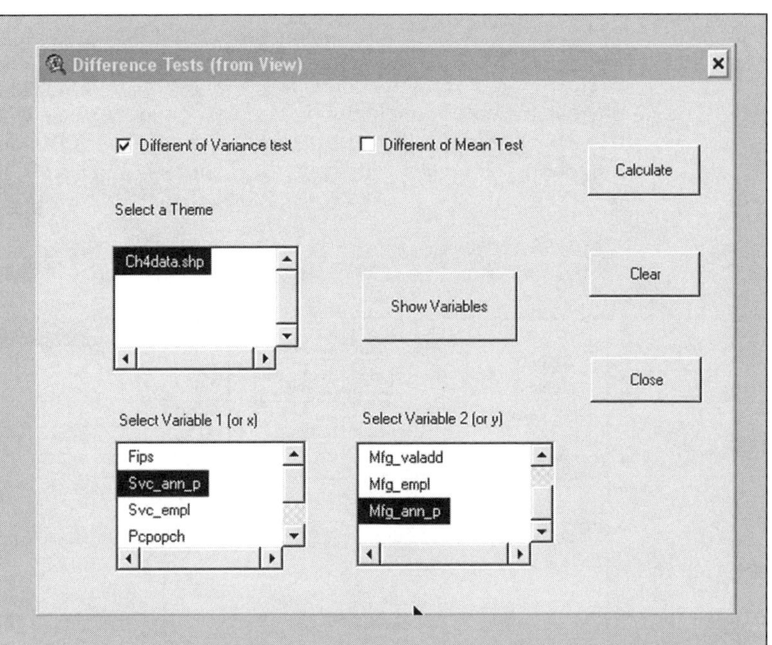

Click the **Calculate** button to proceed.

Step 6 Determine action
As shown in the resulting dialog box, we see that the *F*-ratio = 6.6511, which has a very, very small probability that is close to 0.

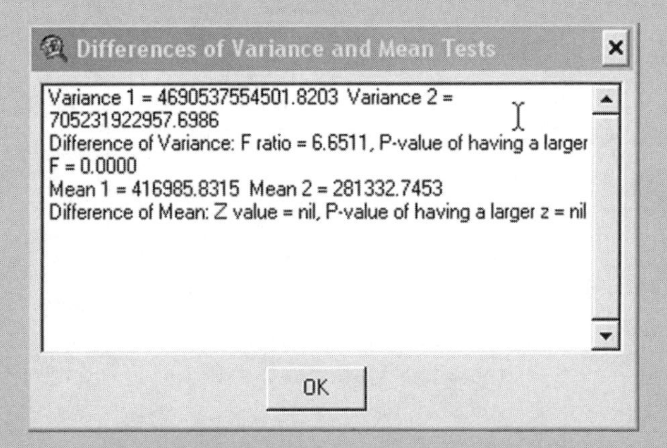

Thus, we reject the null hypothesis and claim that there is evidence to support the conclusion that the alternate hypothesis may be true.

Step 7 End

4.6 DIFFERENCE IN MEANS

The mean is the most common statistical parameter we can obtain from a sample because it is a good representative of the sample. If the purpose of studying the sample is to estimate the population, a common method is to test whether the sample mean is different from the population mean. In this case, we have a one-sample situation.

Sometimes we may have two samples and therefore two means to compare. We want to find out if the difference between the two sample means is attributable to sampling error or to the fact that the two samples come from different populations. Let's address the two-sample situation first. Afterward, it will be relatively easy to understand the one-sample case.

There are slightly different procedures for testing the difference in means, depending on at least two factors: the sample size and whether the variances of the two samples are statistically different or not. In other words, results from testing the variance difference can affect the way we test the mean difference. Therefore, it is logical to test the variance difference first.

4.6.1 Small Sample Size

In statistical testing, a small sample size usually refers to an n of less than 30. In small sample cases, we use **Student's *t*-statistic** to test the difference in means. Since we are testing whether the means are significantly different, we can set up the null hypothesis as

$$H_0\colon \mu_1 = \mu_2,$$

where μ_1 and μ_2 are the means to be tested for the significance of their difference. The alternate hypothesis to the above null hypothesis is

$$H_a\colon \mu_1 \neq \mu_2.$$

Student's t-statistic, used to test the difference in means, is formulated in the following manner:

$$t = \frac{|\bar{x}_1 - \bar{x}_2|}{S\sqrt{\dfrac{1}{n_1} + \dfrac{1}{n_2}}}, \tag{4.8}$$

where S is the pooled variance of the two samples, S_1^2 and S_2^2, and is defined as

$$S = \sqrt{\frac{S_1^2(n_1 - 1) + S_2^2(n_2 - 1)}{n_1 + n_2 - 2}}. \tag{4.9}$$

This is called the pooled variance because the variances of the two samples are combined or pooled together to derive an estimate of the variance for all the observations. We can pool the variances of the two samples together only when they are not significantly different. Note that the denominator of Equation 4.8 is essentially an estimate of the standard error of the difference in means. Therefore, we can use this estimate of standard error and the pooled variance only when the two variances are tested with no significant difference or when

$$H_0: \sigma_1^2 = \sigma_2^2$$

is not rejected.

If H_0 for the variance test is rejected, it means that the two variances are significantly different from each other. In that case, the pooled variance in Equation 4.9 cannot be used to estimate the standard error of the difference in means. If there is evidence supporting the conclusion that $\sigma_1^2 \neq \sigma_2^2$, then the standard error, *SE*, has to be estimated separately from the two sample variances as follows:

$$SE = \sqrt{\frac{S_1^2}{n_1} + \frac{S_2^2}{n_2}}. \tag{4.10}$$

The standard error estimated in Equation 4.10 can be substituted back into the denominator of Equation 4.8 to derive Student's *t*-statistic when the variances of the two groups are different:

$$t = \frac{|\bar{x}_1 - \bar{x}_2|}{\sqrt{\frac{S_1^2}{n_1} + \frac{S_2^2}{n_2}}}. \tag{4.11}$$

Using our example in the variance test earlier, we add the information about the mean so that we can perform a difference-of-means test:

$$n_1 = 11 \qquad n_2 = 16$$

$$S_1^2 = 6.2 \qquad S_2^2 = 3.9$$

$$\bar{x}_1 = 6.5 \qquad \bar{x}_2 = 9.8$$

Then the calculated *t*-statistic is

$$t = \frac{|6.5 - 9.8|}{S\sqrt{\frac{1}{11} + \frac{1}{16}}},$$

where

$$S = \sqrt{\frac{6.2 \times 10 + 3.9 \times 15}{(11 + 16 - 2)}} = 2.1955,$$

so that

$$t = \frac{3.3}{2.1955 \times \sqrt{0.1534}} = 3.8377.$$

One may look up a t-distribution table to determine what probability level corresponds to the calculated t-statistic. But one parameter is needed to determine the probability level. That is the degrees of freedom (*df*), which is equal to $n_1 + n_2 - 2$. In this example, $df = 11 + 16 - 2 = 25$. With $df = 25$, we find that the probability of having a t-statistic as large as 3.8377 is very low, or unlikely. This probability would be smaller than 0.005, as shown in most t-distribution tables. If one obtains the result from a statistical package, the package will report a probability of `prob < 0.0008`.

If we set the level of significance, α, to the traditional level of 0.05, we should definitely reject the null hypothesis because the probability of the calculated t-statistic is much smaller than the probability of the significance level, or 0.05. However, it should be noted that the t-distribution has two tails, one larger than 0 and the other smaller than 0. With that, α should be divided and allocated to both tails such that each tail has a critical region of size $\alpha/2$, or 0.025 in this example. We will divide α into the two tails when we test if the means are different or not. Note that the "difference" can be larger or smaller. Therefore, we need to cover both the larger tail and the smaller tail with the confidence level. This is the **two-tail test.**

In our formulation of the t-statistic in Equation 4.11, we took the absolute value of the two means in the numerator because we were concerned with whether they are different but not if one is larger than the other. This is consistent with our null hypothesis. Consequently, the size of the critical region in one of the tails should be one-half of α, or 0.025. Getting back to the example, the probability of the calculated t-statistic of 0.0008 is still much smaller than $\alpha/2$, or 0.025. This indicates that having such a t-value of the calculated statistic is very unlikely if the two means are not significantly different. As a result, we should reject the null hypothesis, which states that the two means are not significantly different. In other words, there is evidence showing that the alternate hypothesis (H_a) may be true.

Another approach to conducting this test is to determine the critical value given the α level to determine the critical region. With that probability, we can check if the calculated statistic falls within the critical region. If we use $\alpha/2$ or 0.025, the critical value (t_{cv}) for $t_{(df=25)} = 2.06$. The calculated statistic of 3.8377 is definitely inside the critical region because $3.8377 > 2.06$, or $t_{calculated} > t_{cv}$; therefore, we can reject the null hypothesis.

If our null hypothesis is that one mean is not larger than the other, then it will be a one-tail test. Given that we now are concerned with the directional difference, we no longer can take the absolute value of the difference in means in the numerator of Equation 4.8 or 4.11. In this case, $\alpha = 0.05$, will be allocated entirely to one tail to define the critical region. If our example is a one-tail case, the critical value (t_{cv}) for $t_{(df=25)}$ will be 1.708, which is much smaller than the t_{cv} for the two-tail case. This is because the critical region is now much larger on the right-hand side, or the entire 0.05 is on the right-hand side.

4.6.2 Large Sample Size

In the above discussion of testing the difference in means, we assume that the sample size of one or both groups is smaller than 30. In these small sample cases, the difference-of-means statistic follows the t-distribution. If the sample size gets larger, the t-distribution will approach the standard normal distribution (z). The general guideline suggested by statisticians is that if the sample size is equal to or larger than 30, the difference-of-means statistic should be relatively close to the standard normal distribution. Therefore, if $n \geq 30$, the calculated statistic should be

$$z = \frac{|\bar{x}_1 - \bar{x}_2|}{\sqrt{\dfrac{S_1^2}{n_1} + \dfrac{S_2^2}{n_2}}}. \tag{4.12}$$

Note that this equation for large sample size is identical to Equation 4.11 for the test of difference in means when n is small and the variances of the two groups are significantly different. However, the statistic in Equation 4.12 will follow the z or standard normal distribution. What Equation 4.12 implies is that when the sample size is large, we need to assume that the variances of the two groups will be significantly different. In other words, the difference in means test for large sample size is not dependent upon the result from the difference in variances test.

The example below illustrates how to test the mean with large sample sizes:

$$n_1 = 120 \qquad n_2 = 150$$
$$S_1^2 = 6.20 \qquad S_2^2 = 6.75$$
$$\bar{x}_1 = 62.2 \qquad \bar{x}_2 = 61.3$$

Note that the means of the two samples are relatively close, with a difference of only 0.9. This difference in means seems too small to be statistically significant. However, it cannot be confirmed unless the sample sizes and variances are also considered. To test if the two means are significantly different, we would calculate the z statistic as

$$z = \frac{|62.2 - 61.3|}{\sqrt{\dfrac{6.20}{120} + \dfrac{6.75}{150}}}.$$

That gives us a z-value of $0.9/0.3109 = 2.8948$. This z-value is applicable to both the two-tail test when testing the difference and the one-tail test when testing if the first mean is larger than the second.

Assuming that our significance level, α, is 0.05 in all of our tests, if we perform a two-tail test, then we will divide the size of the critical region, 0.05, into two halves such that the critical region in each tail is only 0.025. With $\alpha/2$, the critical value (z_{cv}) will be 1.96, and we will reject the null hypothesis only if the absolute value of the calculated statistic is larger than the critical value. In this case, the calculated z-score of 2.8948 is larger than the z_{cv} of 1.96, or

$$z_{\text{calculated}} = 2.8948 > z_{cv} = 1.96.$$

Therefore, we can reject the null hypothesis and claim that there is evidence that the two means are significantly different.

If we perform a one-tail test to test if the first mean is larger than the second mean statistically, we can use the same z-score as the calculated statistic. However, because it is a one-tail test, α will be allocated entirely to the right-hand tail, and z_{cv} will be 1.645. With a larger critical region, z_{cv} will be smaller. In this case, the calculated statistic ($z = 2.8948$) is still larger than the z_{cv} of 1.645, or

$$z_{\text{calculated}} = 2.8948 > z_{cv} = 1.645.$$

Therefore, the first mean is significantly larger than the second mean.

4.6.3 ArcView Note

The several menu items of the Ch.4 extension offer functions to perform the difference-of-mean test. First, "Prob > Z" will return the probability of a value larger than a given z-value. This function can be used to test if the z-score from the difference-of-means test for large samples is significant. Using the example above, the calculated z-score was 2.8948. The returned probability from the function is 0.0039, which is much smaller than the standard significance level of 0.05 or 0.025 for a one-tail situation. "Prob > t" is another function for testing the mean difference, but it is used for smaller sample sizes. In the previous example of the difference-in-means test for small samples, the t-value was 3.8377. If this t-value is entered as the input into the "Prob > t" function with 25 df, it will return a probability 0.000375 for a one-tail situation or 0.00075 for a two-tail situation.

It is cumbersome to compute the z- or t-values separately and determine the corresponding probability. The "Difference Tests" function can perform a difference-of-means test from two selected attributes stored in a theme or table in ArcView. As discussed above, the mean test will be different, depending upon the sample size and whether the variances of the two variables are different or not. If the sample size is large ($n \geq 30$), the procedure will assume that the variances are different. If the sample size is small, the difference-of-means test will depend upon the result of the difference-of-variance test if the latter test is also selected. If the test indicates that the variances are not significantly different ($\alpha = 0.05$), then a pooled variance will be estimated to compute the standard error for the t-test. Otherwise, the variance will be estimated separately. If the difference-of-variance test is not selected, and thus will not be performed, the difference-of-means test will ask the user if the different variances assumption should be used or not.

In ArcView Example 4.1, we will use the z-score to test if income changes in individual states are significantly different from the national average.

Arcview Example 4.2: Changes in Per Capita Income between 1997 and 2001

Step 1 Data source

In this example, the required data come from
`C:\Temp\Data\Ch4_data\ch4data.shp.`
These shapefiles contain boundaries of counties of the 48 contiguous states plus Washington, DC, in the United States.

Step 2 Explore downloaded data

- Start ArcView GIS with an empty view.
- Using the **Add Theme** button, , add the shapefile, `ch4data`, to the opened **View** document. Next, use the **Open Theme Table,** , to open the attribute table of ch4data.

Within the attribute table, the following attribute fields are available:

`Shape`: Polygon.
`Name`: County name.

`State_name`: Name of the state for each county.

`FIPS`: FIPS code for each county. For this example, all counties in Ohio have FIPS codes starting with 39, and all counties in Virginia have FIPS codes starting with 51.

`Svc_ann_p`: Annual payroll for service industries in each county (000's) in 1997.

`Svc_empl`: Number of employees in service industries in each county in 1997.

`Pcpopch`: Percentage change in population between 1997 and 2001.

`Pctcpinc`: Percentage change in per capita income between 1997 and 2001.

`Mfg_valadd`: Value added in manufacturing industries in each county in 1997.

`Mfg_empl`: Number of employees in manufacturing industries in each county in 1997.

`Mfg_ann_p`: Annual payroll for manufacturing industries in each county (000's) in 1997.

Step 3 Load ArcView extension, ch4.avx

- As with extensions in previous chapters, choose the Ch4 extension from the menu item of **Files,** then **Extensions** . . .

After the extension is loaded, a new menu item, **Ch.4,** is available from View documents and from Table documents.

Step 4 Calculate national averages

If the Ch4data shapefile has not already been added, use the **Add Theme** button to add it to your View document and then open the attribute table. First, find out the national average of percentage changes in per capita income and population, both at the county level. To do this:

- Open the attribute table if it is not already open.
- Click the title button of pcpopch; choose the **Field** menu, then **Statistics.**

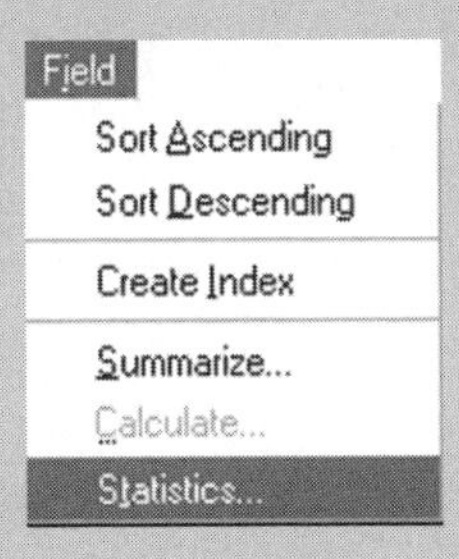

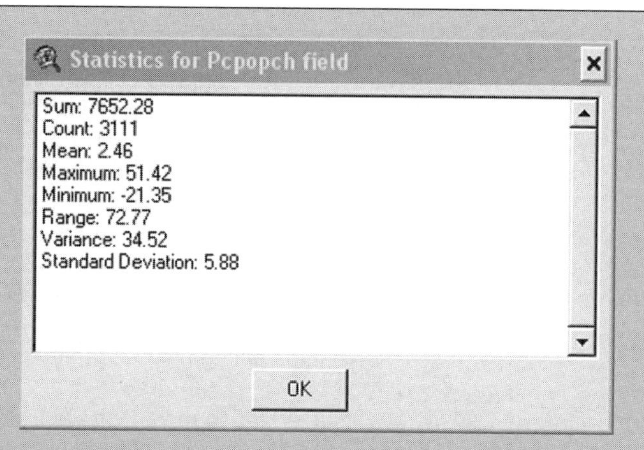

For all counties in the continental U.S., the average percentage change in population between 1997 and 2001 is **2.46%.**
For the percentage change in per capita income:

- Click the title button of `pctpcinc`; then choose the **Field** menu, then **Statistics.**

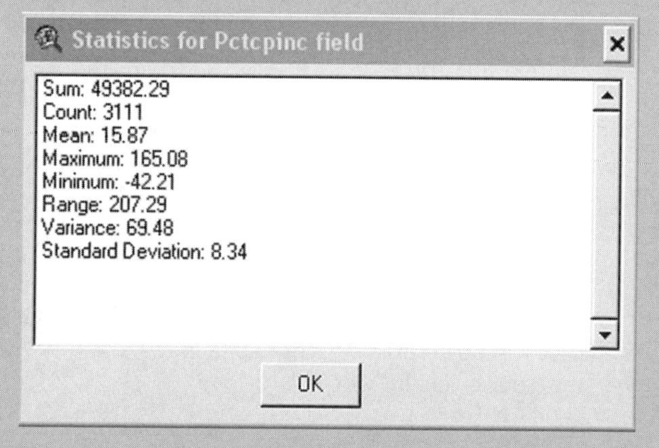

As can be seen from the output, the mean percentage change in per capita income at the county level is **15.87%.**

Step 5 Formulate hypotheses
To see if the changes in population and per capita income in Ohio and Virginia are not significantly different from the average percentage changes in all counties in the 48 states, we can use the *z*-score calculated using each state's counties. Essentially, we are testing

$$H_0\colon \mu_{\text{Ohio}} = \mu; \quad H_1\colon \mu_{\text{Ohio}} \neq \mu$$

and

$$H_0: \mu_{\text{Virginia}} = \mu; \quad H_1: \mu_{\text{Virginia}} \neq \mu$$

Step 6 Calculate the descriptive statistics
First, select only the Ohio counties. To do so:

- Use the **Query Builder** button, . Specify `[State_name] = ''Ohio''` as:

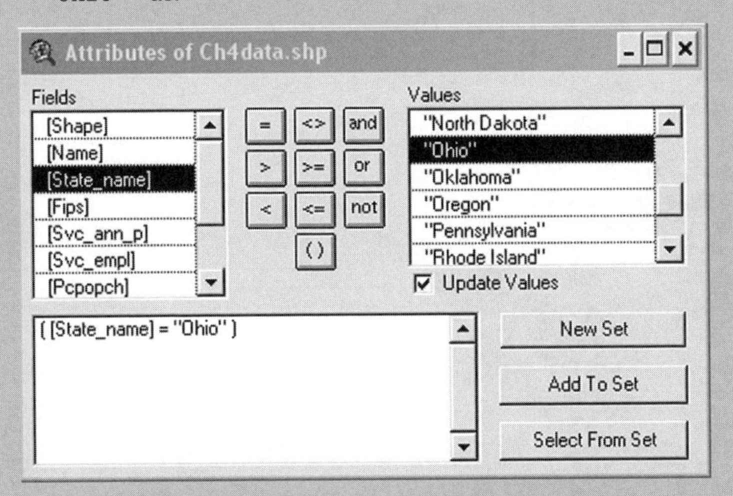

- Now click the title button of `pctcpinc` to indicate that this is the attribute field to be used for calculation. Next, from the **Field** menu, choose **Statistics.** We have:

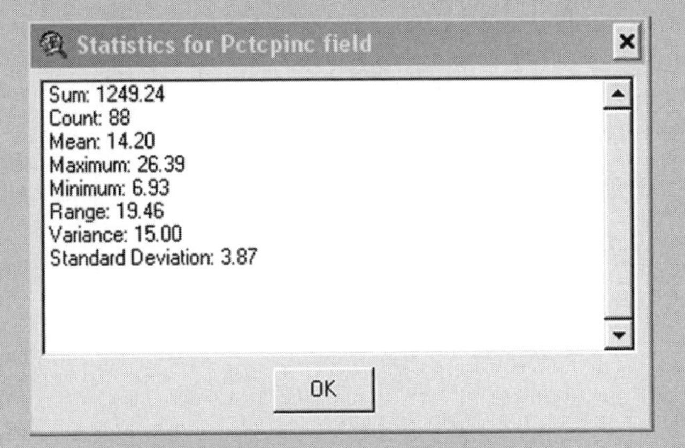

Before proceeding to do the same for Virginia:

- Be sure to clear the selected counties with the **Select None** button, .

- Next, select counties in Virginia by using the **Query Builder** button again and specify the query as `[State_name] = ''Virginia''`.
- Also with the **Field** menu's **Statistics,** we have:

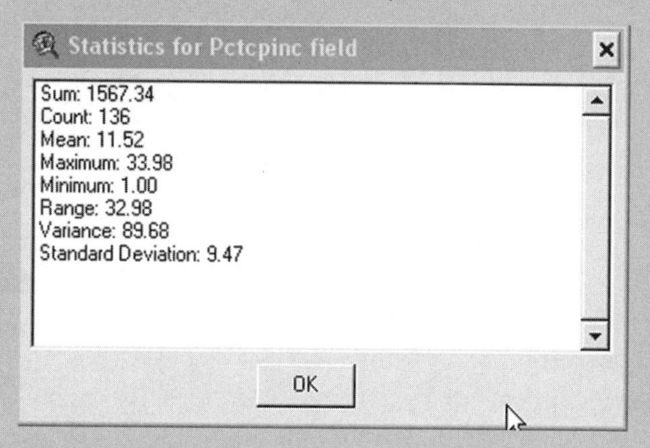

Step 7 Calculate *z*-scores
Recall from Equation 4.12 that we have

$$z = \frac{|\bar{x}_1 - \bar{x}_2|}{\sqrt{\dfrac{S_1^2}{n_1} + \dfrac{S_2^2}{n_2}}}.$$

With the descriptive statistics we have calculated, we can calculate the *z*-scores for Ohio and Virginia.
For Ohio, we have

$$z = \frac{|14.2 - 15.87|}{\sqrt{\dfrac{15}{88} + \dfrac{69.48}{3111}}} = \frac{1.67}{\sqrt{0.1704 + 0.0223}} = \frac{1.67}{\sqrt{0.1927}} = 3.80.$$

For Virginia, we have

$$z = \frac{|11.52 - 15.87|}{\sqrt{\dfrac{89.68}{136} + \dfrac{69.48}{3111}}} = \frac{4.35}{\sqrt{0.6594 + 0.0223}} = \frac{4.35}{\sqrt{0.6817}} = 3.45.$$

Step 8 *z*-test

- From the **Ch.4** menu, use **Prob > *Z*,** with 3.8. We have a probability of 0.0001729273, a value far lower than 0.025 (half of $\alpha = 0.05$) for the two-tail test. We reject H_0 and conclude that, for percentage changes in per capita income between 1997 and 2001, Ohio's changes are statistically significantly different from the national trend.
- For Virginia, use **Prob > *Z*** and plug in 3.45. We have a probability of 0.000380635. Again, this is far below 0.025. Therefore, we also conclude by rejecting H_0.

Step 9 Testing population changes
Can you repeat the test for percent change in population for these two states?

Step 10 End

4.7 DIFFERENCE BETWEEN A MEAN AND A FIXED VALUE

In many cases, we may want to compare the mean calculated from a sample to a fixed value—a value with no variance information. This value could be a known population mean, an observed value, a value with special meaning, or even zero.

Let's call the fixed value μ. Then, when the sample size is small ($n < 30$), the test statistic, which will follow a *t*-distribution, is defined as

$$t = \frac{\bar{x} - \mu}{\frac{s}{\sqrt{n}}}, \tag{4.13}$$

with $(n - 1)$ degrees of freedom (*df*). The denominator is essentially the standard error.

If the sample size is large ($n \geq 30$), the formulation of the test statistic will be the same as for the small sample case:

$$z = \frac{\bar{x} - \mu}{\frac{s}{\sqrt{n}}}, \tag{4.14}$$

but it will follow a *z*-distribution.

4.7.1 ArcView Note

Though there is no specific function in the Chapter 4 extension to test the difference between the mean and a fixed value, if the t-value is computed, we can still use the "Prob > t" function to find the probability in order to decide to reject or not to reject the null hypothesis.

In ArcView Example 4.2, we tested if the difference between the state means and the national mean is statistically significant. That test can be reformulated to test the difference between the means and a fixed value (national mean), using the z-statistic.

4.8 SIGNIFICANCE OF PEARSON'S CORRELATION COEFFICIENT

In the previous chapter, we discussed using Pearson's correlation coefficient, or r, to evaluate the strength and direction of the correlation between two sets of values. The coefficient can indicate the direction through the positive or negative sign and the strength through its coefficient value, which ranges from 0 for no correlation to 1 for the strongest correlation in absolute value.

Occasionally, we may observe a strong correlation between two variables, but based on a small number of observations. In that case, the strong correlation may be due to sampling error. To verify if the correlation is real, we may want to test if the correlation coefficient is significantly different from zero or not. The calculated statistic, which will follow a t-distribution, is defined as

$$t = r\sqrt{\frac{n - 2}{1 - r^2}} \tag{4.15}$$

with $(n - 2)$ degrees of freedom.

In the example in Section 3.4 examining the correlation between median household income and median housing value among the 50 states and D.C., the correlation was 0.79. If $n = 51$, the t-statistic is

$$t_{\text{calculated}} = 0.79 \times \sqrt{\frac{51 - 2}{1 - 0.79^2}} = 0.79 \times \sqrt{130.3538} = 9.0196.$$

The critical value with $t_{(df=49)}$ is approximately 1.677 for a one-tail test (α = 0.05), and 2.011 for a two-tail test ($\alpha/2 = 0.025$). In both cases, the calculated t-statistic is larger than the critical values, and thus it is safe to claim that the observed correlation coefficient of 0.79 is significantly different from zero. In general, the larger the sample size, the more likely the correlation coefficient is to be significant.

4.8.1 ArcView Note

Correlation is part of bivariate analysis. The Chapter 4 extension has a "Bivariate" menu item that computes several bivariate measures. This user interface or the dialog of this menu item is identical to the interface in the Chapter 3 bivariate extension. However, significance testing routines are added so that it reports both the correlation statistics and the significance levels of those statistics. If we check the box for Pearson's correlation coefficient, the correlation and the associated probability (P-value) will be reported. The smaller the probability, the more significant the correlation.

A word of caution: when using this function, one should recognize that different statistics should be used for different types of data. Refer to the discussions of various correlation coefficients suitable for nominal, ordinal, and interval/ratio data in Chapter 3.

Below, ArcView Example 4.3 shows a case in which we test the statistical significance of the correlation between changes in per capita income and economic activities in manufacturing and service industries. While this is not the appropriate place to discuss issues in economic geography in detail, it is interesting to note that statistical analysis often provides great insights in understanding geographic phenomena.

ArcView Example 4.3: Statistical Significance of the Correlation Coefficient

Step 1 Data preparation and extension

As in the previous ArcView example, load the Ch4.avx extension. Add the Ch4Data shapefile to the active View document.

Step 2 Data exploration

In this example, we will be working with

`Pctcpinc`: percentage change in per capita income, 1997–2001

`Pcpopch`: percentage change in population, 1997–2001

`Svc_ann_p`: annual payrolls (000's) in service industries, 1997

`Mfg_ann_p`: annual payrolls (000's) in manufacturing industries, 1997

Step 3 Data query

First, select counties that have valid data:

- Use the **Query Builder** button to select records with `[svc_ann_p] > 0`

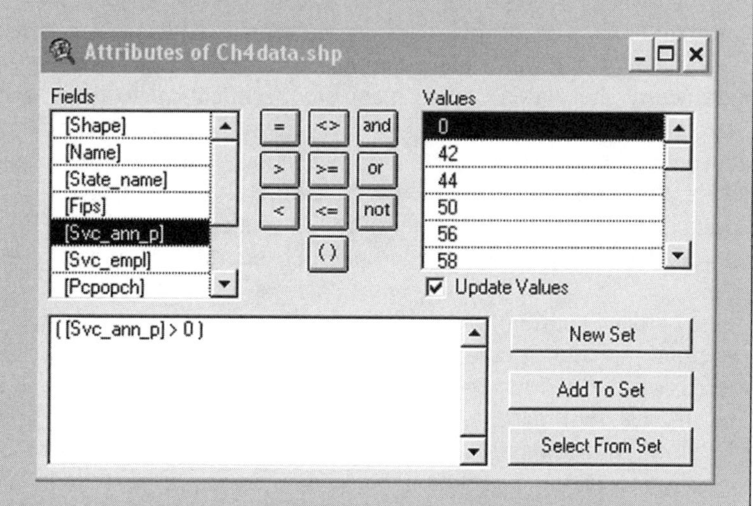

- Click the **New Set** button to proceed.
- Close the **Query Builder** dialog box when you are finished.

Notice that the selected counties are highlighted in yellow.

Step 4 Calculate test statistics

To calculate the correlation coefficient and the test statistics with probabilities:

- From the drop-down menu of **Ch.4,** choose **Bivariate**
- In the dialog box.
- Check the **Pearson Correlation Coefficient.**
- Click the data theme, `Ch4Data`.
- Click the **Show Variable** button to list all attribute fields.
- For Variable 1, choose `pcpopch`, and for Variable 2, choose `svc_ann_p`.
- Click the **Calculate** button to proceed with the calculation.

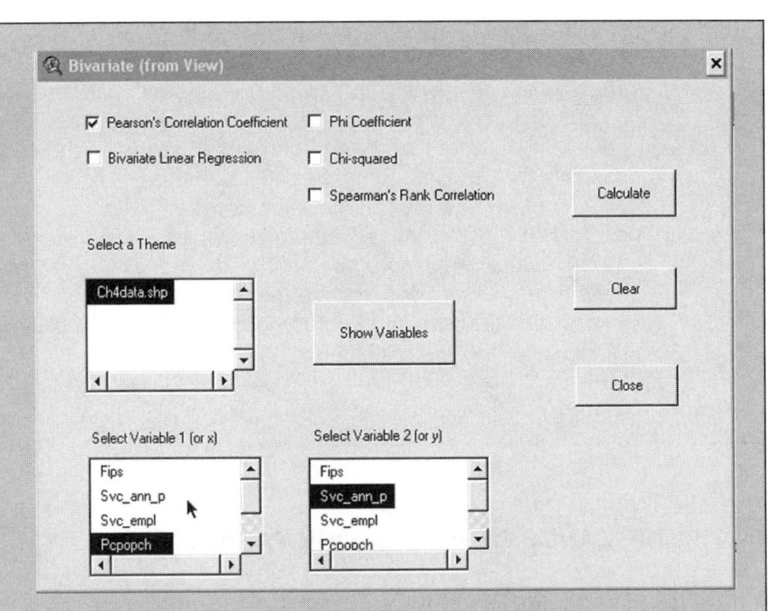

When this is completed, the coefficient, statistic, and probability are displayed in a dialog box, as

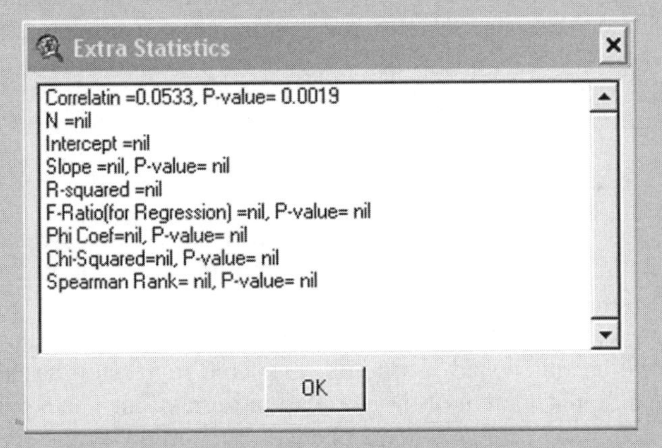

Between pcpopch and svc_ann_p, the correlation is a very low value of 0.0533. It represents a very weak, almost nonexistent, correlation between the two variables.

Step 5 Additional variables
Using the menu item, **Ch.4/Bivariate,** with the option of **Pearson Correlation Coefficient,** we have the following results:

Between pcpopch and svc_ann_p, $R = 0.0533$, $P = 0.0019$
Between pcpopch and mfg_ann_p, $R = 0.0342$, $P = 0.0647$
Between pctcpinc and svc_ann_p, $R = 0.1353$, $P = 0.0000$
Between pctcpinc and mfg_ann_p, $R = 0.1779$, $P = 0.0000$

The results show that manufacturing activities have slightly stronger relationships with changes in percent changes in per capita income and percent changes in population, both at the county level. Note that even though most values of the correlation coefficients are relatively small (<0.18) in this set of comparisons, we can claim that the relationships between pairs of variables are weak, and most of these weak relationships are significant, as their *P*-values are smaller than the traditional 0.05 level; that is, there is less than a 5% chance that we will observe a correlation level at this magnitude. In other words, we can have weak but statistically significant relationships.

Step 6 End

4.9 SIGNIFICANCE OF REGRESSION PARAMETERS

As discussed in Chapter 3, closely related to Pearson's correlation coefficient is the bivariate regression, which offers a mathematical description of the relationship (or trend) between the independent variable (x) and the dependent variable (y). To review briefly, the regression will provide a predicted value, $\hat{y}_i$, of observation i, given the value of the independent variable, x_i, for observation i. Thus, the regression can be written as

$$\hat{y}_i = \hat{a} + \hat{b}x_i, \tag{4.16}$$

where $\hat{a}$ is the estimate of the intercept, that is, the value of y when $x = 0$, and $\hat{b}$ is the estimate of the slope of the regression line parameter β, that is, the ratio of changes in y to a change in x.

As in testing the difference between the sample mean and the population mean, we are interested in finding out if the regression based on the sample is a good representative of the population regression. The b coefficient reflects how much y will change when x changes by 1 unit. Of most interest is whether the relationship is statistically significance or not. In other words, we want to find out if the slope parameter, b, is significantly different from zero or not. To test b, we can formulate a null hypothesis and an alternate hypothesis in the following manner:

$$H_0: \beta = 0$$

$$H_a: \beta \neq 0$$

While testing if the slope parameter, β, is significantly different from zero, we can treat the situation as if we are testing whether the mean is different from a fixed value, as discussed earlier. The test statistic will be a t-statistic as follows:

$$t_{n-1} = \frac{b - \beta}{S_b}, \tag{4.17}$$

where S_b is the standard error of the slope parameter, b, and the t-statistic has $(n - 1)$ degrees of freedom (df). Because the null hypothesis is $\beta = 0$, the t-statistic can be simplified to

$$t_{n-1} = \frac{b - 0}{S_b} = \frac{b}{S_b}, \tag{4.18}$$

and S_b, in turn, is defined as

$$S_b = \sqrt{\frac{S_e^2}{(n - 1)S_x^2}}, \tag{4.19}$$

where S_x^2 is the variance of the independent variable, x, and S_e^2 is the variance of the residuals, which can be defined as

$$S_e^2 = \sum_i^n \frac{(y_i - \hat{y}_i)^2}{n - 2}. \tag{4.20}$$

As the simple regression is closely related to the Pearson correlation coefficient, recall that the **coefficient of determination,** r^2, indicates how well the regression performs. It is literally the square of the Pearson correlation coefficient. We could find out if the regression is successful by testing the significance of r^2 using an F-test, which can be defined as

$$F = \frac{r^2 (n - 2)}{1 - r^2}. \tag{4.21}$$

This F-statistic will have 1 and $(n - 2)$ degrees of freedom.

4.9.1 ArcView Note

The "Bivariate" function in the Ch.4 extension will report not just the regression parameter, but also the F-statistic (and the associated probability) for testing the significance of the regression, as well as the t-statistic (and the associated probability) for testing the significance of the regression parameter when the regression option is checked.

In ArcView Example 4.4, we explore the relationship between the value added in manufacturing industries and the annual payrolls in manufacturing industries at the county level. Assuming that counties engaging in manufacturing activities with higher values added tend to have higher annual payrolls, we can use regression to find out if this relationship is statistically significant or not.

ArcView Example 4.4: Manufacturing Value Added and Annual Payrolls

Step 1 Data preparation and extension loading
As in previous ArcView examples in this chapter, we use Ch4Data.shp for this analysis.
Among the attribute fields, we are concerned with

`mfg_ann_p`: annual payrolls in manufacturing industries
`mfg_valadd`: value added in manufacturing industries,

both at the county level.
Similar to other ArcView examples, load the Ch4.avx using the **File** menu's **Extensions** . . .

Step 2 Regression model
Assuming that annual payrolls are a function of value added, we have the following regression model:

`mfg_ann_p` = *f* (`mfg_valadd`)

In this way, the independent variable is `mfg_valadd` and the dependent variable is `mfg_ann_p`.

Step 3 Calculation of regression parameters
To calculate the regression parameters, we first select valid records in which we have value added and annual payrolls greater than or equal to 0. Next, we use the **Bivariate** option from the **Ch.4** menu to calculate the regression parameters.

- Click the **Query Builder** button and construct the query `[mfg_ann_p] >= 0 and [mfg_valadd] >= 0`, as below

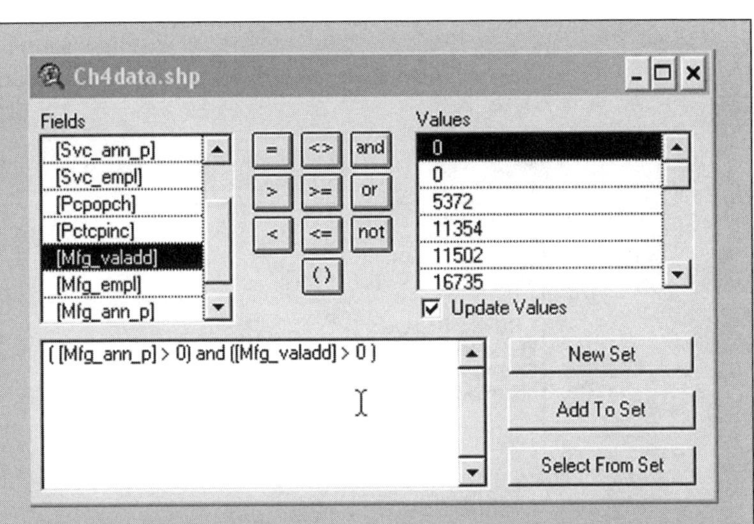

- Next, use the **Bivariate** option from the **Ch.4** menu to bring up the **Bivariate** dialog box.
- Click the `Ch4Data` data theme.
- Click the **Show Variable** button to list variables.
- Check the box by **Bivariate Linear Regression** to choose the regression.
- Choose `mfg_valadd` as the independent variable (variable 1 or *x*) and `mfg_ann_p` as the dependent variable (variable 2 or *y*).

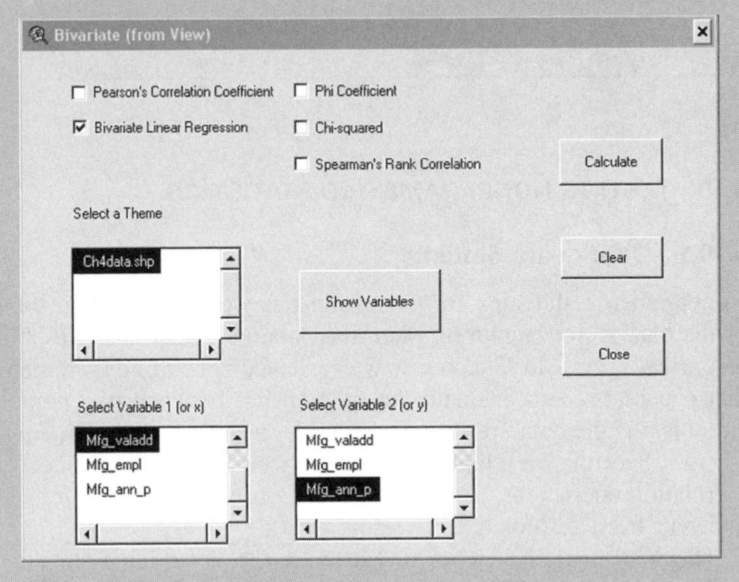

- Click the **Calculate** button to proceed.

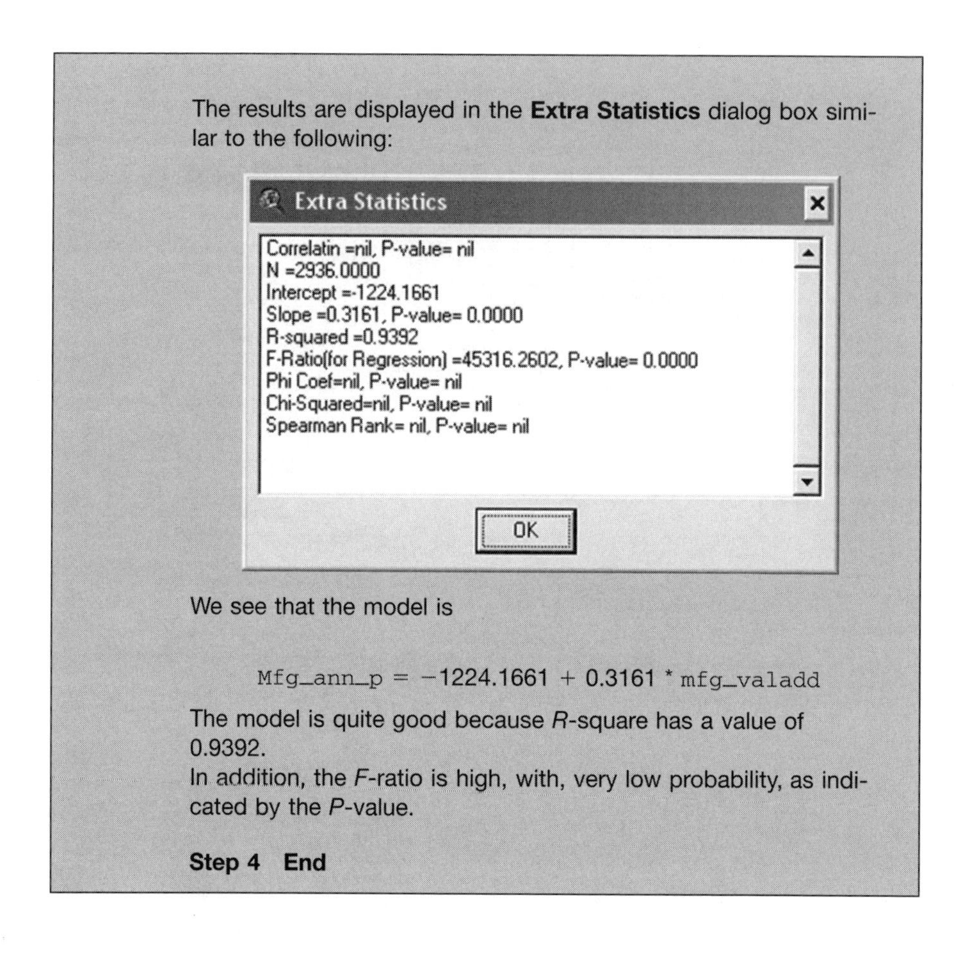

The results are displayed in the **Extra Statistics** dialog box similar to the following:

We see that the model is

`Mfg_ann_p` = −1224.1661 + 0.3161 * `mfg_valadd`

The model is quite good because *R*-square has a value of 0.9392.
In addition, the *F*-ratio is high, with, very low probability, as indicated by the *P*-value.

Step 4 End

4.10 TESTING NONPARAMETRIC STATISTICS

4.10.1 Chi-Square Statistic

Nonparametric statistics are those statistics we can use when the distribution of the data is not known or when the data are categorical variables (nominal or ordinal scale). In Chapter 3, we reviewed several nonparametric statistics that can be used to evaluate the relationship between two noninterval scale variables. If the data are in nominal scale, we can use the chi-square statistic, χ^2, to reflect the correlation. Similar to testing the significance of Pearson's correlation coefficient, we can test if the observed χ^2 statistic is significantly different from zero or not.

The χ^2 statistic has its own distribution. Conventional statistics textbooks often provide such tables. Given a calculated chi-square, or $\chi^2_{\text{calculated}}$, we can compare it with the critical value (χ^2_{cv}) with the corresponding significance level (α) and degrees of freedom. For χ^2, the degree of freedom is equal to

$(r - 1)(c - 1)$, where r and c are the number of rows and the number of columns, respectively, in the frequency table for the χ^2 calculation. To reject the null hypothesis, the calculated statistic has to be larger than χ^2_{cv}.

In the example used in Section 3.2, $\chi^2_{\text{calculated}} = 28.5589$. The frequency table has three rows and three columns. Therefore, $df = (3 - 1)(3 - 1) = 4$. If we use $\alpha = 0.05$ and $df = 4$, the critical value in χ^2 statistic is $\chi^2_{\text{cv}} = 9.488$. In this case, we can reject the null hypothesis because

$$\chi^2_{\text{calculated}} = 25.5589 > \chi^2_{\text{cv}} = 9.488.$$

Since the calculated χ^2 value is larger than the χ^2_{cv}, we can claim that it is very likely that the observed χ^2 is different from zero. Note that the χ^2 test is a one-tail test, as the minimum of χ^2 is zero.

4.10.2 ArcView Note

The "Bivariate" menu item in the Chapter 4 extension reports not just the χ^2 statistic, but also the associated probability.

In ArcView Example 4.5, two attribute fields contain categorical data. These two fields are used in the analysis of χ^2. The two fields were classified from pcpopch and pctcpinc in Ch4Data.shp. The classification is based on the minimum, maximum, mean, and standard deviation in pcpopch and pctcpinc. Specifically:

Attribute	Maximum	Minimum	Mean	Standard Deviation
Pcpopch	51.42	−21.35	2.46	5.88
Pctcpinc	165.08	−42.21	15.87	8.34

Pcpopch Value Range	Category	Pctcpinc Value Range	Category
−21.35 ~ −3.42	1	−42.21 ~ 7.53	1
−3.43 ~ 2.46	2	7.54 ~ 15.87	2
2.47 ~ 8.34	3	15.88 ~ 24.21	3
8.35 ~ 51.42	4	24.22 ~ 165.08	4

ArcView Example 4.5: Nonparametric Tests

Step 1 Data preparation and loading extension

Start ArcView GIS if it is not already started.

Use the **Add Theme** button to add `Ch4Data-C.shp`. This shapefile contains the categorical attribute fields:

`popch_c`, from pcpopch
`inch_c`, from pctcpinc.

Again, load the Ch4.avx extension via the menu item, **File,** then **Extensions** . . .

Step 2 Hypothesis
When using the χ^2 test, the null hypothesis is that there is no statistically significant relationship between the two variables tested. The alternate hypothesis is that the relationship is statistically significant.

Step 3 Statistics calculation
To calculate χ^2 statistics, use the **Bivariate** option from the **Ch.4** menu.

- First, check the box **Chi-squared** to select this statistics for output.
- Click the `Ch4data-c.shp` as the data theme for analysis.
- Click the **Show Variables** button to list all attribute fields.
- For Variable 1, click `popch_c`.
- For Variable 2, click `incch_c`.
- Click the **Calculate** button to proceed.

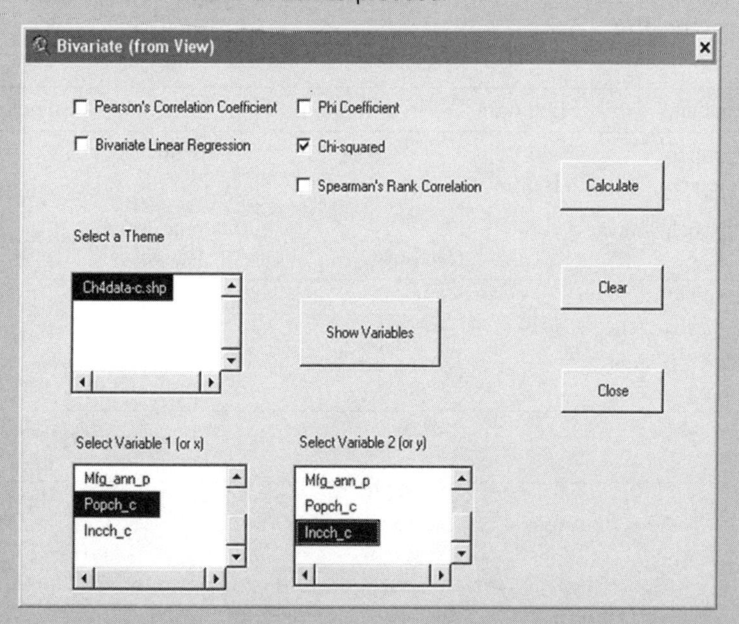

- Click **OK** when you are informed that no features are selected.
- Click **Yes** when you are informed that all features have been selected for calculating the statistic.

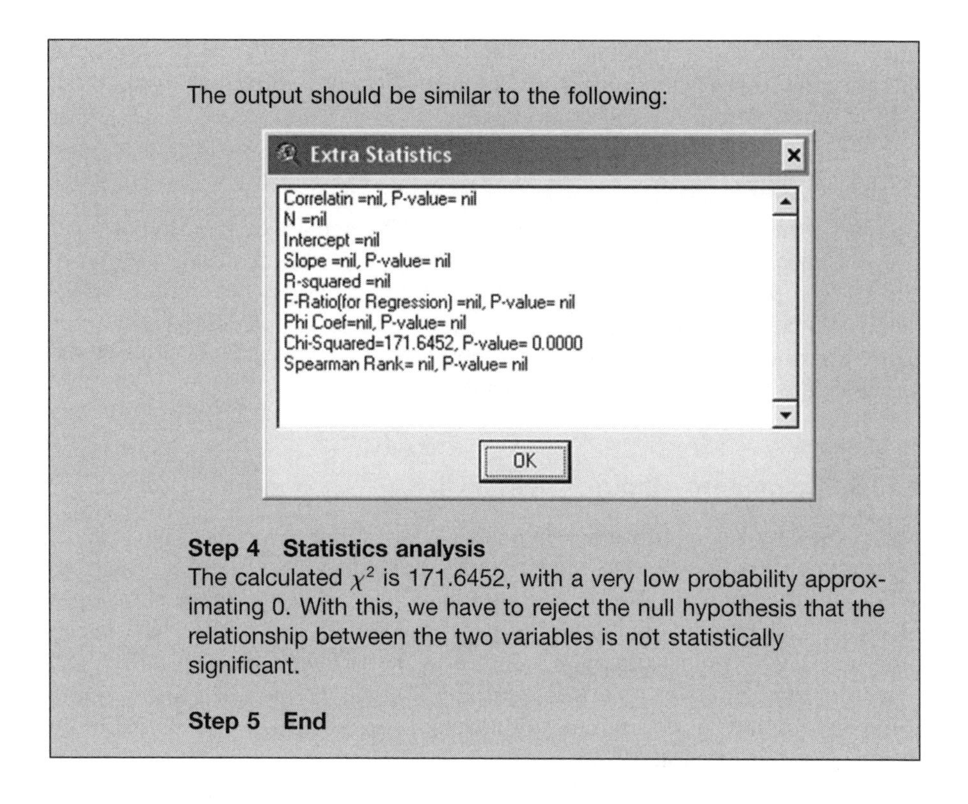
The output should be similar to the following:

Step 4 Statistics analysis
The calculated χ^2 is 171.6452, with a very low probability approximating 0. With this, we have to reject the null hypothesis that the relationship between the two variables is not statistically significant.

Step 5 End

4.10.3 Spearman's Rank Correlation Coefficient

For ordinal data, the correlation between two variables can be evaluated by Spearman's rank correlation coefficient, or r_s, as discussed in Section 3.3. Similar to the above tests, we can also test if the correlation reflected by Spearman's rank correlation coefficient is significantly different from zero or not. The calculated statistic, which also follows a t-distribution, is

$$t = r_s \sqrt{n - 1} \tag{4.22}$$

with $(n - 1)$ degrees of freedom. In the previous example in Section 3.3, the ranks of 50 states plus D.C. by median income and house value are compared using Spearman's rank correlation coefficient. The calculated r_s is 0.8662. To compute the corresponding t-statistic,

$$t_{\text{calculated}} = 0.8662 \sqrt{51 - 1} = 6.1250.$$

Because r_s can be positive or negative, α should be allocated to both tails. Given that the df is $(51 - 1)$, the critical value with $\alpha/2$ (i.e., 0.025) is approximately 2.011, smaller than the calculated statistic. Therefore, we can reject the null hypothesis and conclude that there is evidence to show that the observed Spearman's rank correlation coefficient is different from zero.

4.10.4 ArcView Note

Similarly, the "Bivariate" menu item in the Chapter 4 extension reports not just the Spearman's rank correlation coefficient, but also its associated probability. The procedure for using the Ch4.avx extension is the same as those demonstrated in various ArcView examples in this chapter. However, note that Spearman's rank correlation coefficient should be used only when the variables are rank-ordered (ordinal data).

4.10.5 Kolmogorov-Smirnov Test

Spearman's rank is applicable when each observation is a rank. The scale of measurement of this type of data is ordinal scale, but with strong ordering. When observations are assigned to groups while the groups have some type of order, such as low, medium, and high, those data are said to have weak ordering. Though the Kolmogorov-Smirnov (K-S) D statistic is usually used to test for conformity to a normal distribution, it can be modified to determine if the distributions of two samples of a weak-ordered variable are significantly different or not (Taylor, 1977).

For instance, we have observations in two regions grouped into the following classes of frequencies:

Region	Low	Med-Low	Medium	Med-High	High	Total
A	3	5	7	3	2	20
B	1	3	10	6	4	24

Based on the frequency tabulation above, we can derive the cumulative frequencies (*cf*) for each group for both regions in the following manner:

Region	Low	Med-Low	Medium	Med-High	High	Total
A (cf_a)	3	8	15	18	20	20
B (cf_b)	1	4	14	20	24	24

The idea is that, for instance, for Region A, the cumulative frequency for the Med-Low group will be the frequency of the previous group (Low = 3) plus the frequency of the current group (Med-Low = 5). That will give us a cumulative frequency of 8 for the Med-Low group. For the cumulative frequency for the Medium group, we just have to add the frequencies of all previous groups and the frequency for itself together (3 + 5 + 7 = 15).

From the cumulative frequencies, we can derive the cumulative proportions (*cp*) for each group in each region. The cumulative proportion is simply the cumulative frequency of a group divided by the total frequency in that region.

Region	Low	Med-Low	Medium	Med-High	High	Total
A (cp_a)	3/20	8/20	15/20	18/20	20/20	1
B (cp_b)	1/24	4/24	14/24	20/24	24/24	1

Note that in the last group, the cumulative proportions should be 1, by definition, except for rounding errors. Next, we compare the cumulative proportions between the two regions in the same group to derive a difference in the proportion (*diff*).

Region	Low	Med-Low	Medium	Med-High	High
A (cp_a)	0.15	0.40	0.75	0.90	1
B (cp_b)	0.04	0.167	0.583	0.833	1
Diff	0.11	0.233	0.167	0.067	0

Also by definition, the difference of the cumulative proportion in the last group should be zero. The K-S D statistic is the one with the maximum absolute difference in the cumulative proportions between the two groups. Formally,

$$D = \max |cp_a - cp_b|. \tag{4.23}$$

Note that we take the absolute differences between the two cumulative proportions in each class. In other words, we do not care which cumulative proportion is larger; only the difference is important. Conceptually, we compare the cumulative distributions of the two regions and focus on the differences between the two distributions, as shown in Figure 4.7. In our example, the maximum absolute difference in cumulative proportion is 0.233.

To test if the observed K-S D statistic is really statistically significant, we use the same testing methodology as before. We can either compare the observed D statistic with the critical value (D_{cv}) at a given α level and total sample sizes (*df*) to determine if the observed statistic is in the critical region or not, or compute the probability of having the D statistic with the same magnitude as the observed D statistic, and see if the probability is larger or smaller than the significance level, α, at the given degrees of freedom. In order to reject the null hypothesis that the two distributions are not significantly different, the observed D statistic must be larger than the critical value.

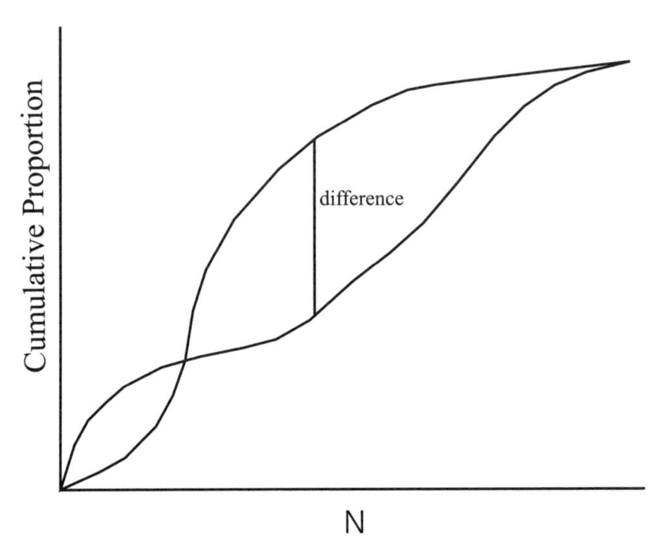

Figure 4.7 Cumulative proportional difference used in deriving the K-S D statistic.

In our example, the calculated D statistic is 0.233. Given the sample sizes of 20 (n_1) and 24 (n_2), and $\alpha = 0.05$, the critical value (D_{cv}) is 0.4118. Because the calculated D statistic of 0.233 is not larger than the D_{cv}, we fail to reject the null hypothesis that the two distributions are significantly different.

4.10.6 ArcView Note

In the Ch4 extension, a menu item, "Prob > D (KS)," is used to report the probability of having a D statistic larger than a given D value. The user needs to provide the calculated D statistic and the sample sizes of the two groups. The function reports the one-tail probability of having a D value larger than the one observed. For instance, by entering 0.233 as the calculated D value, and 20 and 24 as the sample sizes, the probability of observing a D value larger than 0.233 is 0.5355. That is, the observed D statistic is highly likely, and therefore, we fail to reject the null hypothesis.

4.11 SUMMARY

This chapter has emphasized basic probability theories related to hypothesis testing. We reviewed basic probability concepts (sampling space, permutation, and combination), discrete (binomial and Poisson) and continuous (normal or Gaussian) probability functions, the central limit theorem, and the standard error. Based upon these concepts, we discussed how different statistics (variance and mean) can be tested for their differences and the significance of the bivariate regression model. Furthermore, we reviewed how to test for the significance of correlation coefficients for data in various measurement scales.

All of these topics set the stage for the following chapters analyzing spatial data when numerous hypotheses about spatial data will be formulated, and the statistics describing patterns of spatial data will be tested for their significance. The framework for testing the significance of means or regression parameters (e.g., the *t*-test) will be central to the upcoming discussions. Certain probability functions, especially the Poisson function, will be used frequently in analyzing point patterns. With the background knowledge presented in this chapter and previous chapters, we are now ready to analyze spatial data in the remainder of this book.

REFERENCES

Press, W. H., B. P. Flannery, S. A. Teukolsky, W. T. Vetterling. 1992. *Numerical Recipes in C: The Art of Scientific Computing.* 2nd Edition. Cambridge University Press.

Rogerson, P. A. 2001. *Statistical Methods for Geography.* Sage Publications

Taylor, P. J. 1977. *Quantitative Methods in Geography.* Houghton Mifflin Company.

EXERCISES

1. Assume that an event has three outcomes (possible values). If that event happens five times, what is the number of all possible outcomes? If we denote the values of the event outcomes as *a*, *b*, and *c*, list all of the possible outcomes to verify your answer.

2. If we are not concerned about the order of the outcome in Question 1 but only about the number of *a*'s in the outcomes, then how many outcomes will have
   - **a.** one *a*?
   - **b.** three *a*'s?
   - **c.** five *a*'s
   - **d.** at least three *a*'s?

3. An archer can hit a bull's-eye only 3 out of 10 times, and getting the bull's-eye will get 10 points. If the archer makes 10 attempts to shoot, what is the probability that he/she will get
   - **a.** 30 points?
   - **b.** 50 points?
   - **c.** less than 50 points?
   - **d.** into the final if it requires at least 70 points?

4. Assume that after we studied the long-term climatic record, we came to the conclusion that on average, the Mid-Atlantic region on the East Coast of the United States will have 3.5 significant snow storms in a year. Then for a given year, what is the chance that the Mid-Atlantic region will have

a. no snow storms?

b. three snow storms?

c. seven or more snow storms?

New England	Deaths (per 1000)	IMR (per 1000)	Mid-Atlantic	Deaths (per 1000)	IMR (per 1000)
Connecticut	868.4	6	Delaware	892.8	10.6
Maine	967	6.2	DC	1037.1	10.8
Massachusetts	886.6	5	Maryland	813.9	8.1
New Hampshire	779.4	3.8	New Jersey	877.8	6.4
Rhode Island	945.7	6.8	New York	834.4	6.1
Vermont	848.5	5.7	Pennsylvania	1054.4	7.2
			Virginia	782	7.4
			West Virginia	1164.2	7.3

In the table above, we have the same data set used in the exercise of Chapter 3. However, additional information regarding the regional membership of the states is included here. The 13 states and D.C. can be divided into two regions: New England (6 states) and the Mid-Atlantic region (7 states and D.C.). The variables, crude death rate (number of deaths/100,000 people) and infant mortality rate (IMR) (number of deaths per 1000 live births), are the same as before.

5. Calculate the descriptive statistics, especially the mean, variance, and standard deviation, for the two variables but separately for the two regional groups. Based upon these statistics, what can you say about the statistical properties of these two variables in the two regions?

6. Focus on the variable IMR. Treat the two sets of IMR values of the two regions as two samples. Convert the IMR values of each region into z-scores. What are the probabilities of observing an IMR value between 5.5 and 6.5 in the New England region and the Mid-Atlantic region? Similarly, what are the probabilities of observing an IMR value larger than 6 in each of the two regions?

7. Suppose that we want to compare the variances and averages IMRs of the two regions. Formulate the null hypothesis and the alternate hypothesis for testing both the variance and the mean.

8. Test the difference in variances first. Based upon the result, test for the means. Which statistic should be used for the difference-of-means test? Why?

9. Repeat Question 8, but using the general mortality variable to test the variance and the mean. Is the procedure different from that of Question 8? Do the two regions differ in their variances and means? Do the two variables provide consistent results?

10. If the national general mortality rate is 848.5 per 100,000 and the national IMR is 6.8 (per 1000 live births), can you say that the two regions are different from the national averages?

11. In the exercise for Chapter 3, you have already computed Pearson's correlation coefficient between general mortality and IMR for the 13 states and D.C. in Question 4. Test if the correlation coefficient is significant. If we divide the observations into the two geographic regions, as above, will the correlation of the two mortality rates (general mortality and IMR) have different significance levels in each region? Verify your results using the ch4 extension.

12. In the exercise for Chapter 3, you have already computed the regression slope parameter estimates between general mortality and IMR for all 13 states and D.C. in Question 7. First, compute the residuals for each observation based upon the regression results. Also, compute the variance of the residuals. Then test if the slope parameter estimate is significantly different from zero. Also, test for the significance of the regression model (coefficient of determination). Verify your answers using the ch4 extension.

13. In the exercise for Chapter 3, Question 2, you created classification variables for general mortality and IMR variables and computed the χ^2 statistic. Test the significance of the χ^2 statistic.

14. Using the same groupings to compute the χ^2 statistic, perform a K-S test.

15. Test the significance of Spearman's rank correlation coefficient obtained from the exercise for Chapter 3.

16. Based upon the results of various correlation measures and their significance, what can you say about the relationship between the two variables?

PART II

SPATIAL STATISTICS

In Chapter 1, we stated that spatial data consist of cartographic and attribute data. Cartographic data describe the location and geometric characteristics of geographic features. Attribute data provide meanings to these features. In later chapters, we discussed classical statistical measures and methods for testing statistics commonly used in classical statistics. The discussions of previous chapters focused exclusively on the analysis of attribute data, without concern for the location or geographic information that may be incorporated into the analysis. Therefore, the analysis was aspatial in nature.

In Chapter 5 and subsequent chapters, we will discuss statistical methods that have been implemented in GIS (Lee and Wong, 2001) and have been specifically designed for analyzing spatial data. Some of these methods analyze only location information, such as point pattern analyzers considered in Chapter 5, while others analyze location information together with the attribute information of the geographic features. To cover all these methods, Chapter 5 discusses descriptive centrographic measures that may be used to describe point distributions. Chapter 6 will introduce methods for comparing and testing point patterns. Chapter 7 will focus on techniques used to describe and analyze linear geographic features. The last chapter, Chapter 8, will deal with polygon features, but with an emphasis on spatial autocorrelation analysis.

CHAPTER 5

POINT PATTERN DESCRIPTORS

This chapter introduces statistics for describing point patterns. These spatial statistics are used to summarize the distribution of a set of points. Similar to the analysis of attribute data using classical statistics, we will examine these statistics in a progressive manner. Therefore, we will start with measures of (spatial) central tendency. Then we will describe the magnitude of dispersion of a set of locations, followed by measures indicating the directional biases of the set of points. In the next chapter, we will start with lower-ordered statistical properties of point distributions using direct measures of distances between pairs of points. Higher-ordered statistical properties of point distributions will be addressed by the K-function and spatial autocorrelation statistics.

5.1 THE NATURE OF POINT FEATURES

In Chapter 1, we mentioned that geographic features or observations are often represented in abstract forms. Three geometric primitives—points, lines, and polygons—are widely used to represent geographic features. Points are used to represent features with little or no spatial extent. Lines, or arcs, are used to represent linear features, and polygons are used to represent areas or features with significant spatial extent. In this chapter, we will focus on analyzing point features.

As stated above, points are used to represent geographic features with little or no spatial extent. This representation is very sensitive to spatial scales. For example, a house on a city map is only a point, even though it becomes a polygon when the house is shown in a neighborhood map or on a large-scale, detailed tax parcel map. In other situations, points are also used to describe the locations of events or incidences such as disease occurrences or traffic accidents. In these cases, points do not represent real geographic features, but rather the locations of the events. In addition, points may be used simply for symbolic purposes. For example, in transportation modeling, urban analysis, and location-allocation modeling, areal units with significant spatial extents are often abstractly represented by representative points or by the geometric centroids of the polygon features. This is done to accommodate the specific data structures, as required by the analytical algorithms or models.

Points are defined by coordinate pairs. Depending on the coordinate systems and the geographic projections adopted, a point on a map can be defined by a pair of latitude and longitude readings, a pair of x and y coordinates, a pair of easting and northing, and so on. While points on a map are all simple geometric objects defined by their coordinates, the attributes associated with these points provide specifics to differentiate among them. The attribute information associated with points gives meaning to the points. Consider a map showing all residential water wells in a city; the points would all look alike except for their locations. If attributes such as owners' names, well depth, dates dug, or dates of last water testing were added to the database as attributes, more meaningful and information-rich maps could be created to show the spatial variation of the wells according to one or more of these attributes.

The description of any spatial relationship between individual points requires the application of a special type of spatial statistics called **centrographic measures** (Kellerman, 1981), as described in this chapter. Specifically, this chapter covers the methods for determining overall patterns of a given set of points. We will also discuss measures used to describe the magnitude of spatial dispersion of a given set of points. Finally, we will examine how the direction bias, if any, of a set of points can be extracted statistically. Note that these centrographic measures do not include the point attributes in the analysis in general, though occasionally the weights or frequencies of observations assigned to the locations will be used. In other words, in this chapter, points are differentiated mostly by their locations and occasionally by their weights.

The spatial measures discussed in this chapter are appropriate for points that represent various types of geographic features in the real world. A word of caution: the accuracy of location information and its associated attribute values must be considered carefully. This is because the reliability and usefulness of the results obtained by analyzing the set of points are always dependent on the quality of the data, even though data quality is not the only factor. This is true of all types of data analysis, not just spatial data analysis. Results of any analysis are only as good as those the data can provide.

Point data obtained from maps may contain cartographic generalizations or location inaccuracy. On a small-scale map, one point may represent a city whose actual areal extent may be tens or hundreds of square miles, while another point may represent a historic landmark or the location of a specific endangered plant species that occupies only several square meters on the ground. Comparing these points directly, however carefully, would be comparing apples with oranges. Point locations derived from calculating or summarizing other point data are especially sensitive to the quality of the input data because the inaccuracy of the input data may be propagated through the computation process, ending up with results of little value.

For this reason, spatial analysis must be performed with careful consideration of the geographic scale of the spatial database and the quality of the data. These types of information are usually recorded as metadata—"data about data." In the United States, all spatial data provided by federal government agencies must include metadata for users to evaluate the quality and appropriateness of the spatial data when they are released to the public. Statistical methods can be very useful when they are used correctly. But, at the same time, they can be very misleading and deceptive when used inappropriately or carelessly, especially when the data cannot support the analysis.

5.2 CENTRAL TENDENCY OF POINT DISTRIBUTIONS

In classical statistics, a common practice at the beginning of data analysis is to describe the data or variables using descriptive summary statistics, including the measurements of central tendency and dispersion (Chapter 2). This is followed by the formulation of hypotheses and testing of these hypotheses regarding, for instance, the differences between samples.

In analyzing spatial data with only location information, we can adopt a similar approach, starting by describing the spatial data of interest. In fact, we can borrow concepts from classical descriptive statistics in describing spatial data. In classical descriptive statistics, we start by describing the central tendency of the data, or the general location of the set of data on the value line. We can also use this concept to describe spatial data.

Central tendency in classical statistics refers to the average or a representative value of a given set of values. The median house value of a neighborhood can give a house hunter a quick general impression of the housing prices of a neighborhood. Similarly, the most common style of houses in the neighborhood, or the mode—say, the colonial style—can offer a general idea of the housing choices in the area. The mean household income of a neighborhood can provide an informative summary of the socioeconomic condition of a neighborhood to local officials or policy makers to assess the need for social services in the area. If someone plans to visit a foreign country during Christmas, it would be useful to know the average December temperature there in order to bring appropriate clothing. The type of clothing will be dramatically different for northern Europe compared to Australia. We can apply the central tendency concept to evaluate a set of point locations to identify a representative location. Using different spatial measures of central tendency, we can obtain different representative locations for a given set of locations.

When comparing multiple sets of numerical values, the concept of **average** is particularly useful. Educators can use average scores on state proficiency tests among elementary schools to assess how schools compare with one another. Comparing the harvests from farms using a new fertilizer with those from farms not using the fertilizer can provide a good basis for judging the effectiveness of the fertilizer. In many settings similar to these, central tendency measures furnish summary information on a set of values that would otherwise be difficult to comprehend. The same idea can be applied to the analysis of point locations. Measures of spatial central tendency can be applied to multiple sets of points. By comparing their representative locations, we can obtain a better understanding of the differences in their geographic and spatial characteristics.

Let us review some basic concepts of measuring central tendency in classical statistics. Given a set of values, x_i, $i = 1, 2, 3, \ldots, n$, measures of central tendency include the mean, median, and mode. The median is the middle value in the set of values when the values are arranged in ascending or descending order. The mode is the value occurring most frequently; sometime there may be no such value(s) when each observation has a unique value. The mean, $\bar{x}$, is simply the arithmetic average of all values:

$$\bar{x} = \frac{\sum_{i=1}^{n} x_i}{n}. \tag{5.1}$$

What if different observations carry different levels of importance? In this case, the measure of central tendency should not be the simple arithmetic mean. Each observation will be weighted according to its level of importance, assuming that we know the weights for different observations. Then each value, x_i, in the data set will first be multiplied by its associated weight, w_i. The sum of the weighted values is then divided by the sum of the weights to obtain the weighted mean:

$$\bar{x}_w = \frac{\sum_{i=1}^{n} x_i w_i}{\sum_{i=1}^{n} w_i}. \tag{5.2}$$

Note that this formula is very similar to that for the weighted mean in Chapter 2, except that now w_i is used instead of f_i, the frequency for i. This indicates that the two terms, weights and frequencies, are interchangeable in the context of computing weighted mean. When frequencies are used, it is likely to happen when, in grouped data, that the value of x_i has occurred or is assumed to occur f_i times. Thus, in calculating the weighted mean, the value x_i has to be weighted f_i times.

When dealing with data that describe observations distributed over a geographic space, one can extend the central tendency concept to measure spatial observations. For instance, the concept of average in classical statistics can be extended to the concept of geographic center, a measure of spatial central tendency. Because geographic features or observations have their locations spatially referenced in a two-dimensional space, the measure of central tendency needs to incorporate coordinates that define the locations of the features or observations. Central tendency measures in the spatial context will be the **mean center,** assuming that all geographic features have the same weights or frequencies. In turn, the weighted mean center can be calculated when the weights are not uniform. We will also discuss the concept of median in summarizing the locations of a set of points.

The positions of such centers can be calculated in several ways, yielding different results based on how the data are organized in the geographic space. Different definitions of the boundaries of a study area, distortions caused by different map projections, or even different map scales at which the data were collected often lead to different results. It is important to realize that there is no *absolutely correct* way to find the center of a spatial distribution. More than one method can be used in different settings, but there is probably no single correct method suitable for all situations. For this reason, the interpretation of the result in calculating the center of a spatial distribution can only be determined by the nature of the problem.

In this chapter, we will discuss a set of point descriptors. These descriptors provide certain descriptive information on the distribution of a set of points. For central tendency information, we will consider **mean centers, weighted mean centers,** and **median centers.** These provide a good summary of how a set of points distributes in the geographic space. To describe the spatial dispersion characteristics of a set of points, we will discuss the measures of **standard distance** and **standard ellipse.** These measures indicate the spatial variation and orientation of a point distribution.

5.2.1 Mean Center

The **mean center,** or **spatial mean,** is a central or average location of a set of points. The points may represent water wells, houses, power poles in a residential subdivision, locations where landslides occurred in a region in the past, or locations of cities on a continental scale. As long as a location can be defined, even with very little or no areal extent, it can still be represented as a point in a spatial database. Whatever the points represent in a spatial database, each point, p_i, may be defined operationally by a pair of coordinates, (x_i, y_i), for its location in a two-dimensional space.

A variety of coordinate systems may be used to define the locations of points. Geographers have devised various map projections and their associated coordinate systems to reference geographic features on Earth. In adopting any such coordinate system, we should keep in mind that projecting features in a three-dimensional space (the Earth) to a two-dimensional space (maps) always results in some degree of distortion.

Depending on the coordinate system adopted, the locations of points in space can be referred to by their latitude/longitude, easting/northing, or other forms of x-y coordinates. When working with known coordinate systems, the location of a point is relatively easy to define on maps. There are, however, many situations that require the use of coordinate systems with arbitrary origins as the reference points. Arbitrary coordinate systems are often created for studies involving small, localized areas or for quick estimation projects. In these cases, the coordinate system needs to be carefully constructed so that (1) it orients to a proper direction for the project, (2) it situates with a proper origin, and (3) it uses suitable measurement units. For a more detailed discussion on this topic, interested readers may refer to the monograph by Monmonier (1993). All of these issues have to be taken into account in interpreting the mean center and other measures of spatial central tendency.

With the coordinate system defined, the mean center can be found easily by calculating the mean of the x-coordinates (or eastings) and the mean of the y-coordinates (or northings). These two means of the coordinates define the location of the mean center as

$$(\bar{x}_{\mathrm{mc}}, \bar{y}_{\mathrm{mc}}) = \left(\frac{\sum_{i=1}^{n} x_i}{n}, \frac{\sum_{i=1}^{n} y_i}{n} \right), \tag{5.3}$$

where

$\bar{x}_{\mathrm{mc}}$ and $\bar{y}_{\mathrm{mc}}$ are the coordinates of the mean center,
x_i and y_i are the coordinates of point i, and
n is the number of points.

As an example, Table 5.1 lists the coordinates of 29 shelters in Washington, D.C., based on the information gathered by a nonprofit community organization. Their locations are shown in Figure 5.1. Given the coordinates of the 29 shelters in Table 5.1, we can use Equation 5.1 to calculate the mean center of all the shelters. As shown in Table 5.1, we first take the mean of the longitude readings (x's) and the mean of the latitude readings (y's). The mean center is located at (−77.0145, 38.9065), slightly due east of the center of the city. The special symbol in Figure 5.1 represents the location of the mean center calculated above.

Since the mean center is defined by the mean of the x coordinates and the mean of the y coordinates, it is located around the center of the 29 shelters

TABLE 5.1 Calculation of the Mean Center

Shelter ID	Longitude in Degrees (x)	Latitude in Degrees (y)
1	−76.9892	38.9472
2	−77.0049	38.9157
3	−76.9926	38.8979
4	−76.9886	38.9418
5	−76.9926	38.8964
6	−77.0189	38.9020
7	−77.0310	38.9125
8	−77.0275	38.9041
9	−77.0138	38.8956
10	−77.0328	38.9206
11	−77.0734	38.9284
12	−77.0191	38.9374
13	−77.0240	38.9076
14	−77.0091	38.9013
15	−77.0191	38.8999
16	−77.0200	38.9058
17	−77.0047	38.9622
18	−77.0333	38.9028
19	−77.0190	38.8988
20	−77.0314	38.9073
21	−77.0201	38.9070
22	−77.0300	38.9257
23	−77.0342	38.9305
24	−77.0182	38.9094
25	−76.9672	38.8720
26	−76.9960	38.8560
27	−76.9939	38.8423
28	−76.9947	38.8809
29	−77.0201	38.8798
$n = 29$	$\Sigma = -2233.4194$ $\bar{x}_{mc} = -2233.4194/5$ $= -77.01446207$	$\Sigma = 1128.2889$ $\bar{y}_{mc} = 1128.2889/5$ $= 38.90651379$

as expected. What it represents is the average location of a set of points composed of the 29 shelters. Note that although this mean center may be used, to define the centroid of a polygon commonly used in GIS applications and in positing the label point of a polygon, it is not desirable. The more robust and more preferable ways to define the centroid of a polygon will likely identify a location that is different from the mean center based upon principles in computational geometry. Readers interested in finding the centroid of a polygon can refer to the book by Wise (2003) as an introduction. In addition, the coordinate system used in computing the spatial mean of the

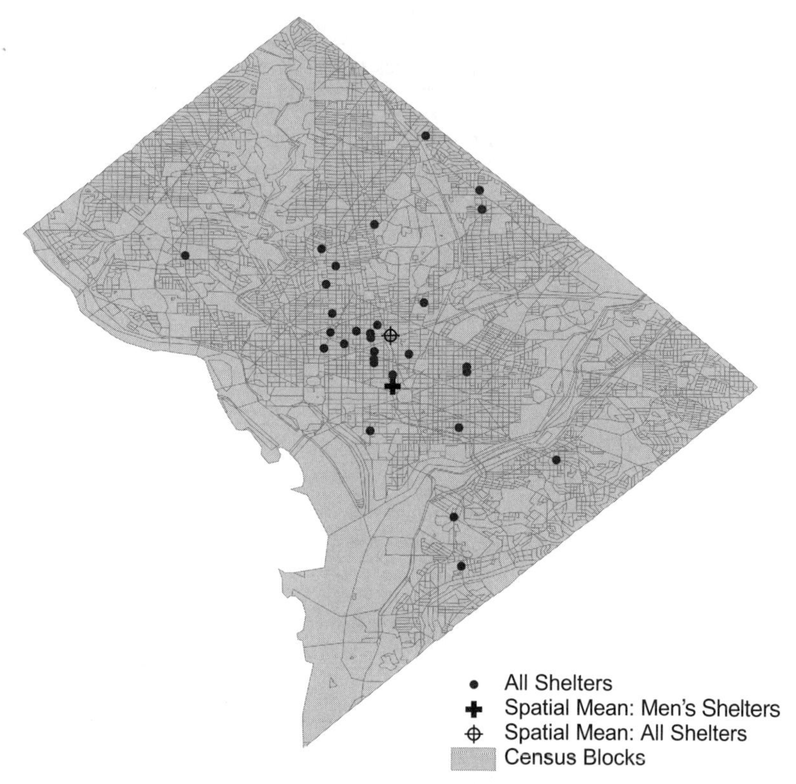

Figure 5.1 Distribution of the 29 shelters in the Washington, D.C., area.

shelters uses the longitude-latitude system, which will create distortion when the area is large. But given the relatively small area of Washington, D.C., even using longitude-latitude in calculating the spatial mean will cause only a negligible amount of distortion.

5.2.2 Weighted Mean Center

In the previous section, the procedure for calculating the mean center treated each point location equally. We did not differentiate shelters with different characteristics, such as some being for women only, or some having relatively large capacities while others are relatively small. We were simply concerned about the locations of shelters, and one shelter was no different from another except in location.

If we want to identify a representative location of all shelter beds in Washington, D.C., then we have to take into account the fact that different shelters have different capacity levels, and therefore, these shelters should be treated differently. In other words, there are situations in which the calculation of the mean center needs to consider more than just the location of points. The importance of individual points in a spatial distribution will not always be the same. If the points are different in some way, the calculation of the mean center should consider this difference.

In calculating the spatial mean of shelter beds, the mean center may give a more realistic picture of the central tendency if it is weighted by the number of beds in each shelter. The mean center will be pulled closer to the shelter (or shelters) with relatively large capacities. Therefore, a **weighted mean center** describes the central tendency better than a mean center does when the points or locations have different frequencies or occurrences of the phenomenon being studied. This is also true when points have different degrees of importance reflected by certain attributes of the locations.

The specific variable or attribute one should use to differentiate point locations or to weight point locations depends on the issue being studied. In studying retail activities, the amount of retail floor space (in square footage) or the number of stores in retail centers can be used as the weights. In examining school locations, the weights can consist of the number of students or the capacity of the school classrooms.

Borrowing the concept of the weighted mean from classical descriptive statistics, the weighted mean center of a distribution of points can be found by multiplying the x- and y-coordinates of each point by the weight assigned to each observation or location. The mean of the weighted x-coordinates and the mean of the weighted y-coordinates define the position of the weighted mean center. More specifically, the weighted mean center can be computed by

$$(\bar{x}_{\text{wmc}}, \bar{y}_{\text{wmc}}) = \left(\frac{\sum_{i=1}^{n} w_i x_i}{\sum_{i=1}^{n} w_i}, \frac{\sum_{i=1}^{n} w_i y_i}{\sum_{i=1}^{n} w_i} \right), \tag{5.4}$$

where

$\bar{x}_{\text{wmc}}$ and $\bar{y}_{\text{wmc}}$ define the coordinates of the weighted mean center and
w_{i} is the weight at point i.

In the case of the shelter distribution in Washington, D.C., if we are concerned with the distribution of shelter capacity, we should use the number of beds as the weights. If we limit the shelters to those for men and the general population only, the weighted mean center will be shifted south. This is because the shelter just slightly north of the weighted spatial mean (i.e., Shelter ID 9 in Table 5.2) has the largest number of beds among all shelters in the

TABLE 5.2 Calculation of the Weighted Mean Center

ID	Size (w_i)	Types	Longitude (x)	Latitude (y)	Size*Long (x)	Size*Lat (y)
7	12	Men	−77.0310	38.9125	−924.372	466.95
9	1350	General	−77.0138	38.8956	−103,968.6	52,509.06
10	27	General	−77.0328	38.9206	−2,079.886	1,050.8562
13	2	Adults	−77.0240	38.9076	−154.048	77.8152
14	8	Men	−77.0091	38.9013	−616.0728	311.2104
15	150	Men	−77.0191	38.8999	−11,552.87	5,834.985
17	10	General	−77.0047	38.9622	−770.047	389.622
26	144	Men	−76.9960	38.8560	−11,087.42	5,595.264
29	170	Men	−77.0201	38.8798	−13,093.42	6,609.566
	$\Sigma w_i = 1873$				$\Sigma w_i x_i = -144{,}246$	$\Sigma w_i y_i = 72{,}845.33$
					$\bar{x}_{\text{wmc}} = \frac{\Sigma w_i x_i}{\Sigma w_i} = \frac{-144{,}246}{1873} = -77.0138$	$\bar{y}_{\text{wmc}} = \frac{\Sigma w_i y_i}{\Sigma w_i} = \frac{72{,}845.33}{1873} = 38.8923$

region. Table 5.2 lists the coordinates of the shelters, as in Table 5.1, and the number of beds in the shelters, but is limited to shelters for men and the general population.

To calculate this weighted mean center, the sizes of shelters are treated as weights (w_i). Then the longitude readings (x) are multiplied by the corresponding weights (w_i), and so are the latitude readings (y). The weighted coordinates are summed separately for the longitude ($\Sigma\ w_i x_i$) and latitude ($\Sigma\ w_i y_i$). These sums are then divided by the total weights ($\Sigma\ w_i = 1873$) or the total capacity of these types of shelters. The result shown in Table 5.2 gives the location of the weighted mean center as (−77.0138, 38.8923), representing a shift in the center to the south, as shown by the symbol (Spatial Mean: Men's Shelters) in Figure 5.1.

5.2.3 Median Center

Similar to the classical descriptive statistics of central tendency, the concept of the median of a set of values can be extended to the **median center** of a set of points, or the **spatial median.** But the spatial median in a geographical space may not be defined precisely. According to Ebdon (1983), the median center of a set of points is defined differently in different parts of the world. For instance, in the British tradition, given a set of points, a median center is the center that divides the study region into four quadrants, each containing the same number of points. However, more than one median center (remember that a point has no areal extent) can serve as the center of the four quadrants with an equal number of points. As this method leaves too much ambiguity, it has not been used widely in geographic studies.

In North America, the median center is more often defined as the center of minimum travel from all points. That is, the sum of all distances from the median center to each of the points in the region is the minimum. In other words, if we select any location other than the median center, and compute the total distance from that location to all locations in the region, the total distance will be larger than the one using the median center. Mathematically, median center, (u, v), satisfies the following objective function:

$$\text{Min} \sum_{i=1}^{n} \sqrt{(x_i - u)^2 + (y_i - v)^2}, \tag{5.5}$$

where x_i and y_i are the x- and y-coordinates of location i. If there are weights attached to the points, a weighted median center can be derived accordingly as

$$\text{Min} \sum_{i=1}^{n} w_i \sqrt{(x_i - u)^2 + (y_i - v)^2}. \tag{5.6}$$

Note that the weights, w_i for i, can be positive or negative values to reflect pull or push effects that points may have on the location of the median center.

To derive the median center, an iterative procedure can be used to guide the search for the location that satisfies the above objective function. The procedure is as follows:

1. Use the mean center (spatial mean) as the initial location in the search for the median center. This first step essentially sets $(u_{\text{init}}, v_{\text{init}}) = (x_{\text{mc}}, y_{\text{mc}})$. An initial location has to be provided in this iterative search process. Theoretically, any location can be chosen as the initial location. But in practice, the starting location should not be too far away from the median center to reduce the search time.
2. In each iteration, t, a new location, (u_t, v_t), is computed to approximate the median center. The approximated location is found by

$$u_t = \frac{\sum_i \frac{w_i x_i}{\sqrt{(x_i - u_{t-1})^2 + (y_i - v_{t-1})^2}}}{\sum_i \frac{w_i}{\sqrt{(x_i - u_{t-1})^2 + (y_i - v_{t-1})^2}}} \tag{5.7}$$

and

$$v_t = \frac{\sum_i \frac{w_i y_i}{\sqrt{(x_i - u_{t-1})^2 + (y_i - v_{t-1})^2}}}{\sum_i \frac{w_i}{\sqrt{(x_i - u_{t-1})^2 + (y_i - v_{t-1})^2}}}. \tag{5.8}$$

The starting point is apparently $(u_0, v_0) = (x_{\text{mc}}, y_{\text{mc}})$.
3. Repeat Step 2 to derive new approximated locations for the median center until the distance between (u_t, v_t) and (u_{t-1}, v_{t-1}) is less than a threshold defined a priori.

Using the Washington, D.C., shelter data again, we are concerned with the central location of the overall shelter capacity. Therefore, we compute the weighted spatial mean of all shelters using the number of beds as the weights. Figure 5.2 is the zoom-in view of the area near the weighted spatial mean, which has a longitude-latitude reading of (−77.0145, 38.8979). Using this longitude-latitude reading as the initial location, $(u_{\text{init}}, v_{\text{init}})$ to search for the spatial median, several intermediate locations are computed for different iterations until the change in location of subsequent iterations becomes too small. Then the search process is terminated. The intermediate spatial median locations are also shown in Figure 5.2.

It is interesting to note that the geographic leap from the initial location (spatial mean) to the location from the first iteration is relatively large, while

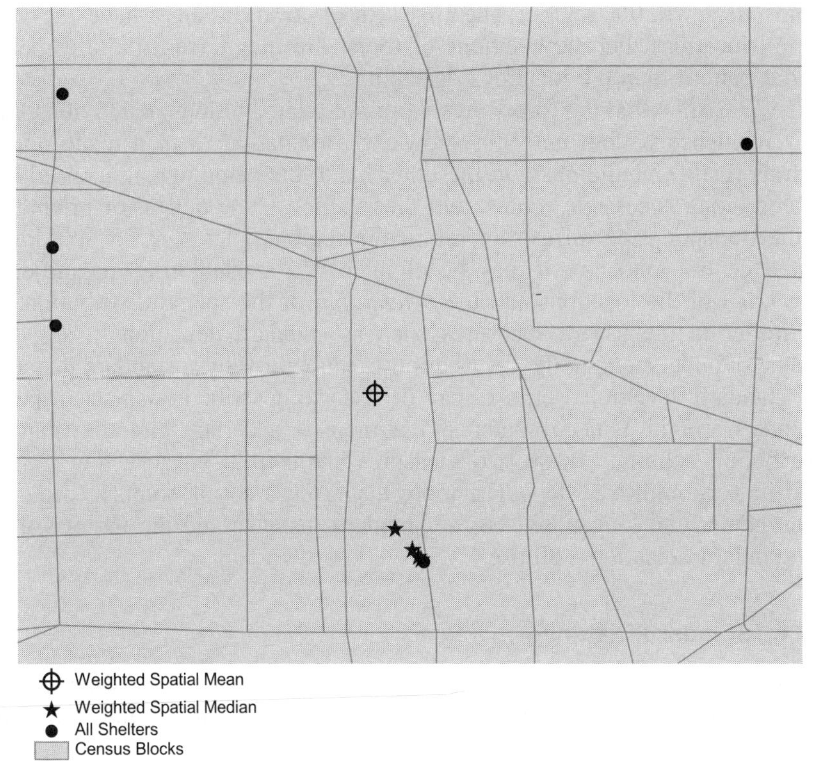

Figure 5.2 Spatial median of the shelters.

the changes in locations for subsequent iterations become smaller and smaller as the computed spatial medians approach the theoretical spatial median, which can never be identified precisely. As in this example, the spatial median ends up being very close to one of the shelters. This shelter has the largest number of bed (1350), thus pulling the spatial median (−77.0138, 38.8956) toward that location.

5.3 DISPERSION AND ORIENTATION OF POINT DISTRIBUTIONS

In the first two sections of this chapter, we are concerned with only one spatial characteristic of a set of points—the general or overall locations of the points. However, there are other spatial characteristics of the set of points that may be of interest to investigators. Two sets of points may occupy the same geographic space and may be interrelated. For instance, one set of points may represent the locations of forest fires and the other the locations of camping cabins in a wildlife region. The two sets of locations may have the same overall locations, but the locations of forest fire may have a more dispersed spatial pattern than the locations of cabins.

If we assume that the forest fires were not related to human activities, then their incidence pattern may not show any orientation to man-made objects such as roads or buildings. On the other hand, the camping cabins are likely located along accessible routes, and thus exhibit some degree of orientation to the transportation infrastructures in the region. Therefore, in addition to spatial central tendency, it may be of interest to evaluate the magnitude of *dispersion* of the locations and the *orientation* of the spatial distribution.

Similar to the use of measures such as standard deviation to assist an analyst in understanding the distribution of numeric values, standard distances and standard deviational ellipses can be used to describe how a set of points disperses around a mean center and if there is a certain direction that the distribution exhibits. These two measures are helpful because they can be used in very intuitive ways. The more dispersed a set of points is around a mean center, the longer will be the standard distance and the larger will be the standard deviational ellipse.

5.3.1 Standard Distance

Similar to those in classical statistics, given a set of n data values, x_i, $i = 1, \ldots, n$, the **population standard deviation,** σ, or the **sample standard deviation,** S, can be computed as

$$\sigma = \sqrt{\frac{\sum_{i=1}^{n} (x_i - \mu)^2}{n}} \tag{5.9}$$

or

$$S = \sqrt{\frac{\sum_{i=1}^{n} (x_i - \bar{x})^2}{n - 1}}, \tag{5.10}$$

where μ is the population mean and $\bar{x}$ is the sample mean. The standard deviation is literally the square root of the average squared deviation from the mean.

Standard distance is the spatial analogy of standard deviation in classical statistics. While standard deviation indicates how observations deviate from the mean, standard distance indicates how locations or points deviate from the mean center or the spatial mean. Standard deviation is expressed in units of the observation values, such as the prices of a commodity or the temperatures of a place, but standard distance is expressed in distance units, which is a function of the coordinate system or projection adopted.

Specifically, standard distance of a point distribution can be calculated by the following equation:

$$SD = \sqrt{\frac{\sum_{i=1}^{n} (x_i - x_{\mathrm{mc}})^2 + \sum_{i=1}^{n} (y_i - y_{\mathrm{mc}})^2}{n}}, \tag{5.11}$$

where $(x_{\mathrm{mc}}, y_{\mathrm{mc}})$ is the mean center of the point distribution.

Equation 5.11 compares the location of each point with the spatial mean along the x-coordinate and the y-coordinate. The differences in distance are then squared separately by the x- and y-coordinates. These differences are then summed and averaged by the number of points. Finally, the square root is taken to obtain the measure in the original distance unit. Thus, the calculated standard distance is essentially the average deviation of a point location from the spatial mean.

Since points in a distribution may have different attribute values that reflect the relative importance of different point observations, these values can be used as weights when calculating their mean center or even their median center, as discussed earlier. It is also possible to weight the points with specific attribute values when calculating standard distance. For the **weighted standard distance,** the following equation can be used:

$$SD = \sqrt{\frac{\sum_{i=1}^{n} w_i(x_i - x_{\mathrm{wmc}})^2 + \sum_{i=1}^{n} w_i(y_i - y_{\mathrm{wmc}})^2}{\sum_{i=1}^{n} w_i}}, \tag{5.12}$$

where

w_i is the weight for point i, and

$(x_{\mathrm{wmc}}, y_{\mathrm{wmc}})$ is the weighed spatial mean, as discussed earlier. Note that the weighted spatial mean is used here instead of the unweighted one.

Different standard distances can be computed for different phenomena within the same region. The distances for the corresponding phenomena can be compared to evaluate the relative dispersion of the locations around the spatial mean. To provide effective visual depiction of the dispersion, the standard distance is often used as a radius to draw a circle centered at the spatial mean. Thus, circles of different sizes based on their standard distances and centered at their spatial means allow simple but effective comparisons of the general locations and spatial dispersions of locations among different sets of point locations.

Using the shelter data in Washington, D.C., again to illustrate the application of standard distance, the data can be divided into at least two groups.

Some shelters serve the general population, but others serve only a specific group in the population. Five shelters in our database with capacity information serve only men, and six serve exclusively women. Their locations are shown in Figure 5.3. To better understand the spatial distribution characteristics of these two groups of shelters, we can compute their standard distances as in Tables 5.3 and 5.4.

Table 5.3 shows the data on the shelter locations and the procedure for computing the standard distance for men-only shelters. In this case, the capability (size) of each shelter was taken into account as the weight in the computation. Because we want to compute a weighted standard distance, we must first find the weighted spatial mean using the sizes of shelters as the weights instead of using the unweighted spatial mean. For this purpose, the first step in Table 5.3 shows that the weighted spatial mean was computed as (−77.0127, 38.8801) among the five men shelters.

Given the location of the weighted spatial mean, we can compare the location of each shelter with the weighted spatial mean. The differences are then squared, summed, and averaged to obtain the standard distance for the men-only shelters. The standard distance for the five men-only shelters is 0.021242 degree. Using this distance, we can draw a standard distance circle centered at the spatial mean to visually depict both the dispersion of the locations and the general location of the shelters. These are shown in Figure 5.3.

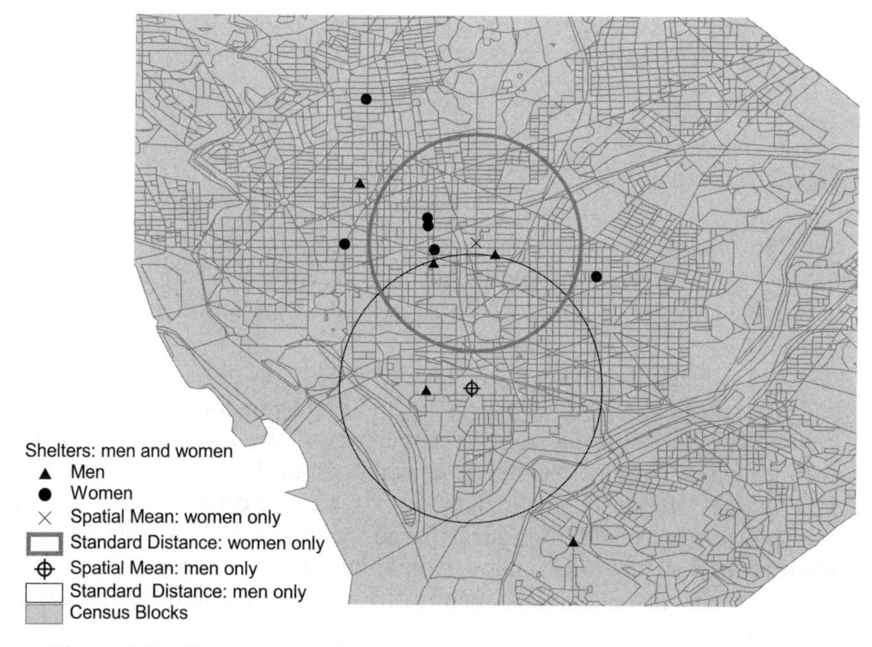

Figure 5.3 Circular areas defined by standard distances and mean centers.

TABLE 5.3 Calculation of Standard Distances

ID	Types	Longitude (x)	Latitude (y)	Size (f)	Long*Size	Lat*Size
7	Men	−77.0310	38.9125	12	−924.372	466.95
14	Men	−77.0091	38.9013	8	−616.073	311.2104
15	Men	−77.0191	38.8999	150	−11,552.9	5,834.985
26	Men	−76.9960	38.8560	144	−11,087.4	5,595.264
29	Men	−77.0201	38.8798	170	−13,093.4	6,609.566
				Σ 484	−37,274.2	18,817.98
					$\frac{-37{,}274.2}{484}$	$\frac{18{,}817.98}{484}$
				$(x_{\text{wmc}}, y_{\text{wmc}})$	= −77.0127	= 38.8801

ID	$x - x_{\text{wmc}}$	$y - y_{\text{wmc}}$	$(x - x_{\text{wmc}})^2$	$(y - y_{\text{wmc}})^2$	$f(x - x_{\text{wmc}})^2$	$f(y - y_{\text{wmc}})^2$
7	−0.0183	+0.0324	0.000335	0.001050	0.004019	0.012597
14	+0.0036	+0.0212	0.000013	0.000449	0.000104	0.003596
15	−0.0064	+0.0198	0.000041	0.000392	0.006144	0.058806
26	+0.0167	−0.0241	0.000279	0.000581	0.040160	0.083637
29	−0.0074	−0.0003	0.000055	0.000000	0.009309	0.000015
			Σ		0.059736	0.158651

$$SD_{\text{men}} = \sqrt{\frac{0.059736 + 0.158651}{484}} = 0.021242$$

Calculating standard distance the same way, the information on the six women-only shelters is shown in Table 5.4. Using the same procedure, the weighted spatial mean (−77.0122, 38.9031) based upon the size of the shelters was calculated first. Then the standard distance, 0.017111 degree, for these shelters was used to plot a standard distance circle centered at the weighted spatial mean. The weighted spatial mean and the circle are shown in Figure 5.3. Note that the circle of the standard distance may appear squashed. This happens when the map display is not in a projection that supports equal distance or equal area representation. Nevertheless, the circle derived from the standard distance has the correct distance measure as the diameter.

Comparing the results from the men-only and women-only shelters offers some insights into the spatial distribution characteristics of the two types of shelters. The overall location of the men-only shelters is more southerly than that of the women-only shelters. We can then compare the sizes of the standard distance circles. It is apparent that the circle for men-only shelters is bigger than the one for women-only shelters. This difference indicates that the shelters for men are more spatially dispersed than those for women. This information can shed light on issues related to access to social services.

TABLE 5.4 Calculation of Standard Distances for Women-Only Shelters

ID	Types	Longitude (x)	Latitude (y)	Size (f)	Long*Size	Lat*Size
3	Women	−76.9926	38.8979	74	−5,697.45	2878.445
6	Women	−77.0189	38.9020	25	−1,925.47	972.55
16	Women	−77.0200	38.9058	20	−1,540.4	778.116
18	Women	−77.0333	38.9028	31	−2,388.03	1205.987
21	Women	−77.0201	38.9070	38	−2,926.76	1478.466
22	Women	−77.0300	38.9257	10	−770.3	389.257
				Σ 198	−15,248.4	7702.82
					$\frac{-15,248.4}{198}$	$\frac{7702.82}{198}$
				(x_{wmc}, y_{wmc})	= −77.0122	= 38.9031

ID	$x - x_{wmc}$	$y - y_{wmc}$	$(x - x_{wmc})^2$	$(y - y_{wmc})^2$	$f(x - x_{wmc})^2$	$f(y - y_{wmc})^2$
3	0.0196	−0.0052	0.000384	0.000027	0.028428	0.002001
6	−0.0067	−0.0011	0.000045	0.000001	0.001122	0.000030
16	−0.0078	0.0027	0.000061	0.000007	0.001217	0.000146
18	−0.0211	−0.0003	0.000445	0.000000	0.013802	0.000003
21	−0.0079	0.0039	0.000062	0.000015	0.002372	0.000578
22	−0.0178	0.0226	0.000317	0.000511	0.003168	0.005108
			Σ		0.050108	0.007865

$$SD_{women} = \sqrt{\frac{0.050108 + 0.007865}{198}}$$

$$= 0.017111$$

Different standard distance circles can be drawn for different types of events or incidences in the same area. Figure 5.3 shows the standard distance circles based upon the calculated standard distances for men-only and women-only shelters. With standard distances, circles can also be computed and drawn for subsets of incidences to be compared. There are other examples in the literature, such as the standard distance circles drawn for different types of traffic accidents in Levine et al. (1995) for the economically disadvantaged populations in Greene (1991).

Even the same types of events/incidences can be drawn in different areas for comparison, both visually and numerically. Between two neighborhoods that have similar numbers of houses, the neighborhood with the larger standard distance based upon houses obviously spreads out more geographically than the other neighborhood.

While similar applications of standard distance to analyze geographic phenomena can be made easily, it is important to understand that comparisons of standard distances between different point distributions may or may not be meaningful. For instance, the standard distance of Japan's largest cities weighted by population counts is calculated as 3.27746 decimal degrees, while it is 8.84955 decimal degrees for Brazil's largest cities based on 1990 data. If the two standard distances are used alone, they indicate that Brazil has a much more dispersed interurban structure than Japan. However, given that the two countries have very different territorial sizes and geographic shapes, the absolute standard distances may not be meaningful in comparing the geographic spread of the largest cities in the two countries.

To adjust for this possible bias due to the different frames of reference, the standard distance should be scaled, for instance, by the average distance between cities in each country or by the area of each country or region. Alternatively, the standard distances can be scaled or weighted by a factor that accounts for the territorial differences of the two countries. In general, standard distances can be scaled by a variable that is a function of the area or size of the study areas. In this example, the weighted standard distances are 0.2379 and 0.0272 for Japan and Brazil, respectively, when their standard distances are scaled by their respective areas, indicating that Japan's largest cities are in fact more disperse than Brazil's cities in respect to the sizes of the two countries.

5.3.2 Standard Deviational Ellipses

The standard distance circle is a very effective visualization tool to show the spatial spread of a set of point locations. Often the set of point locations describes or represents a particular geographic phenomenon that has a directional bias. For instance, accidents along a section of a highway probably will not form a circular shape. Rather, they will exhibit a linear pattern dictated by the shape of that highway section. Similarly, occurrences of algae on the surface of a lake will form patterns that are limited by the shape of the lake, especially if the lake has an elongated shape. Under these circumstances, using standard distance circles will not fully reveal the directional bias of the spatial process.

A logical extension of the standard distance circle is the **standard deviational ellipse** (Furfey, 1927). It can capture the directional bias in a point distribution. Three components are needed to describe and define a standard deviational ellipse: an angle of rotation, deviation along the major axis (the longer axis), and deviation along the minor axis (the shorter axis). If the set of points exhibits a certain directional bias (i.e., the points do not disperse around the center equally in all directions), then we can identify the direction with the maximum spread of the points. Perpendicular to this direction is the direction with the minimum spread of the points. The two axes can be thought of as the x and y axes in the Cartesian coordinate system but rotated to an angle corresponding to the geographic orientation of the point distribution. This *angle of rotation*, θ, is the angle between the north and the y-axis rotated clockwise. Note that the rotated y-axis can be either the major or the minor axis. Figure 5.4 illustrates these terms related to the ellipse.

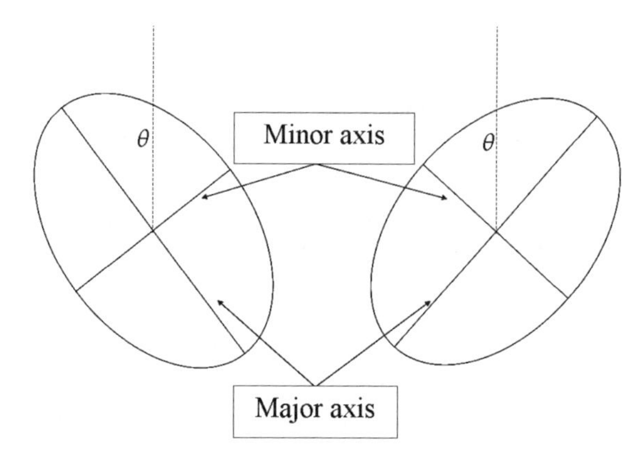

Figure 5.4 Elements defining a standard deviational ellipse.

To derive the standard deviational ellipse for a set of point locations, we can do the following:

1. Calculate the averages of coordinates of the set of points to derive the coordinates of the mean center, $(x_{\mathrm{mc}}, y_{\mathrm{mc}})$. These will be used as the origin in the next step to transform the coordinates.

2. For each point, p_i, distributed in the study region, transform each of its coordinates by the following algebraic operations:

$$x'_i = x_i - x_{\mathrm{mc}}$$

$$y'_i = y_i - y_{\mathrm{mc}}$$

These two operations essentially "move" all the points in the region to center at the mean center $(x_{\mathrm{mc}}, y_{\mathrm{mc}})$. That is, the coordinates of each point will be in reference to the mean center. Thus, the center of the transformed coordinate will be (0, 0) and the transformed coordinates will have positive and negative values, indicating that the observations are on different sides of the origin (0, 0) or the original mean center.

3. Calculate the angle of rotation, θ, such that

$$\tan\theta = \frac{\left(\sum_{i=1}^{n} x_i'^2 - \sum_{i=1}^{n} y_i'^2\right) + \sqrt{\left(\sum_{i=1}^{n} x_i'^2 - \sum_{i=1}^{n} y_i'^2\right)^2 + 4\left(\sum_{i=1}^{n} x'_i y'_i\right)^2}}{2\sum_{i=1}^{n} x'_i\, y'_i}, \quad (5.13)$$

where $\tan\theta$ can be positive or negative.

If the tangent is positive, it means that the rotated y-axis is the longer or major axis, and the longer axis rotates clockwise from north. If the tangent is negative, it means that the major axis rotates counter-clockwise from the north by an angle, θ. If the tangent is positive, we can simply take the inverse of tangent θ (arctan) to obtain θ, the angle of rotation for subsequent steps. If the tangent is negative, taking the inverse of its value will yield a negative angle, say $-x$. The negative tangent implies rotation of the angle counter-clockwise. But because the angle of the rotation is defined as the angle rotating clockwise from north to the y-axis, 90 degrees have to be added to the negative angle (i.e., $90 - x$) to derive θ, the angle of rotation clockwise from north.

4. With θ from Step 3, we can calculate the deviation along the x- and y-axes as follows:

$$\delta_x = \sqrt{\frac{\sum_{i=1}^{n} (x_i' \cos\theta - y_i' \sin\theta)^2}{n}} \tag{5.14}$$

and

$$\delta_y = \sqrt{\frac{\sum_{i=1}^{n} (x_i' \sin\theta + y_i' \cos\theta)^2}{n}}. \tag{5.15}$$

Using the data on men-only and women-only shelters in the previous section, we can compute a deviational ellipse for each type of shelter. Table 5.5 shows the procedure for computing the ellipse for the men-only and women-only shelters separately.

With $x'y'$, x'^2, and y'^2, $\tan\theta$ is equal to -0.519481 for men-only shelters. If we ignore the negative sign of $\tan\theta$ and take the inverse of the tangent or arctan of 0.519481, θ will be equal to 27.451019 degrees. But because $\tan\theta$ is negative, it implies that the angle provided is in the second quadrant or pointing toward northwest, while the definition of the angle of rotation is the angle rotated clockwise from the y-axis or the north. Therefore, the actual angle of rotation should be $90 - 27.451017 = 62.548980$ degrees. With θ of the ellipse, δ_x, and δ_y, the standard deviations along the two axes, which are equal to 0.022238 and 0.006476, respectively, can be derived accordingly. Using these elements, the ellipse can be derived to fit the spatial distribution of men-only shelters, and the ellipse is plotted in Figure 5.5. Similarly, we can compute the standard deviation ellipse for women-only shelters.

With $x'y'$, x'^2, and y'^2, $\tan\theta$ is equal to -1.9402 for the women-only shelters. Similar to the situation of men-only shelters, the negative tangent indicates a northwest-southeast ellipse. Thus, the angle of rotation, or θ, is equal

TABLE 5.5 Calculation of Standard Deviational Ellipses

ID	Types	Longitude (x)	Latitude (y)	x'	y'	$x'y'$	x'^2	y'^2
7	Men	−77.0310	38.9125	−0.0159	0.0226	−0.000359	0.000253	0.000511
14	Men	−77.0091	38.9013	0.0060	0.0114	0.000068	0.000036	0.000130
15	Men	−77.0191	38.8999	−0.0040	0.0100	−0.000040	0.000016	0.000100
26	Men	−76.9960	38.8560	0.0191	−0.0339	−0.000647	0.000365	0.001149
29	Men	−77.0201	38.8798	−0.0050	−0.0101	0.000050	0.000025	0.000102
		Mean −77.0151	38.8899			Σ −0.000928	0.000695	0.001992

ID	Types	Longitude (x)	Latitude (y)	x'	y'	$x'y'$	x'^2	y'^2
3	Women	−76.9926	38.8979	0.0266	−0.0090	−0.000239	0.000708	0.000081
6	Women	−77.0189	38.9020	0.0003	−0.0049	−0.000001	0.000000	0.000024
16	Women	−77.0200	38.9058	−0.0008	−0.0011	0.000001	0.000001	0.000001
18	Women	−77.0333	38.9028	−0.0141	−0.0041	0.000058	0.000199	0.000017
21	Women	−77.0201	38.9070	−0.0009	0.0001	0.000000	0.000001	0.000000
22	Women	−77.0300	38.9257	−0.0108	0.0188	−0.000203	0.000117	0.000353
		Mean −77.0192	38.9069			Σ −0.000385	0.001025	0.000476

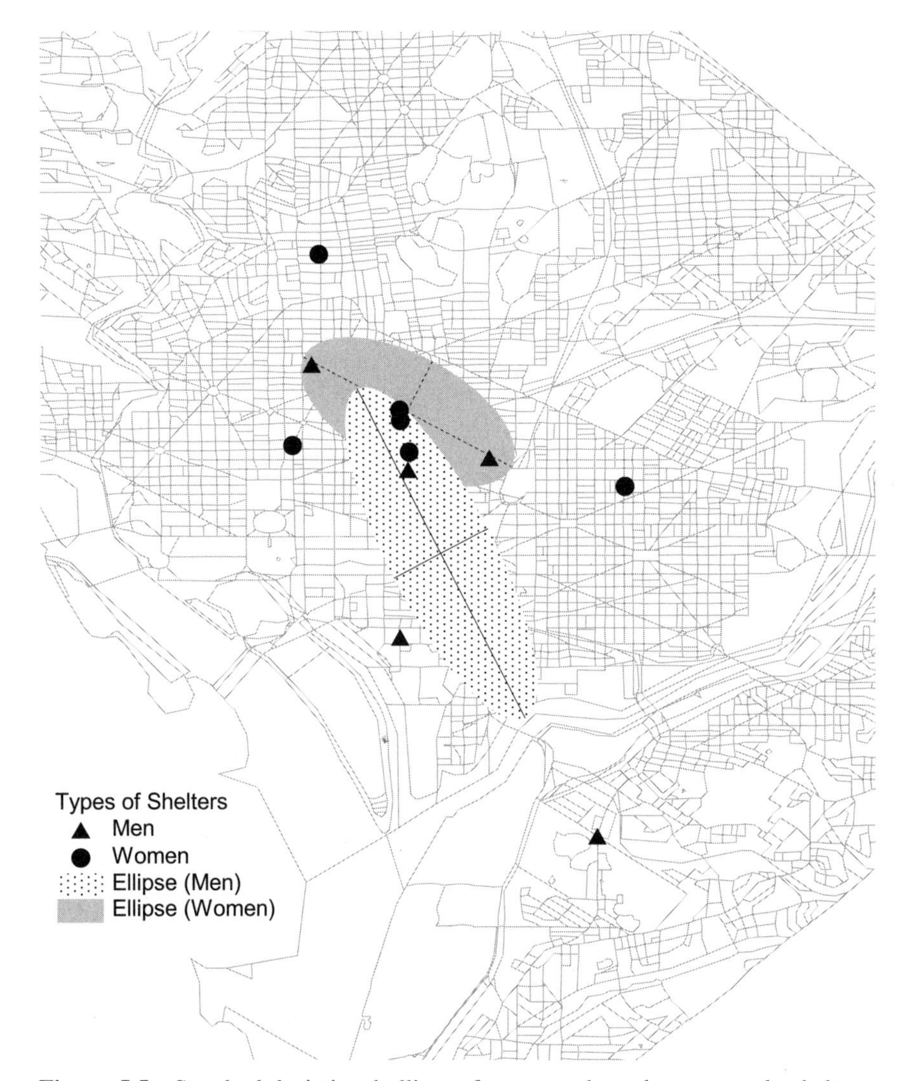

Figure 5.5 Standard deviational ellipses for men-only and women-only shelters.

to 90-arctan(1.9402). Therefore, the result is 27.267. When θ is known, the deviation along the x-axis and the deviation along the y-axis are also computed. The resultant ellipse for women-only shelters is also plotted in Figure 5.5.

In the above discussion of the standard deviational ellipse, we assume that in each location there is only one observation. Using notation adopted earlier, $w_i = 1$ for all i's, or each observation has an equal weight of 1. However, in many situations, each location may have more than one observation. For instance, as in the example discussed earlier in computing the spatial mean of the population, the centroid of a census unit is used to represent the location of all people residing in that areal unit, even though people distribute over the unit in reality. In this situation, the location is weighted by the number of people in the areal unit. Similarly, in deriving the standard deviational ellipse, locations may have different weights. In that case, we just need to assume multiple observations in the same coordinates and multiply the corresponding coordinate-based expressions in the formulas for $\tan\theta$, δ_x, and δ_y by w_i time for location i.

Note that in all of the above examples, we used the latitude-longitude coordinate system and assumed that the distance measures are in degrees. Obviously, this practice is problematic in general because the actual distance for one degree latitude or longitude is not a constant—it varies by location. Though the tools developed for these statistics in this book can use the latitude-longitude coordinate system in computing these centrographic measures, it is desirable to use data in projected coordinates more robust in distance and area calculations.

5.4 ARCVIEW NOTES

As in previous chapters, an ArcView extension was developed for calculating spatial mean, spatial median, standard distance, and standard deviational ellipse. The Ch5.avx extension should have been pasted into the EXT32 folder inside the ArcView installation folder on the computer's hard disk. Now, it can be loaded via the File menu's extensions, as for other extensions.

Here, we will demonstrate how the spatial distribution of car thefts changed over time from 2000 to 2003. The data for this example include three shapefiles: Carthefts.shp, Streets.shp, and Tract2000.shp. Carthefts.shp is a point shapefile that has years of car theft incidents as an attribute field. The other two fields are for references to provide the geographic context of Carthefts.shp.

For Ch5.avx, note that the calculation of a spatial median requires users to specify a convergence distance. As described earlier, the true spatial median cannot be identified. The iterative procedure discussed and adopted in the extension is to gradually get closer to the true spatial median using an iterative procedure. In each iteration, a new location (u_t, v_t) closer to the true spatial median is derived, and a distance can be computed between the new location and the previous location (u_{t-1}, v_{t-1}). This distance can be regarded as an improvement, and it should get smaller as the number of iteration increases. The convergence distance should be a small distance so that the users believe that such a small improvement in getting closer to the true spatial median is enough to terminate the procedure.

ArcView Example 5.1: Centrographic Measures

Step 1 Preparing data

The data set used in this ArcView example contains the point-based locations of motor vehicle thefts in Cleveland, Ohio, from 2000 to 2003. Three shapefiles are found under C:\Temp\Data\Ch5_data:

- `Carthefts.shp:` a point shapefile that contains locations of car thefts from 2000 to 2003
- `Streets.shp:` a line shapefile that contains street center lines of the area
- `Tracts2000.shp:` a polygon shapefile that contains 2000 Census tracts of the area

The data we will analyze are in carthefts.shp. The other two shapefiles are displayed to provide the geographic context of where the incidents occurred.

To prepare for this example, start ArcView, create a new View document, and bring the three shapefiles in using the **Add Theme** button.

Once the three themes are added, use the **View/Properties** menu to bring out the **View Properties** dialog.

- Specify `feet` as the **Map Unit.**
- Specify `feet` as the **Distance Unit.**
- Click **OK** to close the **View Properties** dialog.

View Properties

Name: View1 | OK

Creation Date: Tuesday, January 04, 2005 9:27:21 AM | Cancel

Creator:

Map Units: feet

Distance Units: feet

Projection... | Area Of Interest...

Background Color: | Select Color...

Comments:

If preferred, use the **Open Theme Table** button to open the attribute table of CarThefts.shp. You would see that there is an attribute field Year indicating in which year the incidents occurred. Now use the **File/Extensions** . . . menu to load the Ch5.avx extension.

Step 2 Calculate spatial means and standard distances
First, compute the spatial means and standard distances for each year. To do so, we will need to select car thefts in each year—2000, 2001, 2002, and 2003—and then calculate the spatial means and the standard distances.
So, for 2000 data:

- Use the **Query Builder** button to bring up the query dialog.
- Specify the query to be [(Year) = ``2000''], as shown below:

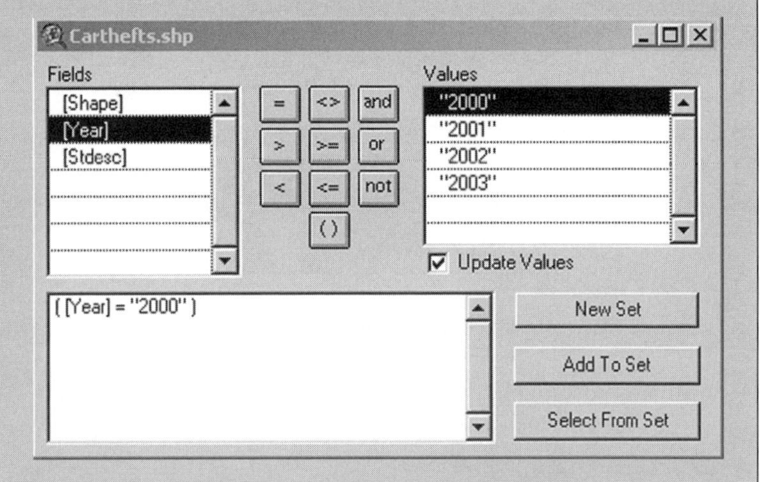

- Click **New Set** to proceed with the selection.
- Close the query dialog when you are done.

Now that the car thefts occurring in 2000 are marked selected, we can calculate the spatial mean and the standard distance. To do so, use the **Ch5** menu to select **Centrographic Measures.**

- In the **Centrographic Measures** dialog, check the boxes besides **Mean** and **Standard Distance.**

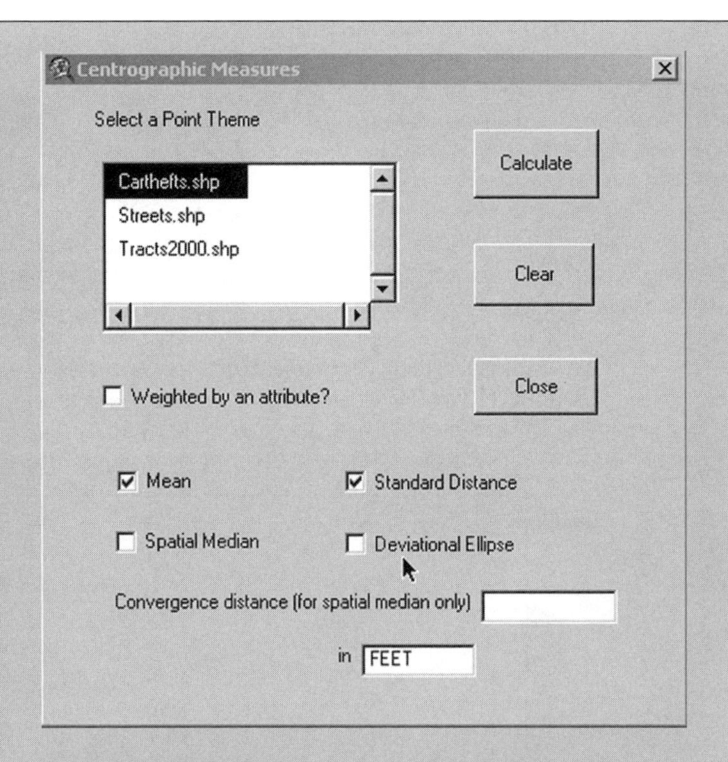

- Click the **Calculate** button to proceed with the calculation.
- When prompted, click the **Yes** button to create a shapefile for the spatial mean. This allows us to bring it back into the View document if needed.

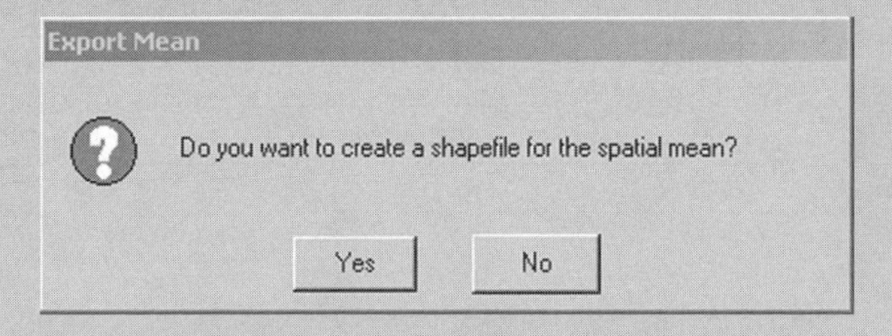

- Select a location to save the shapefile.
- When prompted, click the **Yes** button to create a shapefile for the standard distance circle. Again, this allows us to bring it back into the View document if needed.

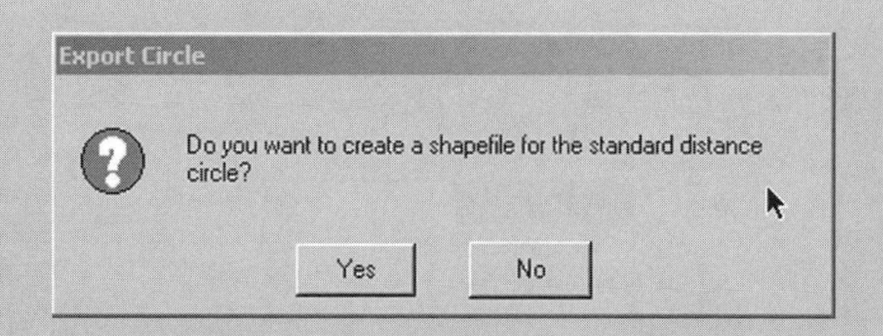

- Select a location to save the shapefile.

When the calculation is complete, the results are reported in an information dialog as:

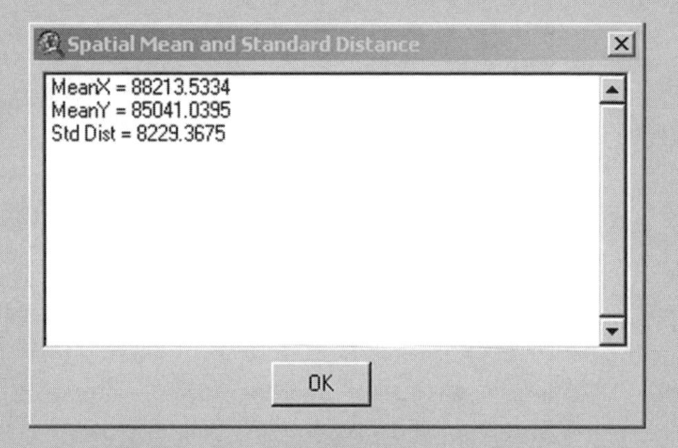

We can see that the spatial mean is (88214, 85041) and the standard distance is 8229.36, or about 1.5 miles.
Repeating the same process for the years 2001, 2002, and 2003, we have:

Year	Spatial Means	Standard Distance
2000	(88214, 85041)	8229
2001	(87958, 84833)	8316
2002	(88109, 84713)	8404
2003	(87654, 84673)	8556

We can see that the standard distances seem to have increased over the studied period, from 8229 to 8556 feet. Geographically, the mean centers seem to have moved westward over the studied period, as indicated on the next page:

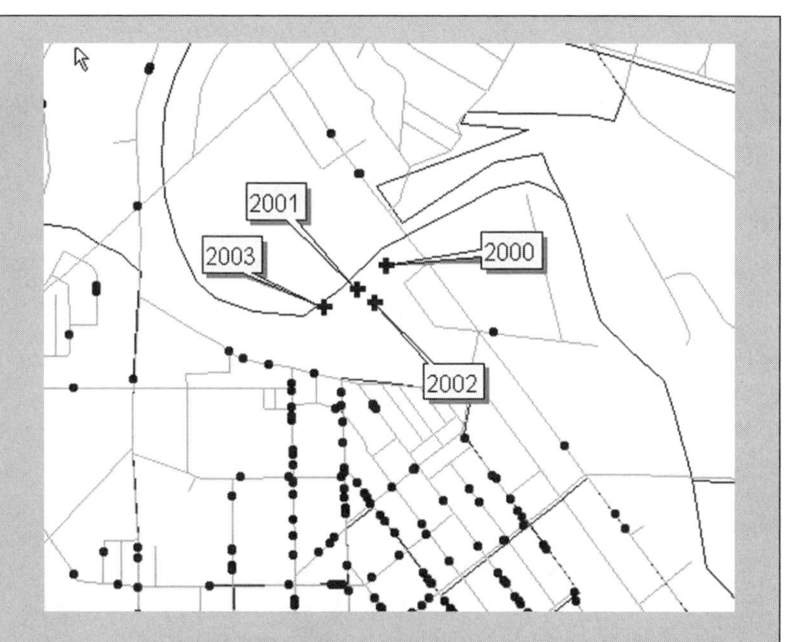

Step 3 Calculate spatial medians and standard deviational ellipses

Because the inner city of Cleveland is oriented to the shore of Lake Erie, running in a northeast-southwest direction, it is probably a good idea to calculate the standard deviational ellipses to show how the locations of the car thefts changed over the studied period.

To calculate the standard deviational ellipses for each year:

1. Select the incidents for each year.
2. Use the **Ch5** menu to select **Centrographic Measures.**
3. Check the box **Standard Deviational Ellipse** in the **Centrographic Measures** dialog.
4. Save each ellipse as a line shapefile to record its long and short axes, as well as a polygon shapefile to record the ellipse.
5. When the four years are processed, use the **Add Theme** button to bring all four polygons back into the view document.

Step 4 Making comparisons

From the results shown below, we see that the change in orientation was most significant in 2001. The standard deviation along the *y*-axis was largest in 2003 and along the *x*-axis was largest in 2001.

Year	Rotation Angle	St Dev Along x-axis	St Dev Along y-axis
2000	54.5	5238	6347
2001	69.6	5420	6307
2002	53.2	5397	6442
2003	48.9	5250	6756

The diagram below shows the four ellipses:

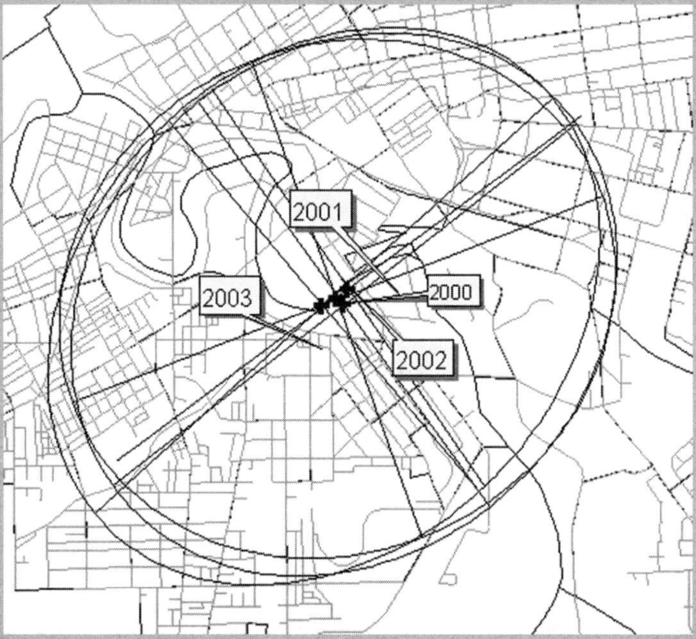

Step 5 End

5.5 APPLICATION EXAMPLES

The descriptive centrographic measures discussed in this chapter have been used widely in a variety of studies, from physical sciences to epidemiology. For instance, the U.S. Bureau of the Census has computed measures of spatial central tendency, namely the spatial mean and spatial median of population counts, after decennial censuses to indicate the geographic change in the overall population location in the United States (*http://www.census.gov/geo/www/cenpop/cntpop2k.html*), though the computation method is more precise and slightly more complicated to account for the curvature of the Earth's surface (Bureau of the Census, 2001). Thapar et al. (1999) conducted a detailed study on the locational changes of the distributions of the U.S. population by comparing centroid locations over censuses at different geographic scales, including the census region, division, and state levels.

By computing the spatial means and standard distances of disadvantaged census tracts in selected metropolitan areas over time, Greene (1991) documented the concentration or dispersion and the locational shifts of the economically disadvantaged population in those areas. Levine et al. (1995) described the directional biases of different types of motor vehicle accidents in Honolulu by fitting a standard deviational ellipse to each type of accident. Based upon a similar concept, Wong (1999) suggested computing a deviational ellipse for each race or ethnic group in population analysis. Subsequently, by overlaying these ellipses, the areas of their intersection and union can be calculated and an index based upon the areas of intersection and union can be formulated to represent the level of spatial segregation. Later, Wong (2000) applied this procedure based upon deviational ellipses to evaluate the level of segregation of several major Chinese ethnic groups.

REFERENCES

Bureau of the Census. 2001. *Centers of Population Computation for 1950, 1960, 1970, 1980, 1990 and 2000* (http://www.census.gov/geo/www/cenpop/calculate2k.pdf).

Ebdon, D. 1988. Statistics in geography. New York: Basil Blackwell.

Furfey, P. H. 1927. A note on Lefever's "Standard deviational ellipse." *American Journal of Sociology* 23: 94–98.

Greene, R. 1991. Poverty concentration measures and the urban underclass. *Economic Geography,* 67(3): 240–252.

Kellerman, A. 1981. *Centrographic Measures in Geography: Concepts and Techniques in Modern Geography (CATMOG) No. 32.* Norwich: Geo Books, University of East Angolia.

Lee, J. and D. W. Wong. 2001. *Statistical Analysis with ArcView GIS.* New York: John Wiley & Sons.

Levine, N., K. E. Kim, and L. H. Nitz. 1995. Spatial analysis of Honolulu motor vehicle crashes: I. Spatial patterns. *Accident Analysis and Prevention* 27(5): 675–685.

Monmonier, M. 1993. *Mapping It Out.* Chicago: The University of Chicago Press.

Thapar, N., D. Wong, and J. Lee. 1999. The changing geography of population centroids in the United States between 1970 and 1990. *The Geographical Bulletin,* 41: 45–56.

Wise, S. 2002. *GIS Basics.* New York: Taylor & Francis.

Wong, D. W. S. 1999. Geostatistics as measures of spatial segregation. *Urban Geography,* 20(7): 635–647.

Wong, D. W. S., 2000. Ethnic integration and spatial segregation of the Chinese population. *Asian Ethnicity,* 1: 53–72.

EXERCISES

Facility	Latitude	Longitude	Fugitive
90	41.41333	−81.78528	43,827
57	41.53889	−81.52917	43,468
33	41.51833	−81.66139	29,581
58	41.48333	−81.66699	16,457
48	41.41833	−81.80833	15,000
88	41.48278	−81.72667	13,345
68	41.50556	−81.64528	11,200
78	41.45364	−81.72456	9,776
94	41.41860	−81.49550	6,161
35	41.47472	−81.76694	5,200
73	41.37222	−81.52000	4,098

Facility	Latitude	Longitude	Point Source
46	41.41667	−81.82194	180,520
83	41.45080	−81.79370	161,248
12	41.45417	−81.78972	133,000
107	41.44167	−81.67500	61,060
19	41.43387	−81.54423	50,163
66	41.58028	−81.53056	48,400
74	41.46139	−81.62500	47,879
28	41.48056	−81.63194	38,533
94	41.41860	−81.49550	18,958
76	41.42690	−81.81390	17,000
65	41.59222	−81.52139	15,900
48	41.41833	−81.80833	15,000

The two tables above were extracted from the Toxic Release Inventory (TRI) database compiled by the U.S. Environmental Protection Agency (EPA) (*http://www.epa.gov*). The TRI data were gathered from local government agencies (counties and states) on various types of toxic substances released to the environment from industries. The database can be tabulated into various formats. One format lists the facilities with their location information (latitude and longitude) and the substances and amounts released to the environment, in the form of air, water, or solid wastes.

Both tables were based upon pollutants released to the air as fugitive emissions (first table) or point source releases (second table) in Cuyahoga County, Ohio, in 2002. They include only facilities with amounts (fugitive and point source) larger than the averages of both types of releases. Note that some facilities are on both lists, and others are on one only.

1. Compute the spatial means for both types of facilities based upon their locations alone. Do the two types of facilities have different general locations?

2. Because some facilities have extremely large amounts of release and others have much smaller amounts, it is desirable to include the amounts in determining the general locations of the two types of facilities. Compute their weighted spatial means using the release amounts. Are the results very different from the results from Question 1? If so, why?

   Transform the two tables into a spreadsheet program and save the two tables as dbf files. Then you can bring the data into ArcView using the following procedure:

   a. Launch ArcView and open up a new project.

   b. In the Project window, click the Table icon once and select Add.

   c. Select one of the dbf files created earlier to be added to the Table document.

   d. Open up a new View document, then select View/Add Event Theme.

   e. Select the field Longitude as the x-coordinate and Latitude as the y-coordinate. The point locations will be added to the View.

   f. Then choose Theme/Convert to Shapefile to save the data in a new shapefile.

   g. Use the same procedure to bring the other tables into ArcView and create another point theme/shapefile.

3. Use the ArcView Ch5.avx extension. Compute the spatial medians of the two types of facilities. Also compute their spatial means. Do they have very different locations?

4. Manually, calculate the standard distances of the two types of facilities. Do they have different magnitudes of dispersion?

5. Use the Ch5.avx extension to verify your answers. Do the standard distance circles generated from the extension provide any insight into the spatial distribution of the two types of facilities?

6. Given that Cuyahoga County is located along Lake Erie and is part of the Cleveland metropolitan area, it is likely that industrial facilities with specific location patterns have a specific orientation. Calculate the three parameters (angle of rotation, deviation along the x-axis, deviation along the y-axis) required to construct a standard deviational ellipse for each type of facility. Do they exhibit different orientations?

7. Use the Ch5.avx extension to verify the ellipse calculations. Bring the two ellipses back to the View document. To what extent do the ellipses for the two types of facilities overlap? What does the overlap indicate?

CHAPTER 6

POINT PATTERN ANALYZERS

The descriptive spatial statistics or centrographic measures introduced in Chapter 5 are useful in summarizing point distributions or comparing different point distributions with or without an attribute as the weight. But the use of these descriptive statistics is only the first step in what spatial analysis can do to explore patterns of point distributions. The next step following the use of summary spatial statistics is to determine if a point distribution corresponds to a recognizable pattern, namely, clustered, dispersed, or random. In other words, this step of geographic analysis consists of evaluating to what extent the observed pattern resembles a pattern generated from a random process, a clustered process, or a dispersed process. To perform this evaluation, however, requires additional tools, which are discussed in this chapter.

Every point distribution is the result of a certain spatial process at a given time and in a given space. To fully understand the various states and dynamics of a particular geographic phenomenon, an analyst must be able to detect spatial patterns from the point distributions and changes in point patterns at different times. The recognition and measurement of patterns from point distributions is therefore a very important step in analyzing geographic information.

In considering how cities distribute over a region, one can easily find situations in which cities distribute unevenly over space. This is because the landscape, transportation network, available natural resources, and many other factors may have influenced the decision to choose different locations for the settlements to start with and the different growth rates these settlements may have had afterward. Consider, for example, the locations of trees affected by a disease in a forest. If we find that these locations exhibited a cluster pattern, this would warrant additional investigation into the origin of the source of infection. Alternatively, if the locations of trees affected by the disease were found to be randomly distributed across the forest area, this suggests that the disease may have had more than one infectious origin or that other factors may have contributed to the spread of the disease.

At a global or continental scale, cities are often represented as points on a map. At a local scale, incidences of a disease, crime in a city or incidences of fire in a forest may exhibit a similar pattern. When studying point distributions such as these, analysts may try to relate them to particular patterns based on their experience or the knowledge generated from previous studies, particularly studies that have developed theories and models. One procedure is to examine how cities are distributed in a region vis-à-vis the theoretical pattern (hexagonal pattern) of city distribution in central place theory (Berry and Pred, 1961). To do so, analysts need tools that do more than summarize the statistical properties of point distributions.

The techniques discussed in this chapter, although having limitations, are useful in detecting spatial patterns of point distributions (Boots and Getis, 1988). First, we will introduce Quadrat Analysis, which allows analysts to determine if a point distribution is similar to a random pattern using a spatial sampling framework. Next, Nearest Neighbor Analysis, which compares the average distance between nearest neighbors in a set of points to that of a theoretical pattern (quite often a random pattern), will be illustrated. Then we will discuss the use of the spatial autocorrelation coefficients to measure how similar or dissimilar neighboring points are in terms of a given attribute. Finally, we will consider how to identify and evaluate the clustering of points at different spatial scales, or extents, using K-function analysis.

But before we discuss the specific spatial statistical techniques, we need to clarify some fundamental spatial issues related to scale and spatial extent. These issues are relevant to the analysis of any given point pattern.

6.1 SCALE AND EXTENT

The methods covered in this chapter are used mainly for detecting or measuring spatial patterns of point distributions. We need to pay special attention to two, and possibly three, critical issues when using these methods. These issues are spatial scale, extent, and projection. Chapter 1 has discussed these issues briefly. Here we will reemphasize the importance of scale and extent, especially in the context of analyzing point patterns.

First, we need to choose a proper geographic scale to work with when analyzing a point pattern in which points represent some geographic objects. This is because geographic objects may be represented differently at different scales, depending on how they are treated. Whether to hold the scale constant throughout the entire study or allow it to vary is an important issue when working with sets of points scattered over a geographic space.

As pointed out earlier, cities are often represented by points at a global or continental scale. The City of Cleveland, for example, appears to be only a point when a map shows it with other major cities in the United States. This same city, however, becomes a polygon object that occupies an entire map sheet when a larger-scale map (say, 1:24,000) shows the city with all of its streets, rivers, and other details. Similarly, a river may be represented in a small-scale map (say, 1:100,000) as a linear feature, but it is an ecosystem by itself if an analyst intends to study its water, riverbeds, banks, and all of its biological objects.

Unfortunately, there is no universal "one-size-fits-all" rule that we can apply everywhere at every scale. In many cases, the selection of a geographic scale for a particular study has to be modified as the study proceeds. The selection of a scale may or may not be apparent or easy, but recognizing its importance and how it may impact the results of your analysis is often the key to a successful and meaningful analysis.

The second issue is the extent of geographic areas in the study. Analysts often need to determine to what extent the area surrounding the geographic objects of interest should be included in the analysis. Let's assume that we are working on a study that examines intercity activities among Akron, Cincinnati, Cleveland, Columbus, and Dayton, Ohio. When only the geographic extent of the State of Ohio is used, the five cities seem to scatter quite far apart from each other, as shown in Figure 6.1. However, these cities seem to be very closely clustered if we define the entire United States as the study area. To increase this difference further, the five cities cluster nearly at one location if they are considered from a global perspective.

Properly delimiting the study area for a project is never easy. There are cases in which political boundaries are appropriate choices, but there are also many cases in which they are not. Many spatial analytical techniques assume that the spatial processes or phenomena under investigation continue over space. In practice, however, we must always define a boundary for the study area so that the study will be feasible. Once the boundary is defined, observations beyond it will not be considered. Therefore, results produced by these spatial analytical techniques could be biased because of their assumption of spatial continuity. The bias will likely be more significant when one evaluates the spatial relationship of geographic features or observations. The relationship will be "truncated" at the edge or boundary of the study area. This truncation or boundary can sometimes be defined arbitrarily or based on reasons not related to how observations are distributed. This issue of bias has been explored in the geographic literature and is called the *edge* or *boundary effect* (Griffith and Amrhein, 1983). While no single, simple solution can be offered for all cases, analysts are urged to consider this issue carefully.

The last issue is the projection used in maps when displaying the distribution of geographic objects. Objects can be distorted in many ways, including *area, shape, direction,* and *distance.* In detecting the spatial patterns of point distributions, it is important to consider the impacts of different map

Figure 6.1 Changes in clustering of selected Ohio cities in regard to the spatial extent of the state or the entire country.

projections. This is because both area and distance are used intensively in the analysis of point patterns. In Quadrat Analysis, the size of the study area affects the density of points, while in Nearest Neighbor Analysis and spatial autocorrelation coefficients, distances between points play a critical role in the calculation of these two types of statistics.

6.2 QUADRAT ANALYSIS

The first method used for detecting point patterns is **Quadrat Analysis.** This method evaluates a point distribution by examining how its density (expressed as the number of points per quadrat) changes over space. The density measured by Quadrat Analysis is then compared with the density of a theoretically constructed random pattern to see if the point distribution in question is more clustered or more dispersed than the random pattern.

6.2.1 General Concepts in Quadrat Analysis

The concept of and procedures for Quadrat Analysis are relatively straightforward. We first overlay the study area with a regular square grid and count the number of points falling in each square. Some squares will contain no points, while others will contain one, two, or more points. We prefer to count the number of squares according to the number of points in the square. That is, we will count squares with zero points, then one point, two points, and so on so that we can construct a frequency distribution of the number of squares based on the numbers of points within them. Quadrat Analysis compares this frequency distribution with that of a known pattern, generally a theoretically constructed random pattern.

The squares in this type of analysis are referred to as **quadrats,** which are essentially sampling units in spatial statistical jargon. That is, quadrats are used as sampling units to determine how dense the points are in different parts of the study region. But quadrats do not have to be squares. Analysts can use other geometric forms, such as circles or hexagons, as long as they are appropriate for the geographic phenomenon being studied. The selection of the form of the quadrat should be based on previous successful experience or on the characteristics of the phenomenon. However, a square is chosen most often because this shape can cover the study area completely. But a square is not geometrically compact. Therefore, some studies also use the circle—the most compact shape. However, circles will not be able to cover the entire geographic space unless they overlap each other. Another choice is the hexagon, which is preferred in the geographic literature, specifically in relation to the central place theory. The hexagon is more compact than the square, and it can also cover the entire region without overlapping. Use of the hexagon was computationally more intensive prior to the GIS era. Currently, implementing Quadrat Analysis using the hexagon is as easy as using other shapes in a GIS environment. There is another important issue regarding quadrat shape. In each analysis, the shape and size of the quadrats have to be constant. That is, if a square is selected, it should be used throughout the entire study, and the size of the square has to be kept the same.

In an extremely clustered point pattern, one can expect all or most of the points to fall inside one or a few squares only. On the other hand, in an extremely dispersed pattern, sometimes also referred to as a *uniform pattern or a triangular lattice,* one would expect all squares to contain a relatively similar number of points. Based upon the distribution of points across the quadrats and the frequency capturing the distribution, one can determine if the point distribution under study is closer to a clustered, a random, or a dispersed pattern. As examples of clustered and dispersed patterns, Figure 6.2 shows the 164 cities in the State of Ohio, along with hypothetical clustered and dispersed patterns with the same number of points.

Overlaying the study area with a regular grid partitions the study area in a systematic manner to avoid over- or undersampling of the points anywhere.

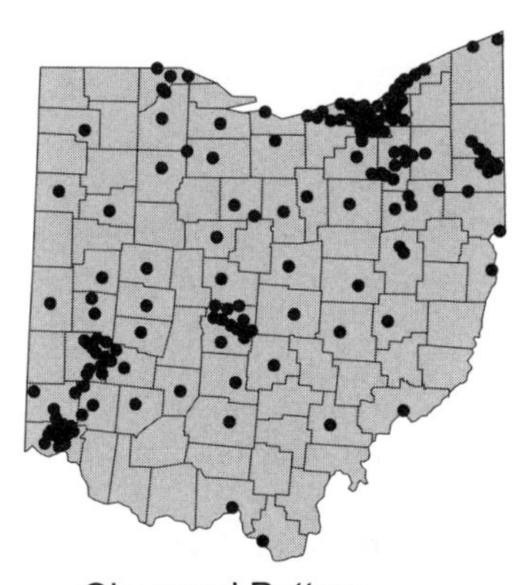

Figure 6.2 Observed pattern of Ohio cities and hypothetical clustering and dispersed patterns.

Since Quadrat Analysis evaluates changes in density over space, it is important to keep the sampling interval uniform across the study area. However, another way to achieve the same effect is to randomly place quadrats of a fixed size over the study area (Figure 6.3). Statistically, Quadrat Analysis will achieve a fair evaluation of the density across the study area if it applies a large enough number of randomly generated quadrats.

The last issue that requires careful consideration when applying Quadrat Analysis is the size of quadrats. According to the Greig-Smith experiment (Greig-Smith, 1952) and subsequent discussions by Taylor (1977, pp. 146–147) and Griffith and Amrhein (1991, pp. 131), an optimal size of quadrat can be calculated by

$$\text{Quadrat size} = \frac{2A}{r},$$

where

A is the area of the study area, and
r is the number of points in the distribution.

This suggests that a quadrat of the appropriate size has a width of $\sqrt{2A/r}$ in the case of square quadrats or a radius of $\sqrt{2A/\pi r}$ in the case of circular quadrats. Based on these relationships, it can be estimated that, with the optimal size for square quadrats, the number of quadrats, n, should be approximately $r/2$.

Once the quadrat size for a point distribution is determined, Quadrat Analysis can proceed to establish the frequency distribution of the number of points for all quadrats, either with a square grid or a hexagon grid that covers the entire study area systematically, or with a set of randomly generated quadrats. The observed frequency distribution derived by tabulating the number of quadrats containing different numbers of points must be compared to the expected frequency distribution describing a random point pattern. The observed and expected frequency distributions can be compared using the Kolmogorov-Smirnov (K-S) statistical test, discussed in Chapter 3. The K-S statistics can be used to test if the observed frequency distribution and the frequency distribution of a theoretical pattern are statistically different or not. This test is simple and straightforward, both conceptually and computationally.

As an example, let's take the 164 Ohio cities and use 81 squares to construct the frequency distribution for a Quadrat Analysis. The frequency distribution of the cities falling into quadrats or squares is listed in Table 6.1. The leftmost column of the table lists the number of cities in each quadrat. The second column shows that there are 36 quadrats with no city at all, 21 quadrats with only one city, 6 quadrats with two cities, and so on. For an

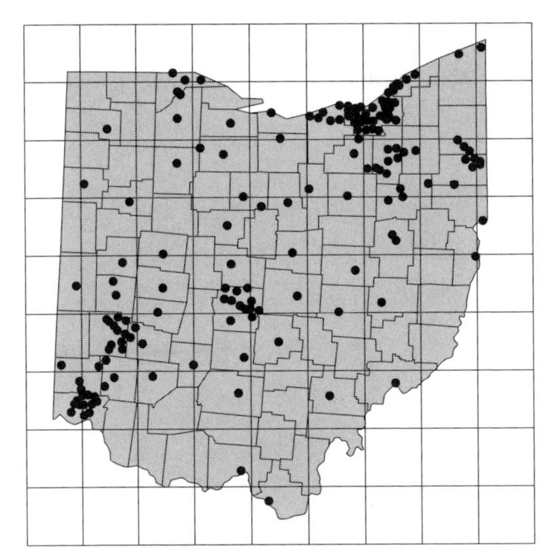

Completed Coverage with Regular Quadrats

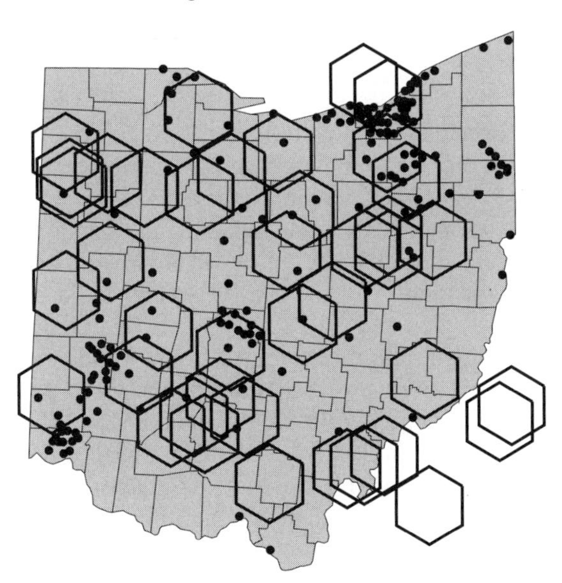

Random Placement with Hexagonal Quadrats

Figure 6.3 Examples of systematic and random quadrats.

TABLE 6.1 Frequency Distributions: 164 Ohio Cities with the Observed and Hypothetical Uniform/Dispersed and Clustered Distributions

Number of Cities in Each Quadrat	Observed	Uniform/Dispersed	Clustered
0	36	16	72
1	21	14	—
2	6	20	3
3	8	16	—
4	1	13	1
5	2	2	1
6	—	—	1
7	2	—	—
11	1	—	—
12	1	—	—
13	1	—	—
15	1	—	—
20	—	—	1
28	1	—	—
40	—	—	1
83	—	—	1
Total Number of Quadrats	81	81	81

example of a uniform/dispersed pattern, the third column lists frequencies of quadrats with cities that are somewhat evenly distributed across all quadrats. The right-most column depicts a highly clustered pattern. It shows that of the 164 cities, 83 are located within one quadrat and another 40 are found in another quadrat, indicating that the cities are highly clustered. On the other hand, there are 72 quadrats with no cities, another indication of a clustered pattern.

The differences among the three frequency distributions clearly show that the observed pattern is more clustered than the dispersed pattern, but it is not as clustered as the distribution of the right-hand column. While the differences among the columns can be visually evaluated, we need some way of measuring them quantitatively. At this stage of our analysis, we can apply the K-S test to determine if the distributions are significantly different or not.

6.2.2 Comparing Observed and Expected Distributions Using the K-S Test

The K-S test allows us to test a pair of frequency distributions at one time. Let's start with the observed pattern and the dispersed pattern. If we know the spatial process generating the dispersed pattern, we can treat this pattern as the theoretical one, though in most cases we prefer to use a random pattern as the theoretical pattern. In this case, we have a one-sample (observed distribution) situation. If we do not know the process generating the dispersed pattern, then we can test if the two distributions are significantly different or not. To test if the observed pattern is significantly different from the dispersed pattern, we are in a two-sample situation.

In the testing process, first we assume that the two frequency distributions are similar enough that we cannot detect any difference between them that is statistically significant. This assumption is in fact our null hypothesis, H_0, that we would like to reject. If the difference between the two frequency distributions is indeed very small, then the difference might have happened simply by chance, or due to sampling error, and therefore is not of concern. The larger the difference is, the less likely that it is created by chance. If we can reject the null hypothesis, we can then claim that there is evidence to support the conclusion that the alternate hypothesis, H_1 may be true, where H_1 states that the two distributions are significantly different.

Following the standard statistical testing procedure described in Chapter 4 and the one used to perform a K-S test, we could compare an observed distribution with a known or theoretical distribution (one-sample case) or with another unknown distribution (two-sample case). We would use the following steps:

1. Formulate the null hypothesis (H_0), which claims or assumes that the difference between the two frequency distributions is not statistically significant. Thus, H_1 should state that the difference between two frequency distributions is significant.
2. Decide on a level of statistical significance, say allowing only 5 out of 100 times ($\alpha = 0.05$) that the result showing the difference is true, while in fact there is no difference. In other word, one wants to be quite conservative in performing this test so that it will be difficult to reject the null hypothesis when it is true (the so-called Type I error).
3. Convert all frequencies to cumulative proportions in both the observed and the theoretical or other distributions.
4. Calculate the D statistic for the K-S test:

$$D = Max|O_i - E_i|, \qquad (6.1)$$

where O_i and E_i are the observed and expected cumulative proportions, respectively, of the ith category in the two distributions. O_i and E_i can also be treated as the cumulative proportions of any two distributions. The Max|.| term indicates that we do not care which one is larger than the other; we are concerned only with the magnitude of the difference. D is then the maximum absolute difference among all pairwise comparisons.

5. Calculate a critical value as the basis for comparison. In the one-sample case,

$$D_{\alpha=0.05} = \frac{1.36}{\sqrt{n}}, \tag{6.2}$$

where n is the number of points. In a two-sample case where we want to test if the two distributions are different or not, the critical value is calculated in the following manner:

$$D_{\alpha=0.05} = 1.36 \sqrt{\frac{n_1 + n_2}{n_1 n_2}}, \tag{6.3}$$

where n_1 and n_2 are the number of quadrats (or observations) in the two sets of distributions.

6. If the calculated D is greater than the critical value of $D_{\alpha=0.05}$, then we can reject the null hypothesis and conclude that the observed and theoretical distributions are statistically significantly different in the one-sample case. In the two-sample case, we can conclude that the two observed distributions are significantly different.

Taking the example of the 164 Ohio cities and the tabulations reported in Table 6.1, we focus only on comparing the observed and dispersed patterns. Table 6.2 shows the original frequency distributions and their corresponding cumulative proportions. The rightmost column lists the absolute differences

TABLE 6.2 Cumulative Proportions from Observed and Dispersed Patterns to Derive the Absolute Differences for the K-S Statistic, *D*

Number of Cities in Each Quadrat	Observed	Cumulated Observed Proportion	Dispersed	Cumulated Dispersed Proportion	Absolute Differences in Proportion
0	36	0.44	16	0.20	0.25
1	21	0.70	14	0.37	**0.33**
2	6	0.78	20	0.62	0.16
3	8	0.88	16	0.81	0.06
4	1	0.89	13	0.98	0.09
5	2	0.91	2	1.00	0.09
6	0	0.91	0	1.00	0.09
7	2	0.94	0	1.00	0.06
>7	5	1.00	0	1.00	0.00
Total	81		81		

between the two columns of cumulative proportions. The largest absolute difference, in this case, is 0.33. Therefore,

$$D = Max|O_i - E_i| = 0.33.$$

The critical value, $D_{\alpha=0.05}$, can be calculated as

$$D_{\alpha=0.05} = \frac{1.36}{\sqrt{81}} = \frac{1.36}{9} = 0.151 < 0.33$$

if we treat this example as a one-sample case where we know the process generating the dispersed pattern. If we just compare the two point distributions without knowing the spatial processes generating the two spatial distributions, then the critical value, $D_{\alpha=0.05}$, in this two-sample case, will be

$$D_{\alpha=0.05} = 1.36 \sqrt{\frac{81 + 81}{81 * 81}} = 0.214 < 0.33.$$

In both cases, the K-S D statistic is larger than the critical values, indicating that the difference between the two frequency distributions is statistically significant at the 0.05 level. We can then reject the null hypothesis that there is no significant difference between the observed pattern of the 164 Ohio cities and the dispersed point pattern. In other words, the 164 Ohio cities do not exhibit a significant similarity to a dispersed point pattern, or the difference between the spatial distribution of the original 164 cities and the hypothetical dispersed pattern is not due to chance, but rather to some systematic processes.

In the above example, we examined the difference between an observed point pattern and a dispersed pattern. However, it is more common to compare an observed point pattern to a point pattern generated by a random process. A well-documented random process in generating point patterns is the Poisson process. The Poisson random process is appropriate for generating count data or frequency distributions. Quadrat Analysis often compares an observed point pattern frequency distribution to a frequency distribution generated by the Poisson random process.

A parameter in the Poisson distribution is the average number of occurrences, λ. In the context of Quadrat Analysis, λ is defined as the average number of points in a quadrat. Assume that we have n quadrats and r points in the entire study area. Then an estimate of λ is $\lambda = r/n$, the average number of points in a quadrat.

Let x be the number of points that can be found in any quadrat. Using the Poisson distribution, the probability of having x points in a quadrat is then defined as

$$p(x) = \frac{e^{-\lambda}\lambda^{x}}{x!}, \tag{6.4}$$

where e is the natural logarithm and $x!$ is the factorial of x, which can be defined as

$$x! = x \times (x - 1) \times (x - 2) \times \ldots \times 1.$$

By definition, $0! = 1$. Using various values for x (e.g., 0, 1, 2, . . . , r) in the above equation computing the Possion distribution, the probabilities of having x points in a quadrat can be derived. We can then multiply the probabilities by the total number of quadrats to generate a theoretical frequency distribution in the same format shown in Table 6.1, but now for a random point distribution.

Generating a probability value from the Poisson distribution equation is rather simple. But if a set of probabilities for a range of x values is required, it becomes quite tedious, as the factorial and natural logarithm have to be applied every time. Fortunately, there is a shortcut to generating a set of Poisson probabilities, as discussed in Chapter 4. Let us review this shortcut method briefly, as it is critical in deriving the expected frequency and probability distributions. We know that if $x = 0$, the Poisson probability is reduced to

$$p(0) = e^{-\lambda}.$$

Probabilities for $x > 0$ can be derived based upon $p(0)$. In general, $p(x) = p(x - 1)\ \lambda/x$. If $x = 1$, then $p(x - 1) = p(0)$. This shortcut formula can efficiently derive an entire set of Poisson probability with a known λ.

In the example of the Ohio cities, there are 164 points (cities). In the simplest situation, we can generate 81 quadrats as a grid layer superimposed on the entire region. Then $\lambda = r/n = 164/81 = 2.0247$. Using this λ value, a set of Poisson probabilities can be derived using the shortcut formula for x from 0 to 7 and beyond. Table 6.3 shows the derivations of the probabilities. The first two columns are identical to those in Table 6.2. The third and the fourth columns are the observed and cumulative probabilities based upon the observed pattern. In the fifth column, a Poisson distribution was generated based upon $\lambda = 2.0247$. This probability distribution indicates the probability of a quadrat receiving different numbers of points based upon a random process. Column six shows the frequencies of a quadrat with the expected number of points (x). The cumulative probabilities of the Poisson are derived in column seven. The last column reports the absolute differences between the two sets of cumulative probabilities. The largest of these differences is the K-S D statistic, which is 0.31, much greater than the critical value of 0.151 based upon the one-sample test framework using an 0.05 level of significance.

TABLE 6.3 Derivation of the K-S D Statistic by Comparing the Observed Pattern with a Random Pattern Generated from a Poisson Distribution

Number of Cities in Each Quadrat (x)	Observed Frequency	Observed Proportion	Cumulated Observed Proportion	Expected (Poisson) Proportion	Expected Frequency	Cumulated Expected (Poisson) Proportion	Absolute Differences in Proportion
0	36	0.44	0.44	0.13	16	0.13	**0.31**
1	21	0.26	0.70	0.27	14	0.40	0.30
2	6	0.07	0.78	0.27	20	0.67	0.11
3	8	0.10	0.88	0.18	16	0.85	0.02
4	1	0.01	0.89	0.09	13	0.95	0.06
5	2	0.02	0.91	0.04	2	0.98	0.07
6	0	0.00	0.91	0.01	0	1.00	0.08
7	2	0.02	0.94	0.00	0	1.00	0.06
>7	5	0.06	1.00	0.00	0	1.00	0.00
Total	81				81		

In the above example, we compared the observed proportions to the proportions generated by a Poisson distribution, which is a random process. In other words, the point locations that a Poisson process generates should exhibit a random pattern. This framework allows us to test if the observed pattern is similar to a random pattern. In theory, we can compare the observed pattern with any pattern of known characteristics. For instance, after we compare the observed pattern with a random pattern and the result indicates that they are significantly different, the next logical step is to test if the observed pattern is similar to a clustered pattern or to a dispersed pattern. Very often, through visual inspection, an analyst can hypothesize what the observed pattern may resemble. In that case, one may perform a process similar to the one discussed above to test if the observed pattern is different from, for instance, a clustered pattern—the expected distribution. Or the null hypothesis will be that the observed pattern is not different from a clustered pattern. In the discussion above, we used the Poisson process to generate the random or expected distribution. If the expected pattern is known to be a clustered pattern, we may want to consider using another probability distribution that will generate a theoretically clustered pattern. A theoretical distribution one may use to generate such a clustered distribution can be the negative binomial or gamma function. The theoretical pattern can then be compared with the observed pattern to test if they are different using the framework discussed above.

Another aspect of all of the previous examples deserves our attention. In all the examples, we computed the probabilities up to the category of 7 points and then >7. The last category ensures inclusiveness—that is, all possible values are included in the classification. We can compute the probabilities for each possible number up to the total number of points, r. But it is obvious that the expected probability gets smaller and approaches zero as x increases. Therefore, one may stop calculating the probability when it gets too small to be meaningful.

6.2.3 Comparing Observed and Expected Patterns Using the Variance-Mean Ratio

Besides using K-S statistics to test if the observed pattern is different from a random pattern (generated from a Poisson distribution), one may perform the **Variance-Mean Ratio Test** by taking advantage of a specific statistical property of the Poisson distribution.

We know that λ is the mean of a Poisson distribution. A unique and useful property of the Poisson distribution is that its variance is also λ. In other words, if a distribution is potentially generated by a random process like the Poisson process, its mean should be equal to its variance. If we compute a ratio by dividing the variance by the mean, which is the variance-mean ratio, this ratio should be very close to 1.

With this ratio, given an observed point pattern and the frequency distribution of points by quadrats, we can compare the pattern's observed variance-mean ratio to 1, the expected variance-mean ratio, to see if they are significantly different or not. This is because the variance-mean ratio is expected to be 1 if the observed pattern conforms to the theoretical random pattern.

The first two columns in Table 6.4 are identical to the information in Table 6.1. They list the number of points in each quadrat and the corresponding number of quadrats with the given number of points. Given that the mean of the observed distribution is $\lambda = 2.0247$, the variance of the observed distribution is defined as

$$\sigma = \frac{\Sigma\, n_i(x_i - \lambda)^2}{n}, \tag{6.5}$$

where x_i is the number of points (cities) in a quadrat, n_i is the number of quadrats with x_i points, and n is the total number of quadrats. Therefore, the third column in Table 6.4 is used to derive part of the numerator of the variance, that is, $(x - \lambda)^2$. In the fourth column, the frequencies (n_i) are multiplied by the corresponding squared values of the mean deviations. The sum of the fourth column gives us the numerator of the variance calculation, and it is 1391.95. Dividing this sum by n (i.e., 81) gives us the variance. The variance based upon the observed point pattern is therefore 17.1845.

This calculation of the variance is essentially the same process of computing the variance for grouped data (Chapter 2). The mean, λ, is compared with the number of points (x) in each quadrat. The difference is then squared and multiplied by the quadrat frequencies. The sum of these products is then divided by the number of quadrats to produce the observed variance. Using

TABLE 6.4 Derivation of the Variance-Mean Ratio

Number of Cities in Each Quadrat (x_i)	Observed Frequency(n_i)	$(x_i - \lambda)^2$	$n_i(x_i - \lambda)^2$
0	36	4.09941	147.5788
1	21	1.05001	22.0502
2	6	0.00061	0.0037
3	8	0.95121	7.6097
4	1	3.90181	3.9018
5	2	8.85241	17.7048
7	2	24.75361	49.5072
11	1	80.55601	80.5560
12	1	99.50661	99.5066
13	1	120.4572	120.4572
15	1	168.3584	168.3584
28	1	674.7162	674.7162
Total	81		1391.9506
Average	2.0247		17.1846

the computed variance and the mean, we can form an observed variance-mean ratio. According to the results from Table 6.4, the variance-mean ratio for the 164 Ohio cities is 17.1846/2.0247 = 8.4874.

If the observed pattern is not statistically significantly different from a random pattern, then we can expect that the observed variance-mean ratio, σ/λ, will be close to 1, the theoretical variance-mean ratio for a Poisson distribution, where σ and λ are the same; thus, the variance-mean ratio should be 1. If $\sigma/\lambda > 1$, this indicates that the pattern may be more clustered, as the relatively large variance implies that some quadrats have many points and others have very few. This is the situation of the 164 Ohio cities, as σ/λ is 8.4874. On the other hand, $\sigma/\lambda < 1$ is an indication of a dispersed pattern, as σ is quite small and all quadrats have similar numbers of points.

It is likely that our observed variance-mean ratio will be different from 1 in most cases due to uncontrollable factors—even if the pattern is actually generated by a random process. Therefore, we should perform a statistical test to see if the observed variance-mean ratio is really different from the theoretical value of 1. Conceptually, this test is the same as the difference-of-means test discussed in Chapter 4, where the observed statistic is compared with the expected value and standardized by the standard error of the statistic in order to obtain a t-statistic. This is also the testing model we will use throughout the rest of this book to test if the observed statistic is significantly different from the expected value or not.

Formally, the null hypothesis is that the observed variance-mean ratio is not significantly different from the expected variance-mean ratio, which is 1. The test statistic, which is a t-statistic, is defined as

$$t_{(df=n-1)} = \frac{\left|\left(\dfrac{\sigma}{\lambda}\right) - 1\right|}{\sqrt{\dfrac{2}{(n-1)}}}, \tag{6.6}$$

with a degree of freedom $= n - 1$, where n is the number of quadrats. The numerator is the absolute difference between the observed variance-mean ratio and the expected ratio (1). We take the absolute difference because we are not concerned with whether the calculated ratio is larger or smaller than the expected value. The denominator is the estimated standard error of the difference of the mean. If the number of quadrats, n, is 30 or more, the calculated statistic will follow a standard normal distribution, z, such that the critical values will be ± 1.96 with a confidence level $\alpha = 0.05$.

Given the data in Table 6.4 with the variance-mean ratio equal to 8.4875,

$$t_{df=80} = \frac{|8.4875 - 1|}{\sqrt{\dfrac{2}{81 - 1}}} = \frac{7.4874}{0.1581} = 47.3586 >> 1.96.$$

The t-statistic of the variance-mean ratio is 47.35, much larger than the standard critical value of 1.96. Therefore, we may reject the null hypothesis that the observed Ohio cities pattern is different from a random pattern. We may also modify the above testing model to test for clustering or dispersion by removing the absolute sign in the numerator of Equation 6.6. In this case, a positive t-value indicates a clustered pattern and a negative t-value indicates a dispersed pattern.

ArcView Example 6.1: Quadrat Analysis of Northeast Ohio Cities

With all the methods discussed in this chapter, we now have a variety of choices when analyzing point distributions. This example uses a data set of 17 cities in the five counties of northeast Ohio to illustrate the steps in using an ArcView extension, Ch6.avx, developed specifically for this chapter. This example does not attempt to offer any explanation for the locations of the 17 cities examined here. Its focus is on introducing the various ways that Ch6.avx can be used to run the point pattern descriptors incorporated in the extension.

Step 1 Preparing data and loading extension

As in previous ArcView examples, the data sets should be in the C:\Temp\Data\Ch6_data folder.

To load the ch6.avx extension:

- Start ArcView.
- From the **File** menu, choose **Extenstions . . .**.
- In the **Extensions** dialog box, check the box beside **Ch6** and click **OK** to proceed.

To bring the data sets into ArcView:

- Click the **Add Theme** button and navigate to the `C:\Temp\Data\Ch6_data` folder.
- Add both `5counties.shp` and `5ccities.shp`.

The 5ccities.shp will be the data theme used by the point pattern descriptors.

Step 2 Quadrat analysis

First, we will perform Quadrat Analysis to see if the five cities are located in a random pattern or not.

- From the **Ch6** menu, choose **Quadrat Analysis** to bring up the Quadrat Analysis dialog.
- Click to highlight the `5ccities.shp` point theme, and then click the **Choose the Theme** button.

Notice that the number of points has been updated to `17`.

- For this example, check the **Complete** button for a complete placement of quadrats.
- Enter `100` as the number of quadrats.

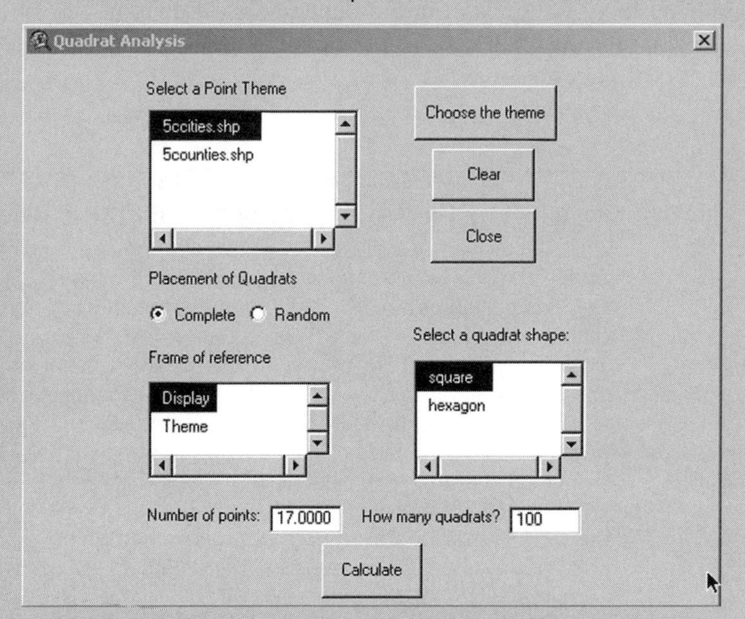

- Click the **Calculate** button to proceed with the calculation.
- When prompted, store the quadrats as a shapefile in a folder with a easily recognizable name.

The first statistic reported is the mean (lamda, λ), which equals 0.163462. Next, the frequency distribution, the number of points in a quadrat denoted x, and the number of quadrats with x points denoted Freq, is reported as

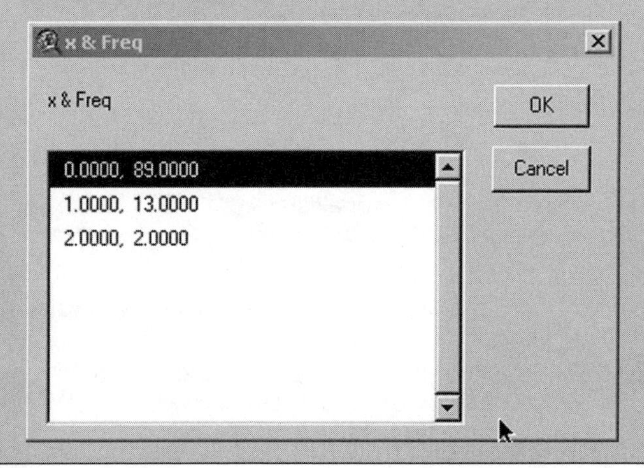

Next, the variance is reported to be 0.175203.
Then the variance-mean ratio is reported to be 0.515496.
Then the K-S *D* statistic is reported to be 0.00724133.
For testing the hypothesis that the point pattern of the 17 cities is not different from a random pattern, we could choose the level of significance as 0.05 in the next dialog box.
With the 0.05 level, we have a critical value of the K-S statistic of 0.133359.

Step 3 Evaluating statistics
Since the critical value of 0.13359 is greater than 0.00724133 as calculated, we cannot reject the hypothesis. Therefore, we conclude that the point pattern formed by the 17 cities is not significantly different from a random pattern. When one carefully examines the distribution of the 17 cities as displayed on the map (the figure below), it is quite obvious that the distribution of the 17 cities does not exhibit any discernible pattern.

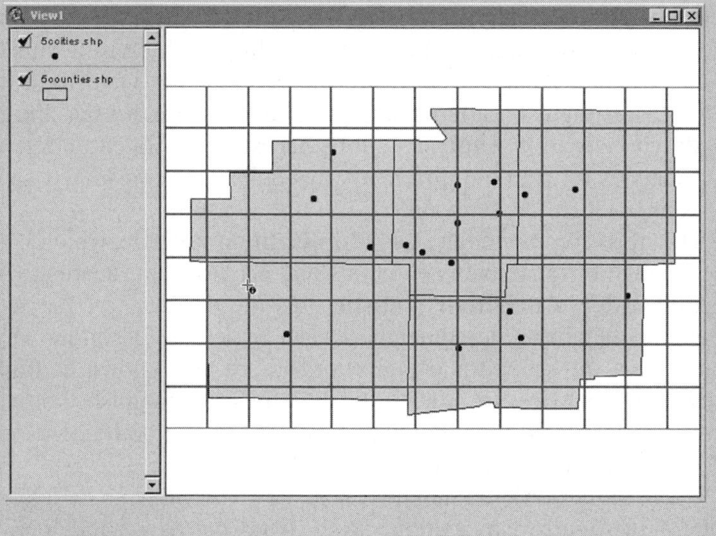

Step 4 End

6.3 ORDERED NEIGHBOR ANALYSIS

Quadrat Analysis is useful in comparing an observed point pattern to a random or theoretically known distribution. However, it has certain limitations. The analysis captures information on the points within each quadrat, but no information on points between quadrats is used in the analysis. As a result, Quadrat Analysis may be insufficient to distinguish between certain point patterns. Figure 6.4 is an example.

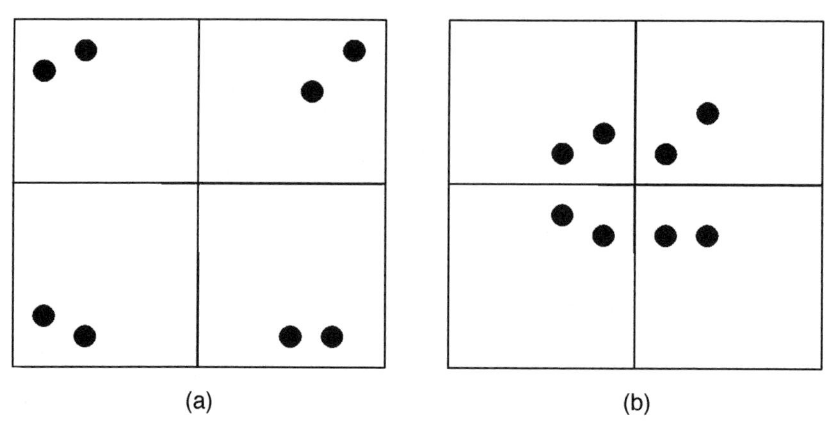

Figure 6.4 Hypothetical patterns illustrating the need for ordered neighbor statistics.

In Figures 6.4a and 6.4b, both spatial configurations have eight points in four quadrats. Visually, the two point patterns are different. Figure 6.4a is a more dispersed pattern, while Figure 6.4b is definitely a clustered pattern. Using Quadrat Analysis, however, the two patterns yield the same result. In order to distinguish patterns similar to those in Figure 6.4 and some others, **Ordered Neighbor Analysis** will be necessary. Different orders in the neighbor analysis refer to the different ways points are considered as neighbors of a given point in the analysis.

The most common is the **nearest neighbor statistic,** which is derived from the average distance between points and each of their nearest neighbors. The **second-ordered neighbor statistic** uses the distances of the second nearest neighbors. Higher-ordered neighbors can be defined in similar ways. Neighbor statistics at the second and higher orders are also known as **higher-ordered neighbor statistics.** In order to quantitatively distinguish the two patterns in Figure 6.4, ordered neighbor statistics, specifically both nearest neighbor and second-order statistics, have to be used.

Using higher-ordered neighbor statistics will allow us to determine that the pattern in Figure 6.4a exhibits a clustered pattern at the local scale but a dispersed pattern at the regional scale. In Figure 6.4b, higher-ordered neighbor statistics reveals that the distribution is a clustered pattern at both local and regional scales. In other words, ordered statistics can evaluate the pattern at different spatial scales.

While both Quadrat Analysis and Nearest Neighbor Analysis test point distributions, they utilize different spatial concepts. Quadrat Analysis tests a point distribution with the *points per area* (or *density*) concept using quadrats as sampling units. Nearest Neighbor Analysis uses the concept of *area per point* (or *spacing*). Quadrat Analysis examines how densities change over space in a point distribution so that the point pattern can be compared with a theoretically constructed random pattern. In Nearest Neighbor Analysis, the test is based on comparing the observed average distances between neighboring points and a known pattern.

Both methods are similar in the sense that the observed pattern, whose spatial distribution characteristics are captured either by quadrat counts with given numbers of points or by the distances to neighbors, is compared with some known distribution, such as a random pattern. For the nearest neighbor statistic, if the observed average distance is greater than that of a random pattern, we can say that the observed point pattern is more dispersed than a random pattern. Similarly, a point pattern is said to be more clustered if its observed average distance between nearest neighbors is found to be less than that of a random pattern.

6.3.1 Nearest Neighbor Statistic

The nearest neighbor statistic was first introduced by two botanists, Clark and Evans (1954). In a homogeneous region, the most uniform pattern formed by a set of points occurs when this region is partitioned into a set of hexagons of identical size and each hexagon has a point at its center (i.e., a triangular lattice). With this setup, the distance between points will be $1.075\sqrt{A/n}$, where A is the area of the region of concern and n is the number of points. This provides a good starting point for us to understand how Nearest Neighbor Analysis works.

In the real world, we rarely see geographic objects distributed in an organized manner such as that partitioned by hexagons of equal size. Realistically, we often see geographic objects, such as population settlements, animal/plant communities, or others distributed in a more irregular fashion. To test if the distribution has any recognizable patterns, let's use a statistic called R, for randomness.

The R statistic, sometimes called the *R scale,* is the ratio of the observed average distance between nearest neighbors of a point distribution and the expected average nearest neighbor distance. It is also the **nearest neighbor statistic.** Specifically, it can be calculated as

$$R = \frac{r_{\text{obs}}}{r_{\text{exp}}}, \tag{6.7}$$

where r_{obs} is the observed average distance between nearest neighbors and the r_{exp} is the expected average distance between nearest neighbors as determined by the theoretical pattern.

To compute r_{obs} from the observed point pattern, we can calculate the distances between each point and all other points. For each point, the shortest distance among all neighboring distances will be associated with the nearest point. This process should be repeated for all points. To formalize this process, we can examine the following example, as depicted in Figure 6.5. In this

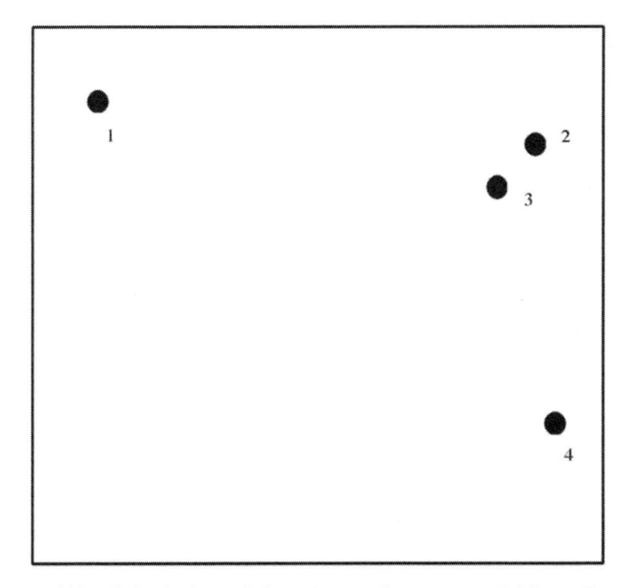

Figure 6.5 Calculation of the observed nearest neighbor distance.

figure, there are four point locations, $i = 1, 2, 3$, and 4. Here d_{ij} is the distance between points i and j, and d_i is the nearest neighbor distance for point i. Then we can derive r_{obs} by the following process:

$$d_1 = d_{13}$$

$$d_2 = d_{23}$$

$$d_3 = d_{32}$$

$$d_4 = d_{43}$$

For point location 1, the nearest neighbor is 3. Therefore, the nearest neighbor distance for point 1 is d_{13}. Similarly, for point location 2, the nearest neighbor is 3. Therefore, the nearest neighbor distance for point 2 is d_{23}. Then r_{obs}, the average nearest neighbor distance, is $(d_1 + d_2 + d_3 + d_4)/4$ or, actually, $(d_{13} + d_{23} + d_{32} + d_{43})/4$. Formally,

$$r_{\text{obs}} = \frac{\Sigma d_i}{n}, \tag{6.8}$$

where d_i is the nearest neighbor distance for point i and n is the number of points.

By selecting the seven largest cities in Ohio, we can compute their nearest neighbor distances and the observed average nearest neighbor distance, r_{obs}. The seven cities are shown in Figure 6.6, and their nearest neighbors (cities) and the corresponding distances are recorded in Table 6.5. By taking the average of the nearest neighbor distances of these seven cities, we can obtain the average nearest neighbor distance, r_{obs}, which is equal to 51.82 miles. Note that the nearest neighbor relationship is not symmetrical. That is, city a is nearest to city b, but this does not imply that the nearest city to city b is city a. This is because the nearest city to city b could be city c on the other side. Specifically, we can see that the city nearest to Toledo is Cleveland, but the city nearest to Cleveland is not Toledo but Akron.

For the theoretical random pattern, we use the following equation to calculate the expected average distance between nearest neighbors:

$$r_{exp} = \frac{1}{2\sqrt{\frac{n}{A}}}, \tag{6.9}$$

Figure 6.6 The seven largest cities in Ohio.

TABLE 6.5 Nearest Cities and Distances for the Seven Cities in Ohio

City	Nearest City	Nearest Distance
Akron	Cleveland	28.73
Cincinnati	Dayton	47.12
Cleveland	Akron	28.73
Columbus	Dayton	65.94
Dayton	Cincinnati	47.12
Toledo	Cleveland	99.43
Youngstown	Akron	45.69
Average Nearest Distance		51.82 (miles)

where

n is the number of points in the distribution, and
A is the area of the study region.

This expression can be rearranged to become

$$r_{exp} = 0.5\sqrt{\frac{A}{n}}. \tag{6.10}$$

In the seven Ohio cities example, n is 7 and A can be the area of the State of Ohio, which is approximately 41,193 square miles. Therefore, the expected average nearest neighbor distance is

$$r_{exp} = 0.5\sqrt{\frac{41{,}193}{7}} = 38.36.$$

With both distances calculated, we can now compute the R statistic:

$$R = \frac{r_{obs}}{r_{exp}} = \frac{51.82}{38.36} = 1.35.$$

Because the R statistic compares the observed with the expected (random) nearest neighbor distances, it indicates that the observed pattern is more dispersed than the random pattern if $r_{obs} > r_{exp}$, or $R > 1$. Therefore, in this example of seven cities with R larger than 1, we can say that the point pattern formed by the seven cities is more dispersed than a random pattern. If $r_{obs} < r_{exp}$, with the observed pattern more clustered than the expected pattern, $R < 1$ indicates a more clustered pattern.

6.3.2 Testing for Pattern Using the Nearest Neighbor Statistic

Now we know how the R statistic is calculated and how to determine whether an observed pattern is more clustered or more dispersed than a random pattern. Many conclusions can be drawn from examining the R statistic with regard to how the seven cities are related to each other. But we are still not sure to what degree this pattern is more clustered than a random pattern. Is the seven-city pattern much more clustered or just slightly more clustered than a random pattern? To appreciate the implications of various values of the R statistic, Figure 6.7 shows a series of hypothetical distributions and their associated R values.

Figure 6.7 shows that patterns with a higher degree of clustering are associated with smaller R values ($r_{obs} < r_{exp}$), while patterns with a higher degree of dispersion often have greater R values ($r_{obs} > r_{exp}$). Figure 6.7 is useful for establishing a general concept of how R values relate to various patterns. It is, however, not sufficient to quantitatively measure the difference between an observed pattern and a random pattern.

The R scale ranges from $R = 0$ (completely clustered) to $R = 1$ (random), and to $R = 2.149$, the theoretical value for the most dispersed pattern—a triangular lattice. However, in this example, we have an empirical R value of 1.35. When $R = 0$, all distances between points are 0, indicating that all points are located at the same location. When $R = 1$ or approximately 1, it means that $r_{obs} = r_{exp}$ or that they are very similar, and the pattern being tested should therefore be a random pattern. When R approaches a value of 2 or more, the patterns display various degrees of dispersions.

When using Nearest Neighbor Analysis, one way to measure the extent to which the observed average distance differs from the expected average distance is to compare their difference with the *standard error* (SE_r) of the average distances among nearest neighbors. The magnitude of this standard error indicates how likely any difference between the observed average nearest neighbor distance and the expected pattern is to occur purely by chance. If the difference is relatively small when compared to its standard error, we say that this difference is not statistically significant. Alternatively, when the difference between the observed and expected distances is relatively large with respect to the standard error, we can claim that the difference is statistically significant; that is, it does not occur by chance.

This concept is identical to the concept of standard error used in the difference-of-means test in Chapter 4. In classical statistical theory, in a normal distribution, there is about a 68% chance that some difference between one negative standard error and one positive standard error will occur by chance when in fact there should not be any difference between two populations being compared. Described in an equation, this means that

$$\text{Probability}(<68\%) = (0 - SE_r, 0 + SE_r).$$

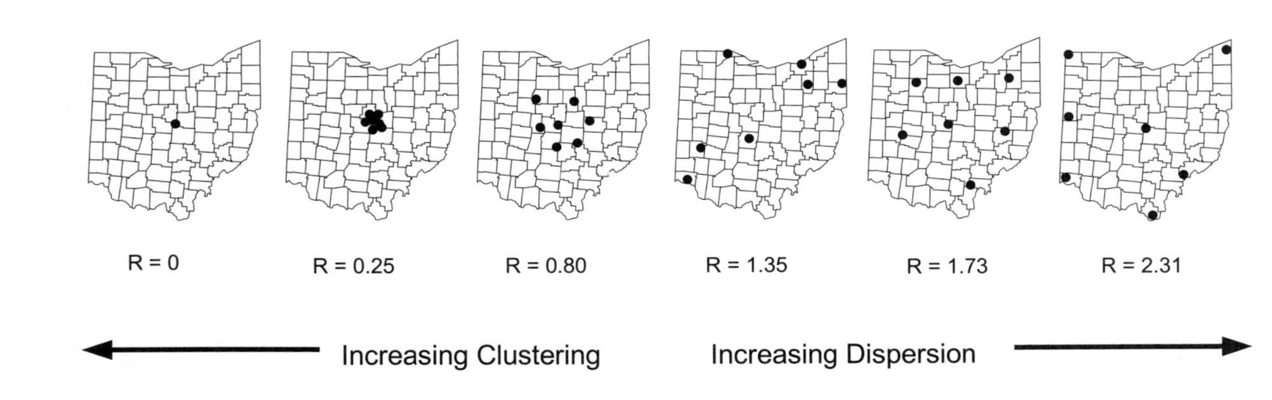

Figure 6.7 The scale of *R* statistics.

We can now define a calculated difference to be statistically significant only when it is smaller than $-SE_r$ or greater than SE_r. Or, if we want to be more rigorous, we can call a difference statistically significant only if it is smaller than $-1.96SE_r$ or greater than $1.96SE_r$. This is because the probability of having a difference of that magnitude is 5 out of 100 times or less:

$$\text{Probability}(<95\%) = (-1.96SE_r + 1.96SE_r).$$

With Probability(<95%), we can say that our level of confidence is 5%. Traditionally, this level of confidence is denoted α. In this case, our α is 0.05.

To calculate the standard error of the difference between the observed and expected average distances for the nearest neighbor statistic, we can use the following equation:

$$SE_r = 0.26136\sqrt{\frac{A}{n^2}}, \tag{6.11}$$

where n and A are as defined previously. We can now test how the difference is compared to its standard error by calculating a standardized Z score:

$$Z_R = \frac{r_{\text{obs}} - r_{\text{exp}}}{SE_r}. \tag{6.12}$$

As mentioned earlier, if $Z_R > 1.96$ or $Z_R < -1.96$, we can conclude that the calculated difference between the observed pattern and the random pattern is statistically significant given that $\alpha = 0.05$. Alternatively, if $-1.96 < Z_R < 1.96$, we can conclude that the observed point pattern, although it may look somewhat clustered or somewhat dispersed visually, is not significantly different from a random pattern, and we will fail to reject the null hypothesis.

In our example of seven Ohio cities, we have

$$SE_r = 0.26136\sqrt{\frac{41{,}193}{7^2}} = 7.580$$

and the Z_R score is

$$Z_R = \frac{51.82 - 38.36}{7.580} = \frac{13.46}{7.580} = 1.7771 < 1.96.$$

Because the standardized score is not larger than 1.96, though the seven cities exhibit a somewhat dispersed pattern, we should conclude that we fail to reject the null hypothesis that the observed seven cities pattern is not statistically different from a random pattern. In other words, a point pattern may seem clustered or dispersed by visual inspection or even indicated by calculating its R value, but we may not be able to reach a conclusion unless a statistical test rejects the null hypothesis that the observed pattern is not significantly different from a random pattern. The meaning of the calculated R value should be confirmed by Z_R to ensure its statistical significance.

Note that in the Ohio cities example Z_R is positive, indicating that the nearest neighbor distance of the observed pattern is larger than expected but statistically insignificant. The sign of the Z-score, however, indicates that the observed pattern has a dispersion tendency. In other words, if the Z-score indicates that the difference between the observed and expected nearest neighbor distances is statistically significant, the sign of the statistic indicates whether the observed pattern is likely clustered or dispersed. Following the logic of hypothesis testing, we can then conduct a one-tail test to see if the Z-score is really negative (smaller than -1.645 using the $\alpha = 0.05$ significance level) or really positive (greater than 1.645). The one-tail test ultimately allows us to decide whether the observed pattern is significantly different from a clustered pattern or a dispersed pattern.

6.3.3 Higher-Order Neighbor Statistics

With its ability to detect patterns in a point distribution, Nearest Neighbor Analysis has been extended to accommodate the second, third, and other higher-order neighbor definitions. When two points are not immediate nearest neighbors but rather the second nearest neighbors, the way distances are computed between them will need to be adjusted accordingly. The extension of Nearest Neighbor Analysis to higher-order neighbors is straightforward and has been used on special occasions when this relationship is important.

Specifically, we can formulate the second-order nearest neighbor statistic, R_2, which is r_{obs}/r_{exp}, with the observed nearest neighbor distance defined as before, $r_{obs} = \Sigma d_i/n$, except that the d_i is the distance between i and its second nearest neighbor. The expected nearest neighbor distance in the denominator of the R_2 statistic is similar to the first-order expected distance, but the constant is changed from 0.5 to 0.75. In other words, the expected second-order nearest neighbor distance is

$$r_{exp} = 0.75\sqrt{\frac{A}{n}}. \tag{6.13}$$

In computing the Z-score and testing the significance of the difference between the observed second-order nearest neighbor distance and the expected distance, the following standard error estimate should be used:

$$SE_r = 0.2722\sqrt{\frac{A}{n^2}}. \tag{6.14}$$

In fact, we can generalize the r_{exp} and SE_r estimates for higher-order neighbor statistics using the following expressions:

$$r_{exp}(k) = \gamma_1(k)\sqrt{\frac{A}{n}} \tag{6.15}$$

and

$$SE_r(k) = \frac{\gamma_2(k)}{\sqrt{\frac{n^2}{A}}}, \tag{6.16}$$

where $\gamma_1(k)$ and $\gamma_2(k)$ are the constants for the expected distance and the standard error estimate for the k-order statistic. The above discussion on second-order statistic can be extended to higher orders, with the constants reported in Table 6.6 to calculate the expected distances and standard error estimates.

The first constant, γ_1, for the expected distance can be generalized (Thompson, 1956) as follows:

$$r_{exp}(k) = \frac{k(2k)!}{(2^k k!)^2}\sqrt{\frac{A}{n}}. \tag{6.17}$$

With the expected distance for the kth order, we can compute the kth order R_k statistic. But according to Levine (2002, p.177), there is no good significance test for the kth-order statistic because of the nonindependence of the different orders. Still, applying higher-order statistics can shed light on the point pattern being investigated.

Using the first-order nearest neighbor statistic, we cannot distinguish the two point patterns shown in Figure 6.4a and Figure 6.4b. This is because, in both patterns, the nearest neighbor of each point is very close. But if the second-order nearest neighbor distances are computed, the results show that

TABLE 6.6 Constants for Calculating Higher-Order Statistics

Order of Neighbors (k)	$\gamma_1(k)$	$\gamma_2(k)$
1	0.5000	0.2613
2	0.7500	0.2722
3	0.9375	0.2757
4	1.0937	0.2775
5	1.2305	0.2784
6	1.3535	0.2789

the pattern in Figure 6.4a is dispersed because the second nearest neighbors are all far away in the other quadrats. On the other hand, the result for the pattern in Figure 6.4b indicates a clustered pattern because all second nearest neighbors are quite close.

By combining the results from the first-order and second-order nearest neighbor statistics, we can conclude that Figure 6.4a has a locally clustered (based upon the first-order statistic) but regionally dispersed pattern (based upon the second-order statistic), while Figure 6.4b has a clustered pattern at both the local and regional scales. To a large extent, using different orders of nearest neighbor statistics can detect spatially heterogeneous processes occurring at different spatial scales.

6.3.4 Boundary Adjustments of the Nearest Neighbor Statistics

One major problem with using ordered neighbor statistics is that the expected distance, r_{exp}, and the standard error estimate are functions of the area of the study region. However, the demarcation of the study region is often arbitrary and may be changed subjectively by different analysts. For instance, Figures 6.8a and 6.8b are based on the same point pattern. But because Figure 6.8b adopts a larger frame, the study area will be larger than the pattern in Figure 6.8a. Therefore, Figure 6.8b exhibits a more clustered pattern than Figure 6.8a.

This issue is only one aspect of the boundary problem. Another aspect is that the derivation of the expected nearest neighbor distance, r_{exp}, is based on the assumption of *complete spatial randomness* (CSR) pattern such that the pattern is true over an infinite surface. When the study area is delineated differently, either artificially or practically, the expected nearest neighbor dis-

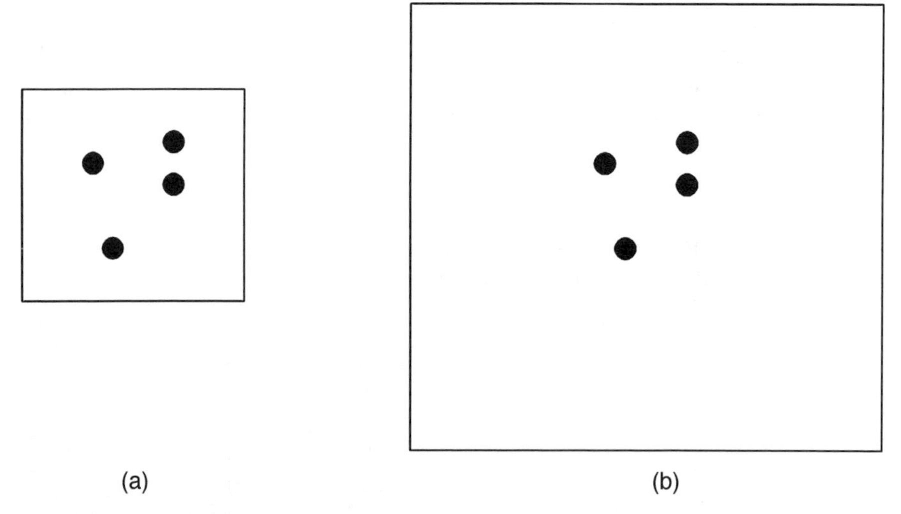

Figure 6.8 Delineation of boundaries affect the degree of clustering or dispersion.

tance must be adjusted. To correct for the distortions imposed by demarcating a boundary of the study region, many solutions have been proposed (Griffith and Amrhein, 1983). One solution is to convert the distribution into a torus of continuous distribution, including a buffer zone around the region, or to create an inset from the original boundary and disregard the points at the edge (Boots and Getis, 1988). Another solution is to adjust the estimation of r_{exp} and the standard error estimate, SE_r, the two statistics affected by the area measure. The most common adjustment method was proposed by Donnelly (1978). This method is appropriate when the number of points is relatively larger and the study region is not highly irregular. Formally, the adjusted r_{exp} is

$$r_{\text{exp}} = 0.5\sqrt{\frac{A}{n}} + \left(0.0514 + \frac{0.041}{\sqrt{n}}\right)\frac{B}{n} \tag{6.18}$$

and the square of the adjusted standard error estimate is

$$SE_r^2 = 0.0683\frac{A}{n^2} + 0.037B\sqrt{\frac{A}{n^5}}. \tag{6.19}$$

In the above two equations, all terms are defined as before, except that B is the perimeter of the study area. Note that Equation 6.19 is for the square of the standard error estimate. One should take the square root of this expression to obtain the standard error. Also, the above adjustments are only for the first-order statistic.

ArcView Example 6.2: Nearest Neighbor Analysis of Northeast Ohio Cities

Following ArcView Example 6.1, we continue the ArcView example with the same 5ccities.shp data theme and the Ch6.avx extension.

Step 1 Nearest Neighbor Analysis

To use Nearest Neighbor Analysis:

- From the **Ch6** menu, choose **Ordered Neighbor Analysis.**
- In the dialog box, click to select the point theme of `5ccities.shp`.
- Click to indicate **Theme** as the study area.
- Click **Nearest** to select it as the order of analysis.
- Check to activate the option of **Donnelly's correction.**
- Finally, click the **Calculate** button to proceed with the calculation

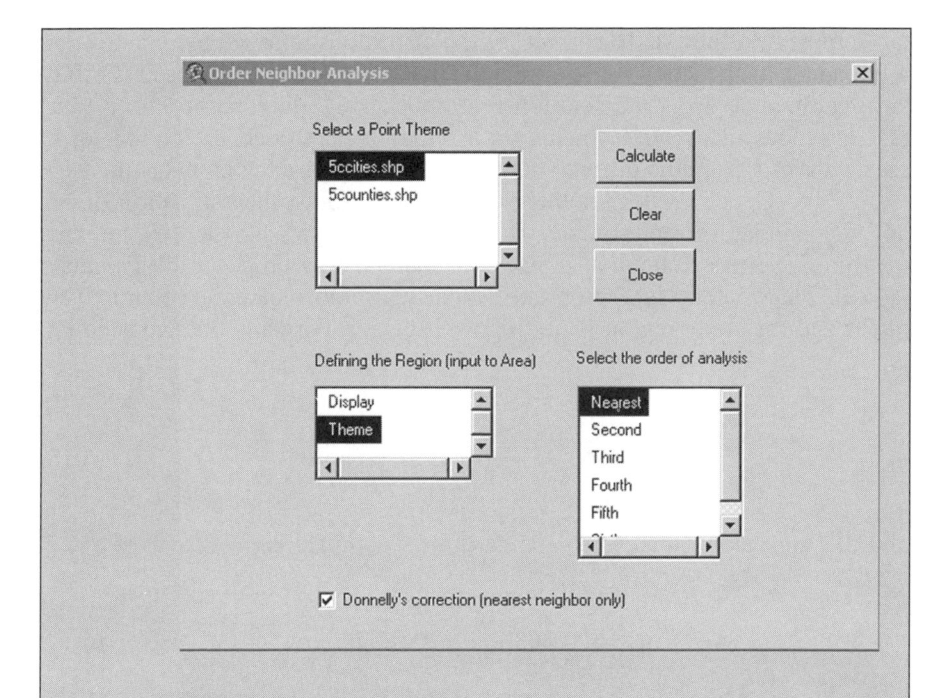

Step 2 Calculation

As discussed in this section, the definition of the study area often affects the results of Nearest Neighbor Analysis. For this example: When prompted, choose `5counties.shp` as the theme defining the study area.

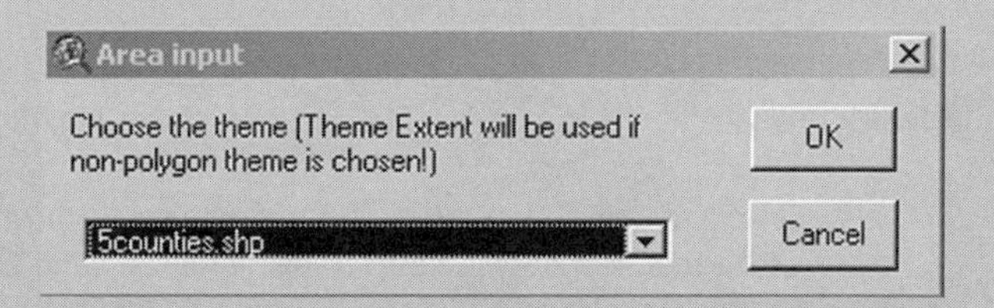

The first reported statistics are the area and perimeter:

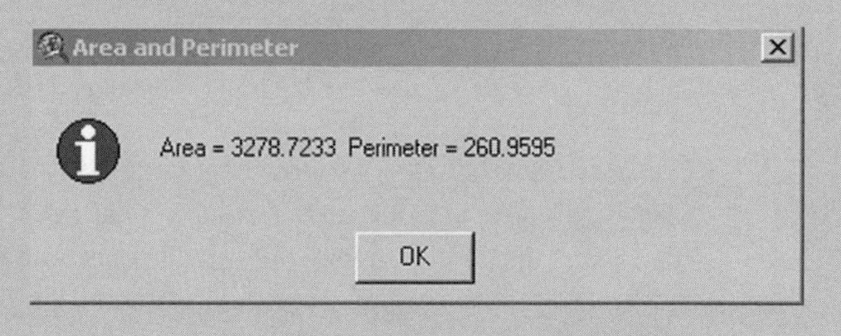

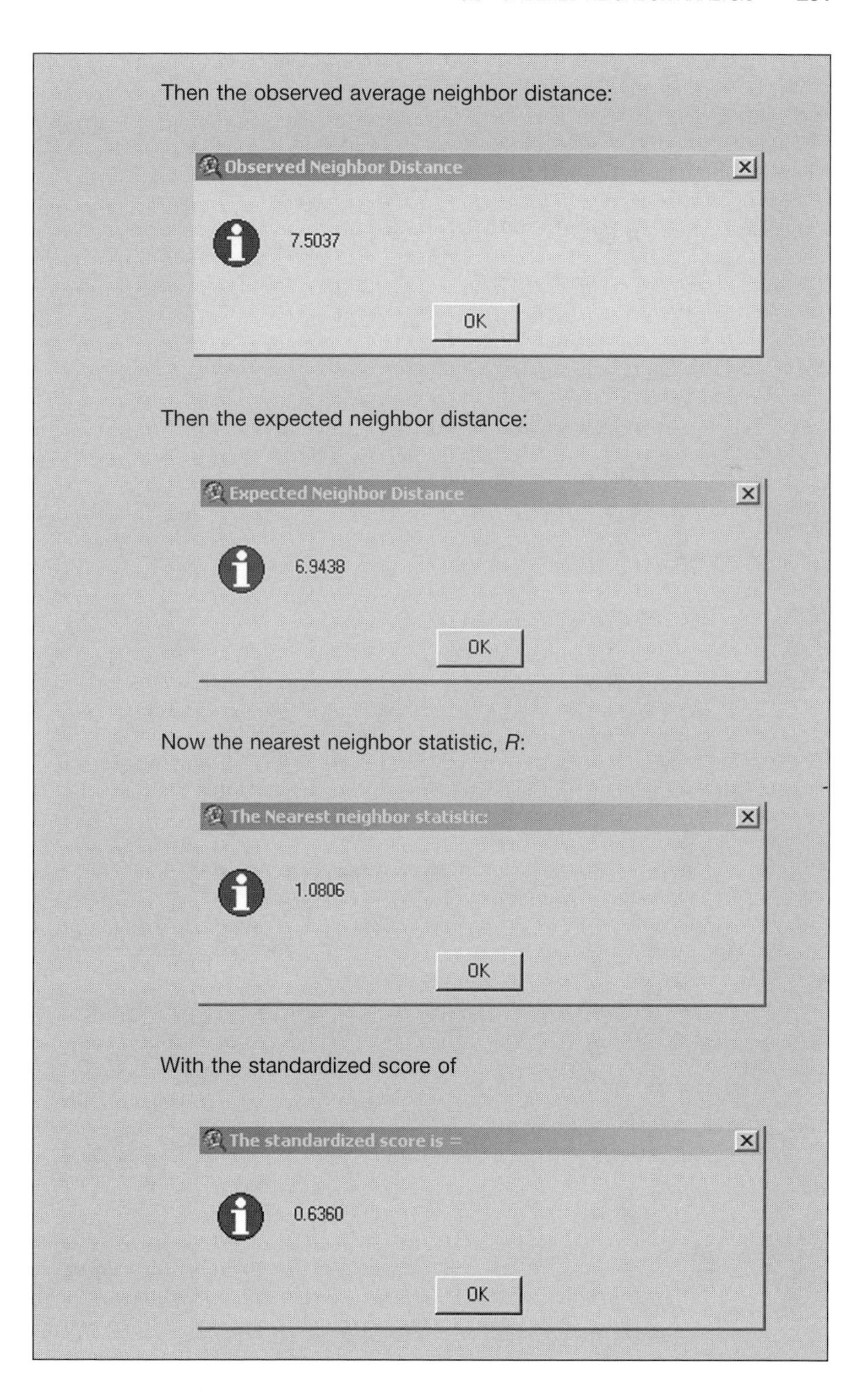

Then the observed average neighbor distance:
Observed Neighbor Distance
7.5037
OK
Then the expected neighbor distance:
Expected Neighbor Distance
6.9438
OK
Now the nearest neighbor statistic, R:
The Nearest neighbor statistic:
1.0806
OK
With the standardized score of
The standardized score is =
0.6360
OK

After the correction, it becomes

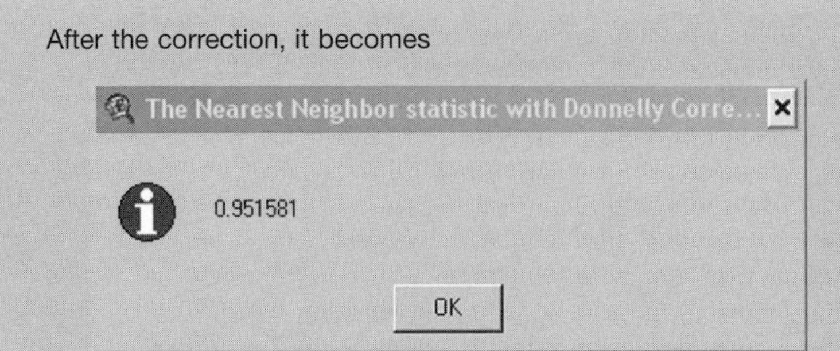

With a standardized score of

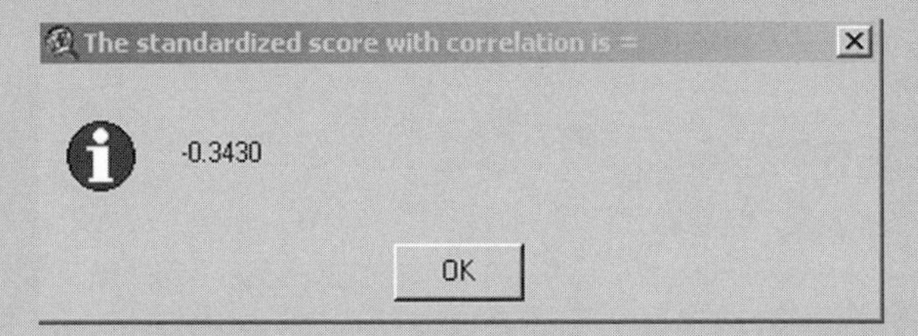

Step 3 Hypothesis testing

As in previous examples, the null hypothesis is that the point pattern formed by the 17 cities is not statistically significantly different from a random pattern.

Given the corrected standardized score of −0.34, we cannot reject the null hypothesis. Therefore, we conclude that the pattern of 17 cities is no difference from a random pattern.

Step 4 Higher ordered nearest neighbor analysis

The higher-order Nearest Neighbor Analysis included in Ch6.avx does not work with Donnelly's correction. Therefore, make sure that it is not checked when running the higher-order Nearest Neighbor Analysis.

For statistics and standardized scores from various ordered Nearest Neighbor Analyses, the table below provides a summary:

Order	R Statistic	Standardized Score	Hypothesis Testing
1st order	0.951581	−0.34303	Fail to reject H_0
2nd order	0.788643	−2.40101	Reject H_0
3rd order	0.778115	−3.11112	Reject H_0

From this example, it is clear that different conclusions may be reached when different orders of neighbor statistics are used.

In this example, the null hypotheses were rejected when second- and third-order nearest neighbor statistics were used, but the result failed to reject the null hypothesis for the first-order nearest neighbor statistic. By taking all these results together, we can conclude that at the local scale the 17 cities do not exhibit any particular pattern, but at the regional scale they exhibit a pattern significantly different from a random pattern. More specifically, at the regional scale, there are indications that the cities may have a clustered pattern, as the R statistics are smaller than 1 with negative standardized scores.

Step 5 End

6.4 *K*-FUNCTION

Both Quadrat Analysis and Nearest Neighbor Analysis attempt to offer overall descriptions of the point patterns being analyzed. If used correctly, they can identify the dominant spatial characteristics or properties of a set of points. These properties sometimes are known as *first-order properties,* as they describe the global pattern. However, the underlying spatial process may not be homogeneous over the study region in regard to the point distribution being analyzed. For example, neighboring units may not cluster at the same magnitude across different parts of the region. To capture this characteristic of local variations, we may have to use other statistics to describe second-order or local properties. The higher-order Nearest Neighbor Analysis can partly serve this purpose, but it is a bit tedious to compute all higher-order statistics.

Another statistic that can offer some insights and is more parsimonious to evaluate if the magnitude of clustering is uniform over different spatial scales is K-function analysis, or Ripley's K statistic, introduced by Ripley (1976). Ripley's K statistic is an extension of the ordered neighbor statistics. It enables us to depict the randomness of a point distribution over different spatial scales. For a set of points in a region, the K-function analysis involves the following steps:

1. Select a distance increment or spatial lag, d, that is analogous to the unit reflecting the change in the spatial scale. If the two farthest points in the region are D units apart, then obviously $d << D$. In this case, r, indicating the number of lags required to include all points in the region, is approximately equal to D/d.
2. Set the iteration number $g = 1$ to begin the process.
3. Around each point i in a region, create a circular buffer with a radius of h, where $h = d \times g$. Therefore, the buffer will have a size d in the first iteration and $2d$ in the second and so on.

4. For each point, count the number of points falling within its buffer of size h and denote that count as $n(h)$.
5. Increase the radius of the buffer by d (i.e., in the second time, the size of the radius will be $d \times 2$ or $2d$).
6. Repeat steps 3, 4, and 5 by increasing h until $g = r$ or $g = D/d$.

The procedure is graphically depicted in Figure 6.9. Though the same procedure should be performed for all points, Figure 6.9 uses only four points to illustrate the concept. Also, only three rings or buffers were created instead of the full range up to D. For a given h, we count the number of points within the buffers centered at all points. In Figure 6.9, point A is rather dispersed from other points, and therefore the counts are relatively low for buffers with small h. For point B, the point is in the middle of the cluster, and therefore the point counts are relatively high with the small buffers, but the increases in point counts are not substantial with larger h's. For points C and D, the points themselves are apart from the clusters. Thus the initial buffers have very low point counts, but the counts increase drastically with larger h's.

With the iterative processes discussed above, if the points form a highly clustered pattern, we will find high count values with small h's, and the counts

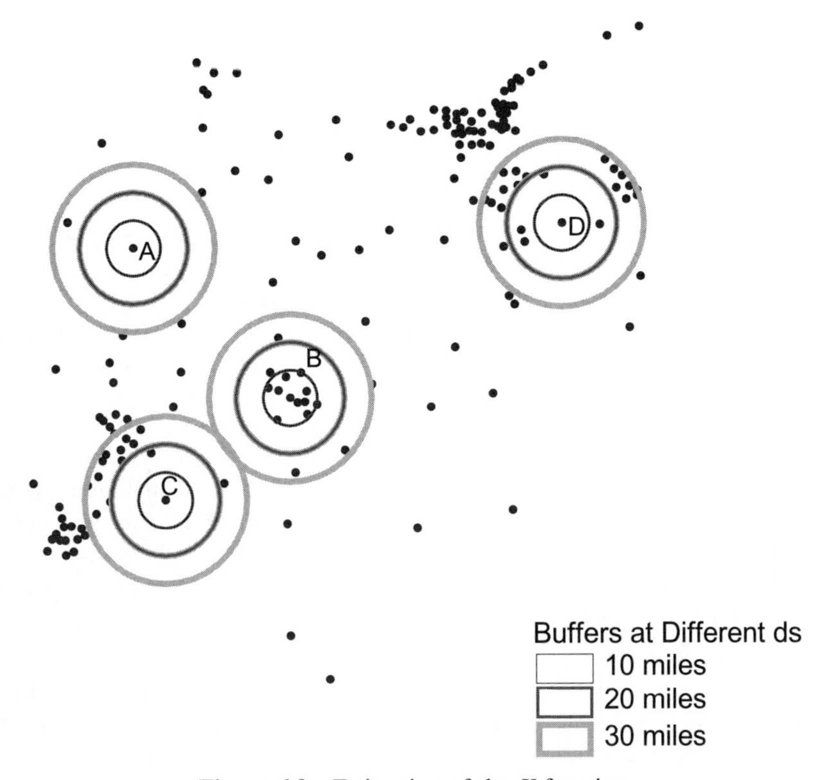

Figure 6.9 Estimation of the K-function.

will not increase much when h is increased to relatively large sizes. Conversely, if the pattern is relatively dispersed, the counts will be relatively low with small h's and will increase quickly when h increases.

The relationship between point counts and the spatial lag, h, from empirical observations can be compared with a known pattern, most likely a random pattern. In a random pattern, we will see the point counts increase with increasing h but in no particular pattern. The most interesting aspect of K-function analysis is to detect clustering at different scales by comparing the relationship between point counts and the size of h to that in a random distribution. Before we can compare the observed K-function with that in an expected pattern, we must discuss the computation of the K-function in more detail.

To formalize the derivation of the K-function, we start by expressing $n(h)$, the number of points within the buffer with a lag h, as follows:

$$n(h) = \sum_i \sum_j I_h(d_{ij}),\ i \neq j, \tag{6.20}$$

where

i and j are the indices of points,
d_{ij} is the distance between the two points, i and j, and
I_h is an indicator function such that $I_h = 1$ if $d_{ij} < h$ and $I_h = 0$ otherwise.

In other words, the expression in Equation 6.20 counts the number of points within the vicinity of d_{ij}. Any point beyond the spatial lag or the vicinity of d_{ij} will be excluded through the binary indicator function. When this point count function is scaled by A/N^2, the inverse of the overall point density in the region with A equals the area of the region. The resulting function is the K-function, which is expressed as

$$K(h) = \frac{A}{N^2} \sum_i \sum_j I_h(d_{ij}). \tag{6.21}$$

Sharing similar problems with other spatial statistical and analytical techniques, the K-function is also subject to the boundary problems. Image that a point is located rather close to the edge of the study region. When buffers are formed around the point, a significant proportion of buffers will be outside of the study area and thus will distort the probability of finding a point within the vicinity of h. Therefore, it is necessary to adjust for this boundary effect (Bailey and Gatrell, 1995). The simplest way to do this is to include a weighting factor, w_i, to accommodate the incomplete buffers centered at i. Formally, w_i is the proportion of the buffer that is centered at i and lies within the study region. With GIS, this proportion is rather easy to compute. With this weight incorporated into the analysis, the K-function can be stated as

$$K(h) = \frac{A}{N^2} \sum_i \sum_j \frac{I_h(d_{ij})}{w_i}. \tag{6.22}$$

This empirical $K(h)$ function is essentially the point density at a given spatial lag, h, and can be compared with the point density function of a random point pattern.

For a random point pattern, a good theoretical estimate of the $K(h)$ function is πh^2. For a dispersed or regular pattern, the empirical $K(h)$ function will be lower than πh^2. For a clustered pattern, the $K(h)$ function will be higher than πh^2. Thus, instead of plotting the empirical K-function and the one for a random point pattern together, we can derive their differences and plot them. The difference function, $L(h)$, can be defined as

$$L(h) = \sqrt{\frac{K(h)}{\pi}} - h. \tag{6.23}$$

High values (positive) in this function indicate clustering at the corresponding spatial lag distance, h, and low values (negative) indicate a dispersion of points at that distance.

To demonstrate the utility of the difference function $L(h)$ to help us understand the spatial patterns, a set of 164 random points, the same number of Ohio cities depicted in Figure 6.2, was generated. K-functions were computed for the two point patterns: the 164 random points and the 164 Ohio cities. The $L(h)$ functions were also calculated for the two sets of points and were plotted against the spatial lag, h, in Figure 6.10.

In Figure 6.10, several interesting observations deserve our attention. First, the $L(h)$ function for the random points has values lower than that for the Ohio cities. As discussed earlier, higher $L(h)$ values indicate a clustering tendency, and the higher values of the L function of the 164 Ohio cities support the conclusion from our visual inspection that the Ohio cities exhibit a somewhat clustered pattern. Second, in the Ohio cities plot, there is a trough at approximately the 50-mile lag and another smaller trough at the 150-mile lag. The spatial lag distances corresponding to these peaks reflect the scales at which clustering of points is most likely. On the other hand, the plot for the 164 random points exhibits a smooth rising trend with a maximum value at about the 150-mile lag. This result implies that the random points do have a somewhat clustered tendency, but not until the 150-mile lag, a relatively large region. The third interesting observation is that the $L(h)$ plot for the Ohio cities increases when the lag is small and then gradually declines when the lag increases. This trend implies that the clustering happens within a short distance, that is, clustering at the local scale. For the 164 random points, values of the $L(h)$ function decline slightly first before rising again, indicating a slightly dispersed pattern at the local scale but more likely due to the random effect.

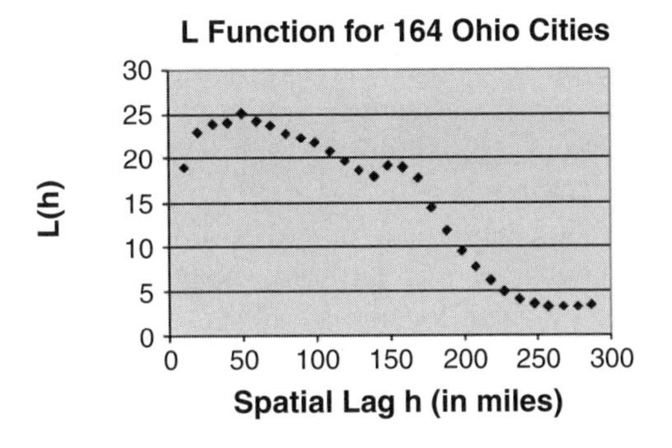

Figure 6.10 $L(h)$ functions for the 164 Ohio cities and 164 random points.

Note that in Figure 6.10, the spatial lag chosen is 10 miles, which is small enough to capture the changes in clustering magnitude over different scales. If the spatial lag is too large, say 100 miles, we may not see the peak at the 50-mile lag for the 164 Ohio cities, and we surely will miss the minor peak at 150 miles. Conversely, we may choose a very small spatial lag to depict the changes at a very fine scale, trying not to miss any small changes in the clustering magnitude over the scale. However, a small lag will increase the demand on computing resources, and the calculation may take much longer to complete. In addition, the *K*-function may pick up a lot of random noises instead of the trend.

We can test whether the clustering or dispersion tendencies reflected by the $L(h)$ function are significance or not (Bailey and Gatrell, 1995). However, the real value of *K*-function analysis is its ability to explore the spatial point pattern over the entire scale of the region. The analysis includes all points and the distances between all points, not just the nearest neighbors. It can pinpoint where the clustering or dispersion is most acute over the geographic scales. The conceptualization is a bit more complicated and less intuitive than that of the ordered neighbor statistics. The applicability of *K*-function analysis to various types of problems is also constrained by the nature of the spatial processes under investigation. If the spatial variability of points is not uniform across the study region or if the spatial process is heterogeneous, then *K*-function analysis may not be effective. Similarly, if the spatial process is not uniform in all directions (anisotropic), then *K*-function analysis will have limited use.

So far, our discussion has been limited to one type of event (i.e., crime events or disease incidence). Furthermore, events of the same type are not distinguished except by their locations. If we want to analyze the patterns across two or more types of events (e.g., robbery and vandalism) using the general framework of *K*-function analysis, cross-*K*-function analysis may be adopted (Bailey and Gatrell, 1995). If events of the same type have different characteristics to distinguish them from each other, then one or more attributes describing the points should be explicitly included in the analysis. When both spatial and attribute information should be included, *K*-function analysis will not be appropriate. In that case, spatial autocorrelation of points with both location and attribute information should be used.

ArcView Example 6.3: *K*-Function Analysis of Northeast Ohio Cities

Following ArcView Examples 6.1 and 6.2, we continue this ArcView example with the same 5ccities.shp data theme and the Ch6.avx extension.

Step 1 *K*-function

The ***K*-Function Analysis** option in the Ch6 menu enables users to create a chart that describes the relationship between various lag sizes, *h*, and *L*(*h*). As discussed in this section, the *L*(*h*) curve provides hints regarding the distances that could be helpful in determining the neighboring relationship between points in the data set.

Note that when using *K*-function analysis in Ch6.avx, it is necessary to first save the results of the calculation to a DBF file before creating a chart that displays the *L*(*h*) curve.

Step 2 Calculation

To calculate the *K*-function:

- From the **Ch6** menu, choose ***K*-Function Analysis.**
- Select `5ccities.shp` as the point theme.

Notice that the distance between the farthest points is calculated and provided here as a reference.

For this data set:

- Enter `10` miles as the **Distance Lag.**
- Click to check both boxes, saving the results to a dbf file and creating a chart.

- Click the **Calculate** button to proceed with the calculation.

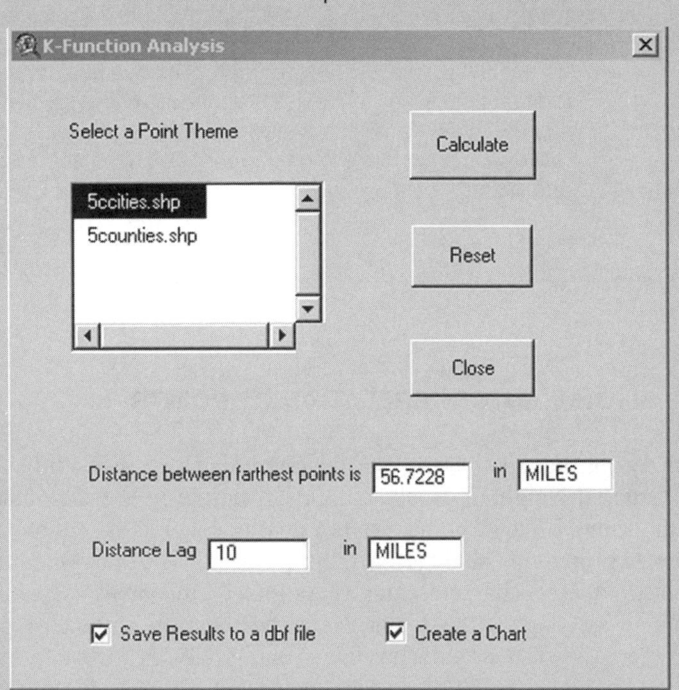

First, the area of the study region is calculated and reported.
Then the number of records is reported.
When prompted, choose a folder and a name to store the calculated results.

Step 3 Plotting the *L* function and *h* lags
The result of the *K*-function analysis is a chart in which the relationship between various lag sizes, *h*, and *L*(*h*) is plotted.

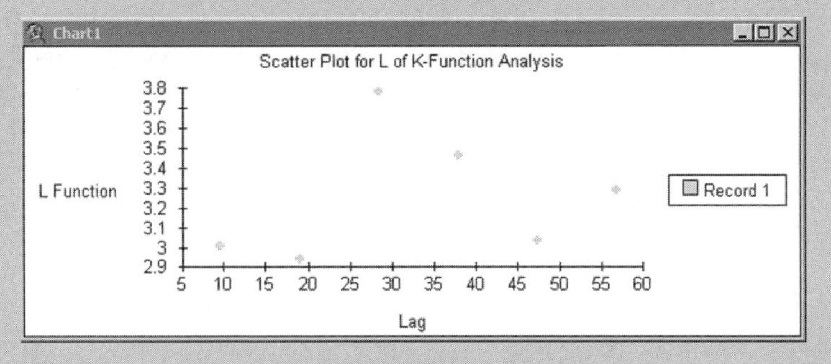

Note that you may need to resize the chart document to obtain a better display of the function values.

From this chart, it appears that 30 miles is a critical distance that deserves further exploration in the neighboring relationship between these cities.

Also note that the *K*-function chart provides useful references to additional analyses in Nearest Neighbor Analysis when distance lags are involved. These analyses are typically carried out in iterations to best explore the neighboring relationships among the data points.

Step 4 End

6.5 SPATIAL AUTOCORRELATION OF POINTS

In analyzing the spatial pattern of a point distribution, both Quadrat Analysis and Ordered Neighbor Analysis treat all points in the distribution as if they are the same. That is, points are not distinguishable except by their locations. These two methods analyze only the locations of points; they do not include the attributes or characteristics of points in the analysis. This approach is adequate as long as the focus is on the location pattern of points and the characteristics of the points are irrelevant. Analyses focusing only on location can be applied to general disease clustering, such as finding clusters of reported West Nile virus cases. An analysis of robbery cases with regard to the amounts involved must consider both the locations and the attributes of the robberies. The analysis may explore if robbery cases involving larger amounts are more clustered than a random pattern. In this case, we need to know both the locations of the robberies and the amounts involved.

In this section, we will discuss a method for detecting spatial patterns of a point distribution by considering both the locations of points and their attributes. This method uses measures known as **spatial autocorrelation coefficients** to measure and test how clustered/dispersed the point locations are with respect to their attribute values. In certain ways, this method may be more powerful and more useful than the methods discussed previously. Different geographic locations rarely have identical characteristics, making it necessary to consider the characteristics of points in addition to their locations. Not only do locations matter; the conditions of these locations or activities are also of great importance.

Spatial autocorrelation of a set of points refers to the degree of similarity between points or events occurring at these points and points or events in nearby locations. If significant positive spatial autocorrelation exists in a point distribution, points with similar characteristics tend to be near each other. Alternatively, if spatial autocorrelation is weak or nonexistent, nearby points do not exhibit any similar or dissimilar pattern or a random pattern exists. This concept corresponds to what was once called the *first law of geography*

(Tobler 1970): *everything is related, but near things are more related* (also cited and discussed in Gould, 1970, pp. 443–444; Cliff and Ord, 1981, p. 8; and Goodchild, 1986, p. 3). This first law has been thoroughly revisited recently (*Annals,* 2004, Vol. 94, No. 2). If the first law is pervasive, then we will rarely observe situations with negative spatial autocorrelation or when nearby points have very dissimilar characteristics (Griffith et al., 2003).

With the spatial autocorrelation coefficient, we can measure

1. the proximity of locations and
2. the similarity of the characteristics of these locations.

For proximity of locations, we evaluate the distances between points. For similarity of the characteristics of these locations, we compare the attributes of points.

6.5.1 Measures for Spatial Autocorrelation

There are two popular indices for measuring spatial autocorrelation applicable to a point distribution: **Geary's Ratio** and **Moran's I Index.** Both indices measure spatial autocorrelation for interval and ratio attribute data. Following some of the notations used in Goodchild (1986, p. 13), we have

s_{ij} representing the similarity of point i's and point j's attributes,
w_{ij} representing the proximity of point i's and point j's locations, with $w_{ii} = 0$ for all points,
x_i representing the value of the attribute of interest for point i, and
n representing the total number of points.

Both spatial autocorrelation measures, Geary's Ratio and Moran's I Index, combine the two measures for attribute similarity and location proximity into a single index, $\sum_{i=1}^{n} \sum_{j=1}^{n} s_{ij} w_{ij}$. This general framework serves as the basis for formulating the two indices. In both cases, the spatial autocorrelation coefficient (SAC) is proportional to the weighted similarity of the point attribute values. Specifically, the equation for the spatial autocorrelation coefficient takes the general form

$$SAC \approx \frac{\sum_{i=1}^{n} \sum_{j=1}^{n} s_{ij} w_{ij}}{\sum_{i=1}^{n} \sum_{j=1}^{n} w_{ij}}. \qquad (6.24)$$

In Geary's Ratio, the similarity of attribute values between two points is defined as

$$s_{ij} = (x_i - x_j)^2.$$

The difference in attribute values between point i and point j is calculated as $(x_i - x_j)$. The difference for each pairs of i and j is squared first and then multiplied by the corresponding spatial weight, w_{ij}. After this process is completed for all i-j pairs, they are summed to obtain the numerator of Geary's Ratio. Following this process, the positive differences will not be offset by negative differences when the attribute values are compared. Specifically, Geary's Ratio C is calculated as follows:

$$C = \frac{\sum_{i=1}^{n}\sum_{j=1}^{n} s_{ij}w_{ij}}{2\sum_{i=1}^{n}\sum_{j=1}^{n} w_{ij}S^2} = \frac{\sum_{i=1}^{n}\sum_{j=1}^{n} w_{ij}(x_i - x_j)^2}{2\sum_{i=1}^{n}\sum_{j=1}^{n} w_{ij}\, S^2}, \qquad (6.25)$$

where S^2 is the sample variance of the attribute x values with a mean of $\bar{x}$, or

$$S^2 = \frac{\sum_{i=1}^{n} (x_i - \bar{x})^2}{(n - 1)}.$$

The computation of Geary's Ratio using Equation 6.25 is quite tedious, as it requires computation of the sample variance first. Therefore, a more efficient formula is

$$C = \frac{(n - 1)\Sigma\,\Sigma\, w_{ij}(x_i - x_j)^2}{2\Sigma\Sigma w_{ij}\Sigma(x_i - \bar{x})^2} \qquad (6.26)$$

with the subscripts for the summation signs omitted.

In Moran's I Index, the similarity of attribute values is defined as the difference between each value and the mean of all attribute values in question. Therefore, for Moran's I,

$$s_{ij} = (x_i - \bar{x})(x_j - \bar{x})$$

and the index can be calculated as

$$I = \frac{\sum_{i=1}^{n}\sum_{j=1}^{n} w_{ij}s_{ij}}{\sigma^2 \sum_{i=1}^{n}\sum_{j=1}^{n} w_{ij}} = \frac{\sum_{i=1}^{n}\sum_{j=1}^{n} w_{ij}\,(x_i - \bar{x})(x_j - \bar{x})}{\sigma^2 \sum_{i=1}^{n}\sum_{j=1}^{n} w_{ij}}, \qquad (6.27)$$

where σ^2 is the population variance, or $\sigma^2 = \Sigma_{i=1}^{n}\,(x_i - \bar{x})^2/n$.

Similar to the derivation of Geary's Ratio formula, the computation of Moran's I using Equation 6.27 involves the calculation of the population variance first. A more efficient formula is

$$I = \frac{n\Sigma\Sigma w_{ij}(x_i - \bar{x})(x_j - \bar{x})}{\Sigma\Sigma w_{ij}\Sigma(x_i - \bar{x})^2} \tag{6.28}$$

with the subscripts for the summation signs omitted.

In Geary Ratio and Moran's I Index, all terms can be calculated directly from the point attribute values. The only value not yet available is w_{ij}, the proximity weight between point i and point j. We often use the inverse of the distance between point i and point j as the weight. This approach assumes that the point attribute values follow the first law of geography, whereby smaller weights are given to points that are farther apart and larger weights are given to points that are closer. For example, w_{ij} can be defined as $1/d_{ij}$ where d_{ij} is the distance between point i and point j. While the simple inverse distance weight format assumes a linear decline in importance as the distance between the two locations increases, there is no reason to apply this linear relationship to all phenomena when a nonlinear relationship in regard to distance may exist. Therefore, occasionally, the inverse distance weight may take the form of a power function such that a parameter, α, is used to reflect how nonlinear the relationship may be. Thus, the inverse distance weight may be rewritten as $1/d_{ij}^{\alpha}$. The distance decay parameter, α, may take any appropriate value based on specific characteristics or empirical evidence associated with the geographic phenomenon in question. Common sense tells us that the distances measured by driving and flying between two airports can have quite different meanings for people's behavior over the geographic space. Many empirical studies indicate that $\alpha = 2$ is widely applicable to many geographic phenomena. Another popular specification of the distance weight function is the exponential function instead of the power function, as discussed above.

Depending on the data models adopted to describe the geographic features, the spatial weights in the computations of the spatial autocorrelation coefficients may take on a form other than a distance-based format. For example, w_{ij} can take a binary form of 1 or 0, depending on whether point i and point j are spatially adjacent. If we use the concept of nodal region in the geographic literature, each point location can be regarded as the centroid of a region. If the two regions share a common boundary (i.e., they are adjacent), the two centroids of these regions can be defined as spatially adjacent. If two points are spatially adjacent, $w_{ij} = 1$; otherwise, $w_{ij} = 0$.

The two spatial autocorrelation indices discussed above are similar in their general formats. The difference between them is whether the differences in attribute values (x_i and x_j) are calculated from direct comparison ($x_i - x_j$) or in reference to their mean $(x_i - \bar{x})(x_j - \bar{x})$. As a result, the two indices yield different numeric ranges, as shown in Table 6.7, and have different statistical properties, with Moran's I Index properties being more desirable. In Table

TABLE 6.7 Numeric Scales of Geary's *C* Ratio and Moran's I

Spatial Patterns	Geary's *C*	Moran's I
Clustered pattern in which adjacent or nearby points show similar characteristics	$0 < C < 1$	$I > E(I)$
Random pattern in which points do not show particular patterns of similarity	$C \sim= 1$	$I \sim= E(I)$
Dispersed pattern in which adjacent or nearby points show different characteristics	$1 < C < 2$	$I < E(I)$

$E(I) = -(1)/(n - 1)$, with n denoting the number of points in the distribution.

6.7, possible values for both indices are listed with respect to the three general spatial patterns: clustered, random, and dispersed.

Note that the index's scale for Geary's Ratio does not correspond to our conventional impression of the correlation coefficient of the $(-1, 1)$ scale, while the scale of Moran's I resembles more closely the scale of the conventional correlation measure:

1. the value for no spatial autocorrelation is not zero but $-1/n - 1$; and
2. the values of Moran's I Index in some empirical studies are not bounded by $(-1, 1)$, especially the upper bound of 1.

The value for a coefficient indicating no spatial autocorrelation is called the *expected value.* The expected value of Geary's Ratio is a constant of 1, indicating a random pattern. For Geary's Ratio, a dissimilar pattern with adjacent or nearby points displaying different characteristics will generate an index value greater than 1. A value of less than 1 suggests a clustered pattern in which adjacent or nearby points show similar attribute values.

It is true that when the number of areal unit (i.e., n) is large, the expected value of Moran's I, $E(I)$, approaches zero. However, the expected value of Moran's I will always be a negative value. When n is small, the expected value will be strongly negative. Therefore, when n is small, a negative Moran's I may not indicate a negative spatial autocorrelation or a dispersed pattern. Only when the calculated Moran's I is smaller than $E(I)$ can we claim the possible existence of negative spatial autocorrelation. At the other end of the scale, Moran's I values that are greater than $E(I)$ typically are from clustered patterns where adjacent or nearby points tend to have similar characteristics.

6.5.2 Significance Testing of Spatial Autocorrelation Measures

When analyzing a point distribution, we may assume that the way the attribute values are distributed to various point locations is only one of many possible arrangements or outcomes with the given set of values. In this case, we adopt the assumption known as *randomization* or *nonfree sampling*. Alternatively, we may assume that the attribute values distributed among the points are only one set of values from an infinite possible set and that each value is independent of others in the set of attribute values. This assumption is sometime called the *normality* or *free sampling* assumption. The difference between these two assumptions affects the way the variances of Geary's Ratio and Moran's I are estimated.

For both indices, we can calculate the variances under free sampling and nonfree sampling assumptions. Free sampling is suitable for cases where sampling is with replacement of observations in sampling possible outcomes, while nonfree sampling is suitable for sampling without replacement.

Let's use R to denote the nonfree sampling assumption (randomization) and N for the free sampling assumption (normality). Following the notations in Goodchild (1986), we can estimate the expected values and variances for a random pattern for Geary's Ratio C as follows:

$$E_N(C) = E_R(C) = 1 \tag{6.29}$$

$$VAR_N(C) = \frac{[(2S_1 + S_2)(n - 1) - 4W^2]}{2(n + 1)W^2} \tag{6.30}$$

$$\begin{aligned} VAR_R(C) = {} & \frac{(n - 1)S_1[n^2 - 3n + 3 - (n - 1)k]}{n(n - 2)(n - 3)W^2} \\ & - \frac{(n - 1)S_2[n^2 + 3n - 6 - (n^2 - n + 2)k]}{4n(n - 2)(n - 3)W^2} \\ & + \frac{W^2[n^2 - 3 - (n - 1)^2 k]}{n(n - 2)(n - 3)W^2} \end{aligned} \tag{6.31}$$

where

$$W = \sum_{i=1}^{n} \sum_{j=1}^{n} w_{ij}, \qquad S_1 = \frac{\sum_{i=1}^{n} \sum_{j=1}^{n} (w_{ij} + w_{ji})^2}{2}, \tag{6.32}$$

$$S_2 = \sum_{i=1}^{n} (w_{i.} + w_{.i})^2, \tag{6.33}$$

and

$$k = \frac{\sum_{i=1}^{n} (x_i - \bar{x})^4}{\left(\sum_{i=1}^{n} (x_i - \bar{x})^2\right)^2}, \tag{6.34}$$

which is the kurtosis of the variable x (see Chapter 2).

The "." in the S_2 expression above denotes the sum of the other subscribe. For instance,

$$w_{i.} = \sum_j w_{ij}. \tag{6.35}$$

For Moran's I, the expected value for a random pattern and the variances are

$$E_N(I) = E_R(I) = \frac{-1}{n-1}, \tag{6.36}$$

$$VAR_N(I) = \frac{(n^2 S_1 - nS_2 + 3W^2)}{W^2(n^2 - 1)} - [E_N(I)]^2, \tag{6.37}$$

and

$$VAR_R(I) = \frac{n[(n^2 - 3n + 3)S_1 - nS_2 + 3W^2]}{(n-1)(n-2)(n-3)W^2} - \frac{k[(n^2 - n)S_1 - nS_2 + 3W^2]}{(n-1)(n-2)(n-3)W^2} - [E_R(I)]^2, \tag{6.38}$$

with W, S_1, S_2, and k similarly defined.

Once the expected values and their variances are calculated, the standardized Z-scores can be calculated:

$$Z = \frac{I - E(I)}{\sqrt{VAR(I)}} \tag{6.39}$$

or

$$Z = \frac{C - E(C)}{\sqrt{VAR(C)}}. \tag{6.40}$$

Note that the same critical values of $-1.96 < Z < 1.96$ can be applied with a statistical significance level of 5%, or $\alpha = 0.05$.

Finally, note that the spatial autocorrelation coefficients discussed here can also be applying in analyzing network characteristics. Specifically, these measures are useful for calculating similarity among network features, such as roads or flight connections (Black, 1992; Lee et al., 1994), and polygon objects, which will be discussed in more detail in Chapter 8. While the previous section focuses on evaluating the level of spatial autocorrelation of given point locations, another theme in spatial statistics or, more precisely, geostatistics is to model the spatial autocorrelation of point events. That approach treats each point event as a spatial sample of the population surface, and thus its purpose is to estimate a value surface based upon the magnitude of spatial autocorrelation from the observed sample locations. This approach falls into the more general category of spatial interpolation; kriging is the most common method. Since kriging has been addressed thoroughly elsewhere (e.g., Isaaks and Srivastava, 1989) and is well integrated into GIS, it will not be discussed further in this book.

ArcView Example 6.4: Spatial Autocorrelation Analysis of Northeast Ohio Cities

Again, this ArcView example uses the 5ccities.shp as the data theme.

Both Geary's Ratio and Moran's I values are useful, but they have different numerical ranges. When interpreting the results after calculating these indices, it is important to keep that in mind in order not to distort the information these indices suggest.

Step 1 Calculate distance matrix

When performing spatial autocorrelation analysis with the Ch6.avx extension, it is necessary to calculate a distance matrix among data points first before using the **Moran_Geary** option in the **Ch6** menu item. The distance matrix contains distances calculated for every possible pairs of points in the ArcView attribute table. The distance information is used as weights in the calculation of Geary's Ratio and Moran's I values.

To calculate the matrix, from the **Ch6** menu item, select the **Create Distance Matrix (Point)** option.

- Click the `5ccities.shp` as the Point Theme.
- Click the **Show Attributes** button to bring up the list of attribute fields, and select `City_fips` as the ID field.
- For this example, click to select using all records.
- Click the **Calculate** button to proceed with the calculation.
- When prompted, give a name for the generated matrix file and indicate the folder to store it.

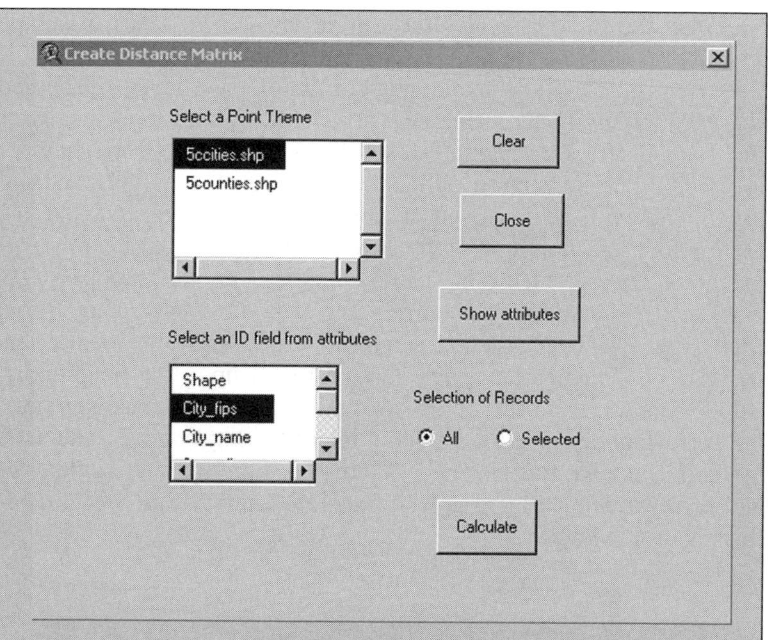

A message should be displayed to inform users that the calculation has been completed. Note that this procedure will create a matrix with $n \times n$ dimension. This matrix dimension also indicates the number of distance computations performed. As n increases, the number of computations will increase exponentially, and the process may take a significant amount of time when n is relatively large.

Step 2 Calculation steps

Note that when computing spatial autocorrelation coefficients, we need to select an attribute from the attribute table of the Point Theme as the variable (x). In this example, we will use the 1990 population of the 17 cities as the point attribute.

- From the **Ch6** menu, select the **Moran_Geary** option.

In the Moran and Geary window,

- First, select `5ccities.shp` as the Point Theme for the analysis.
- Check the box besides "Distance matrix has been created."
- Check the option of using All records.
- Click the **Show Attributes** button to bring up the list of available attributes.
- For ID attributes on the left list, click to select `City_fips` (FIPS codes of cities). Be sure that this ID field matches the one used to create the distance matrix in the previous step.
- On the right list, click to select `Pop1990` as the variable for analysis.

- Finally, enter 1 as the **Distance Decay Parameter (power).**
- Click the **Calculate** button to proceed.

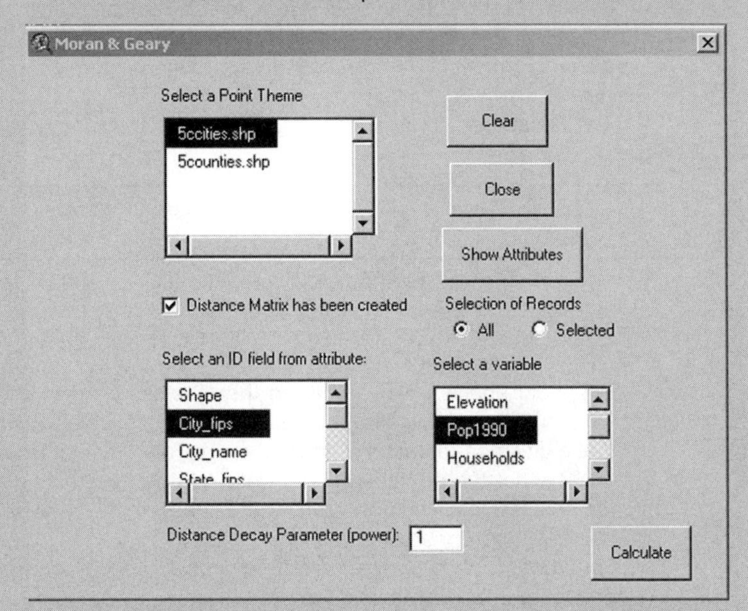

- When prompted for distance matrix input, navigate to the folder where the distance matrix was stored. Click the file and the **OK** button to proceed with the calculation.

Step 3 Reported coefficient and test values
Upon completion of the calculation, a window (titled Report) will display the results of the calculation.

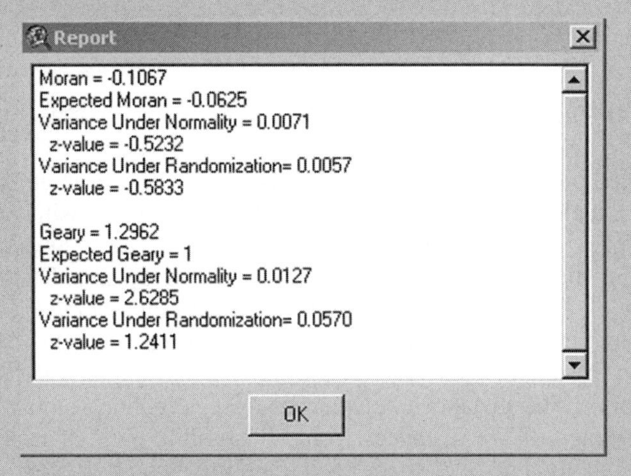

Note that coefficient values and test statistics are reported for both Geary's *C* Ratio and Moran's I.

Step 4 Interpreting the results
As discussed in this section, the coefficient values alone are not useful. They are useful only if they are accompanied by the test statistics. In addition, we need to determine whether the normality assumption or the randomness assumption is more appropriate for the analysis.
The results are summarized in the following table.

Index	Index Value	Expected Value	Variance Normality	*Z*-Score Normality	Variance Random	*Z*-Score Random
Moran's I	−0.1067	−0.0625	0.0071	−0.5232	0.0057	−0.5833
Geary's C	1.2962	1	0.0127	2.6285	0.0569	1.2411

Since our analysis was limited to the 17 cities in northeast Ohio and their population counts, it is not justifiable to treat these cities as the sample from a normal distribution, as we believe that the population count data do not resemble a normal distribution. Thus, the randomness assumption seems to be more appropriate.
First, note that the values of both Moran's I and Geary's Ratio provide a consistent evaluation of the population count data—mild negative spatial autocorrelation (−0.1067 for Moran's I and 1.2962 for Gearys' Ratio). Using the randomness assumption, we see that the *Z*-scores from both Moran's I and Geary's Ratio indicate that we cannot reject the null hypothesis that the spatial autocorrelation is not statistically significant.

Step 5 End

6.6 APPLICATION EXAMPLES

The methods discussed in this chapter are used to evaluate spatial point patterns. Quadrat Analysis is concerned with how the densities of points change over space and, it adopts a spatial sampling approach. Quadrats of consistent size and shape are overlaid on points in the study area. The frequency distribution of the number of quadrats with different points is constructed and compared with the frequency distribution of a theoretical random pattern. Nearest Neighbor Analysis, on the other hand, exploits the spacing between neighbors. The distances between nearest neighboring points are measured. The average of the distances from all possible pairs of nearest neighbors is compared to that of a theoretical random pattern. Beyond using distances between immediate nearest neighbors, the *K*-function provides a way to account for higher-ordered neighbors among points in a distribution.

Both Quadrat Analysis and Nearest Neighbor Analysis are useful in detecting spatial patterns and comparing them with other known patterns. However, only the locations of the points are considered. These two methods do not take into account that different points in a distribution may be different in some ways or may represent different activities or events.

On the other hand, spatial autocorrelation coefficients consider point locations in regard to the similarity of point attributes. These coefficients evaluate how attribute values of point locations are similar or dissimilar over space. Both Moran's I and Geary's Ratio are designed to capture the same spatial property of the point distribution—the level of spatial autocorrelation—but they differ in their numeric scales and statistical properties.

In this section, we will demonstrate how these methods can help us understand the distribution of point data and, to a large extent, how they can be used to detect any distinguishable spatial pattern in the data. We will use two sets of point-based data that document the level of water transparency of monitored lakes to demonstrate the utilities of these point pattern analyzers.

The first data set is the water transparency data of lakes monitored by the Environmental Monitoring and Assessment Program (EMAP). EMAP is a program supported by the U.S. Environmental Protection Agency (EPA). The program collects and analyzes environmental quality data for the United States. The data used here are a subset of the data available from the program's website, which can be accessed via the EPA's main web page (*http://www.epa.gov*). EMAP has developed an intensive monitoring plan that attempted to fully characterize the lake water quality of the northeastern United States. Data collected include water transparency measured by the Secchi disk, water chemical variables, watershed characteristics, fish, zooplankton, and diatom assemblage information. The sampled lakes were selected using a stratified probabilistic approach, which randomly selected lakes based on a criterion that defined the statistical population of lakes in the northeastern United States (Larsen et al., 1994).

Another data set is taken from the Great American Secchi Dip-In program (Dip-In) supported by the North American Lake Management Society. Each year during the week of July 4th, thousands of volunteers across the United States and Canada dip their Secchi disks into lakes of their choice to measure the water transparency (Jenerette et al., 1998). These volunteers then report the finding to the program's office, along with their answers to other questions about the lake's environmental conditions and how the lake is being used. The selection of lakes is entirely determined by the volunteers without any prestructured sampling framework. As a result, the lakes examined in this program are those that are being used, that volunteers care about, and, consequently, that need our attention.

One of the issues discussed recently is the sampling designs used in the two lake monitoring programs. EMAP, using great efforts, selected lakes by what it considered to be a random pattern based on the stratified probabilistic approach. The Dip-In program, on the other hand, let volunteers make the selections. The philosophical and theoretical approaches behind the two programs are entirely different, and it will be interesting to examine how the outcomes differ. To provide a visual impression of how the monitored lakes distribute, Figure 6.11 shows the locations of lakes for EMAP and Dip-In.

On the issue of which program better samples lakes for monitoring, we can use the methods discussed in this chapter to examine how the two data sets differ. We will measure to what degree the lakes selected for examination by the two programs deviate from a random pattern to indicate indirectly how the two programs differ in their sampling outcomes.

Now that the Ch6.avx extension is available, it is just a matter of running the procedures on the two data sets. To examine the data in more detail, we can divide each data set by the state boundaries to create subsets of data in both cases. This data manipulation process allows us to see how the spatial patterns change over scales. When the entire data set is used in the analysis, we test the spatial pattern at a multistate scale. When testing the subsets, we examine the spatial patterns at a more detailed local-regional scale.

Both data sets contain water transparency data (as attributes in the shapefiles). Each of the data sets is divided into eight subsets for the following eight states: Connecticut (CT), Delaware (DE), Massachusetts (MA), Maine (ME), New Hampshire (NH), New York (NY), Rhode Island (RI), and Ver-

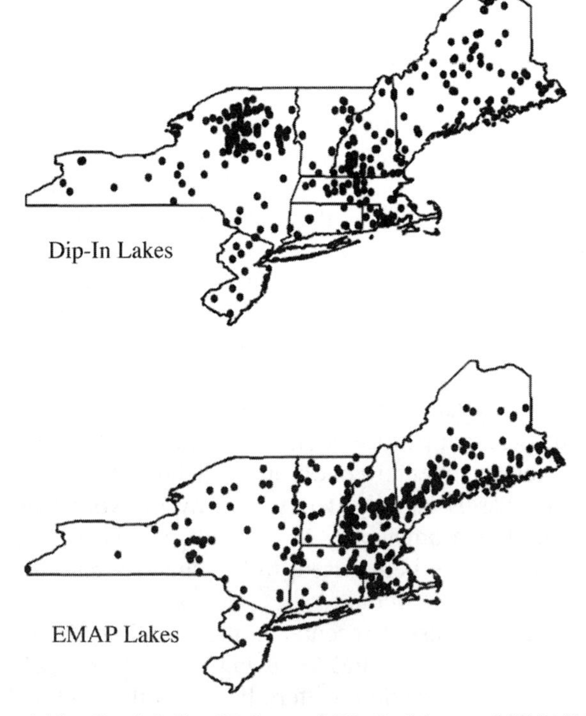

Figure 6.11 Spatial distributions of Dip-In lakes and EMAP lakes.

mont (VT). Including the entire northeastern United States, each program has 9 data sets in this analysis.

Table 6.8 lists the results of using the Ch6.avx extension on each of the data sets and subsets. The table gives the statistics and their Z-scores for the Quadrat Analysis and the Nearest Neighbor Analysis. For Quadrat Analysis, 100 quadrats were used. We highlight the subsets when one of the programs has too few points (lakes). We suggest dropping these subsets (states) from our analysis.

In Table 6.8, the number of data points in each data set is also shown. For example, EMAP has 350 lakes for the entire northeastern (NE) United States and Dip-In has 303 lakes. Some subsets are excluded because one of the programs has fewer than 5 data points. They are shown in gray type. To identify significant results easily, those Z-scores that are greater than 1.96 or less than −1.96 are shown in boldface italics because these Z-scores indicate statistical significance at the level of $\alpha = 0.05$ (5%).

For the entire northeastern United States, both the EMAP and Dip-In programs show spatial patterns statistically different from a random pattern by the two statistics, but with the nearest neighbor statistic indicating a clustering pattern. They deviate from the random pattern with a statistical significance at the level of $\alpha = 0.05$. When data are partitioned into subsets of individual

TABLE 6.8 Quadrat Analysis and Nearest Neighbor Analysis of EMAP Data and Dip-In Data

			Quadrat Analysis		Nearest Neighbor Analysis	
		N	Statistic (*D*)	Critical Value	Corrected *R*	*Z*-Score
NE	**EMAP**	350	***0.6097***	0.0730	0.3284	***−23.2606***
	Dip-In	303	***0.6284***	0.0780	0.4194	***−18.6692***
CT	*EMAP*	*14*	*0.0486*	*0.3490*	*0.2722*	***−4.6841***
	Dip-In	*4*	*0.0007*	*0.6240*	*0.6300*	*−1.2608*
DE	*EMAP*	*13*	*0.0213*	*0.3610*	*0.8597*	*−0.8699*
	Dip-In	*4*	*0.0089*	*0.6240*	*0.4968*	*−1.7159*
MA	**EMAP**	38	0.1114	0.2210	0.4324	***−5.8614***
	Dip-In	32	0.0287	0.2404	0.7649	***−2.3152***
ME	**EMAP**	74	0.1429	0.1581	0.4650	***−8.2803***
	Dip-In	99	***0.3084***	0.1367	0.4893	***−9.2038***
NH	**EMAP**	45	0.0930	0.2027	0.6197	***−4.5287***
	Dip-In	78	***0.2542***	0.1540	0.4741	***−8.3579***
NY	**EMAP**	142	***0.4183***	0.1141	0.3465	***−14.1918***
	Dip-In	44	0.0964	0.2050	0.7224	***−3.2649***
RI	*EMAP*	*4*	*0.0008*	*0.6240*	*1.2043*	*0.6972*
	Dip-In	*18*	*0.0248*	*0.3090*	*0.7768*	*−1.6437*
VT	**EMAP**	18	0.0652	0.3090	0.3068	***−5.1022***
	Dip-In	21	0.0107	0.2700	0.7937	−1.6468

states, they show statistically significant clustering lake patterns in MA, ME, NH, and NY in both programs according to nearest neighbor statistics, except for the Dip-In data for VT, an insignificant level but still a clustering pattern.

On the other hand, the state-level analysis shows that the spatial distributions of sampled lakes of only one of the two programs in ME, NH, and NY are statistically different from a random pattern based upon Quadrat Analysis. We can make several observations here. First, we cannot say which statistical method (Quadrat Analysis vs. Nearest Neighbor Analysis) is more powerful. The two methods will likely provide different results, as they adopt different approaches. Quadrat Analysis adopts the spatial sampling approach in which each quadrat, regardless of its shape, is a spatial sampling unit, while the Nearest Neighbor Analysis is a distance-based approach that focuses only on the spatial relationship of nearest neighbors if we use first-order analysis. Second, volunteers without special instructions selected lakes that show degrees of dispersion similar to those of the lakes selected by EMAP's stratified sampling, at least according to the results of Nearest Neighbor Analysis. Third, the clustering or dispersion pattern of lakes is determined to a large degree by the landscape characteristics of lake formation. Therefore, the analysis results may not reflect sampling biasness.

We also wondered what spatial pattern each data set would display; that is, if lakes in that same region have similar water transparency values. Therefore, spatial autocorrelation analysis is also used here to assist the detection of a pattern, if any. Table 6.9 shows the results of using Ch6.avx in calculating Geary's C and Moran's I. Similar to Table 6.9, states with fewer than 5 lakes in either EMAP or Dip-In are dropped from further analysis and are shown in gray type. Any *Z*-score that is either greater than 1.96 or less than -1.96 is highlighted in boldface italics, as they are statistically significant at the level of $\alpha = 0.05$. *Z*-scores are calculated for the free sampling (or normality) assumption.

First, we need to bear in mind that the numerical ranges of Moran's I and Geary' Ratio have a somewhat inverse relationship. For positive spatial autocorrelation, the *Z*-scores of Moran's I will be positive but those for Geary's Ratio will be negative, and vice versa. Note that positive spatial autocorrelation in this example means that the neighboring lakes tend to show similar values of water transparency. In Table 6.9, the *Z*-scores of the two statistics provide consistent results, with the exception of the Dip-In programs in NY and RI, and both exceptional cases are statistically insignificant.

For the entire Northeast, the two programs indicate some positive autocorrelation. This relationship is significant for both programs based upon Geary's Ratio, but it is significant only for the Dip-In program according to Moran's I. However, the positive autocorrelation at the regional level did not trickle down to the state level. Some state programs, such as the EMAP in MA, have a clearly negative spatial autocorrelation confirmed by both statistics, and the relationship is significant. Another one with a negative spatial

TABLE 6.9 Spatial Autocorrelation in EMAP Data and Dip-In Data

		N	Moran's I	Z-Score	Geary's *C*	Z-Score
NE	**EMAP**	350	0.0105	1.1000	0.9052	***−2.9313***
	Dip-In	303	0.1886	***15.2892***	0.6948	***−8.3505***
CT	*EMAP*	*14*	*−0.3883*	*−1.5328*	*1.4895*	***2.0007***
	Dip-In	*4*	*−0.3017*	*0.0898*	*0.9937*	*−0.0649*
DE	*EMAP*	*13*	*0.0667*	*1.3628*	*0.9548*	*−0.4368*
	Dip-In	*4*	*−0.2298*	*0.2641*	*0.8015*	*−0.8073*
MA	**EMAP**	38	−0.1908	***−2.4342***	1.3320	***2.9784***
	Dip-In	32	0.1301	1.1851	0.8006	−1.1121
ME	**EMAP**	74	−0.0096	0.0944	0.9703	−0.4539
	Dip-In	99	0.0786	***2.9778***	0.7030	***−4.2418***
NH	**EMAP**	45	0.1261	1.8567	0.8835	−0.8667
	Dip-In	78	0.1525	***4.8644***	0.7967	***−3.5792***
NY	**EMAP**	142	0.0081	0.6325	0.7700	***−3.8579***
	Dip-In	44	−0.0357	−0.1924	0.9612	−0.3955
RI	*Dip-In*	*4*	*−0.3051*	*0.0820*	*0.9995*	*−0.0068*
	Dip-In	*18*	*−0.0952*	*−0.2872*	*0.7832*	*−1.3330*
VT	**EMAP**	18	−0.5206	***−2.1993***	1.6218	***2.0342***
	Dip-In	21	−0.0142	0.4250	0.9375	−0.7091

autocorrelation pattern is the Dip-In program in NY, but with an insignificant relationship.

When we compare the two programs at the state level, MA, VT, and, to a lesser extent, NY have contradictory situations, with one program (EMAP in the case of MA and VT) indicating a negative autocorrelation but Dip-In a positive autocorrelation. But in both MA and VT, the relationships for the Dip-In program are not significant. The relationships for both programs in NY are also not significant, except for the EMAP program according to Geary's Ratio. On the other hand, no state with both programs has significant relationships. In most cases, only one of the programs has a significant relationship.

The example discussed in this section shows the different usages of the three methods. Before the availability of computer codes such as those in Ch6.avx, the exploration of various approaches to analyze or detect spatial patterns would have involved a tremendous effort. Now it is feasible and convenient.

REFERENCES

Bailey, T. C. and A. C. Gatrell. 1995. *Interactive Spatial Data Analysis* New York: Longman.

Berry, B. J. L. and A. Pred. 1961. *Central Place Studies: A Bibliography of Theory and Applications.* Bibliography Series Number One. Philadelphia, PA: Regional Science Research Institute.

Black, W. R. 1992. Network autocorrelation in transport network and flow systems. *Geographical Analysis* 24(3): 207–222.

Boots, B. N. and A. Getis. 1988. *Point Pattern Analysis.* Newberry Park, CA: Sage Publications.

Clark, P. J. and F. C. Evans. 1954. Distance to nearest neighbor as a measure of spatial relationships in populations. *Ecology* 35: 445, 453.

Cliff, A. D. and J. K. Ord. 1981. *Spatial Processes: Models and Applications.* London: Pion.

Donnelly, K. P. 1978. Simulations to determine the variance and edge effect of total nearest neighbor distance. In I. Hodder (ed.), *Simulation Methods in Archaeology,* Cambridge: Cambridge University Press, pp. 91–95.

Goodchild, M. F. 1986. *Spatial Autocorrelation.* CATMOG 47. Norwich: GeoBooks, University of East Angolia.

Gould, P. R. 1970. Is statistix inferens the geographical name for a wild goose? *Economic Geography* 46: 439–448.

Greig-Smith, P. 1952. The use of random and contiguous quadrats in the study of the structure of plant communities. *Annals of Botany,* New Series, 16: 312.

Griffith, D. A. and C. G. Amrhein. 1983. An evaluation of correction techniques for boundary effects in spatial statistical analysis: traditional methods. *Geographical Analysis* 15: 352–360.

Griffith, D. A. and C. G. Amrhein. 1991. *Statistical Analysis for Geographers.* Englewood Cliffs, NJ: Prentice Hall.

Griffith, D. A., D. W. S. Wong, and T. Whitfield. 2003. Exploring relationships between the global and regional measures of spatial autocorrelation. *Journal of Regional Science* 43(4): 683–710.

Isaaks, E. H. and R. M. Srivastava. 1989. *An Introduction to Applied Geostatistics.* New York: Oxford University Press.

Jenerette, G. D., J. Lee, D. Waller, and R. C. Carlson. 1998. The effect of spatial dimension on regionalization of lake water quality data. In T. K. Poiker and N. Chrisman (eds.), *Proceedings of the 8th International Symposium on Spatial Data Handling.* Burnaby, B.C., Canada: I.G.U., G.I.S. Study Group, 98: 256–266.

Larsen, D. P., K. W. Thornton, N. S. Urquhart, and S. G. Paulsen. 1994. The role of sample surveys for monitoring the condition of the nation's lakes. *Environmental Monitoring and Assessment* 32: 101–134.

Lee, J., L. G. Chen, and S. L. Shaw. 1994. A method for the exploratory analysis of airline networks. *The Professional Geographer* 46(4): 468–477.

Levine, N. 2002. *CrimeStat: A Spatial Statistics Program for the Analysis of Crime Incident Locations* (*v 2.0*). Houston, TX: Ned Levine & Associates, and Washington, DC: National Institute of Justice.

Ripley, B. D. 1976. The second-order analysis of stationary point process. *Journal of Applied Probability* 13: 255–266.

Taylor, P. J. 1977. *Quantitative Methods in Geography: An Introduction to Spatial Analysis.* Prospect Heights, IL: Waveland Press.

Thompson, H. R. 1956. Distribution of distance to *n*th neighbor in a population of randomly distributed individuals. *Ecology* 37: 391–394.

Tobler, W. R. 1970. A computer movie simulating urban growth in the Detroit region. *Economic Geography,* 46 (Supplement): 234–240.

EXERCISES

Facility	Latitude	Longitude	Fugitive
90	41.41333	−81.78528	43,827
57	41.53889	−81.52917	43,468
33	41.51833	−81.66139	29,581
58	41.48333	−81.66699	16,457
48	41.41833	−81.80833	15,000
88	41.48278	−81.72667	13,345
68	41.50556	−81.64528	11,200
78	41.45364	−81.72456	9,776
94	41.41860	−81.49550	6,161
35	41.47472	−81.76694	5,200
73	41.37222	−81.52000	4,098

Facility	Latitude	Longitude	Point Source
46	41.41667	−81.82194	180,520
83	41.45080	−81.79370	161,248
12	41.45417	−81.78972	133,000
107	41.44167	−81.67500	61,060
19	41.43387	−81.54423	50,163
66	41.58028	−81.53056	48,400
74	41.46139	−81.62500	47,879
28	41.48056	−81.63194	38,533
94	41.41860	−81.49550	18,958
76	41.42690	−81.81390	17,000
65	41.59222	−81.52139	15,900
48	41.41833	−81.80833	15,000

The above two tables were used in the exercise in Chapter 5, which is limited to describing point patterns using centrographic measures. The tables were extracted from the U.S. EPA Toxic Release Inventory (TRI) database for facilities releasing the largest amounts of air pollutants through fugitive release and point source emissions in Cuyahoga County, Ohio, in 2002. In this chapter, we will use the techniques for analyzing point patterns to study the data in these two tables.

1. If you have not done so in Chapter 5, key the tables into two separate spreadsheets. Then save the data in dbf format. Bring the tables into the ArcView Table document. Then select View/Add Event Theme to bring the data into ArcView and save them (Theme/Convert to shapefiles) as two point themes: Fugitive and Point Source. For more detailed instruction, refer to the Exercises in Chapter 5. Also, bring in the Ch6 extension as described earlier in this chapter.

2. Given that the two data sets (Fugitive and Point Source) have similar numbers of observations, what will be the appropriate number of quadrats for both data sets?

3. Use the function in the Ch6 extension to create quadrats for complete coverage of both data sets. Store the quadrat layers as shapefiles and then terminate the operation for now. Bring the quadrat theme back to the View document with the point theme. Tabulate the observed frequency distributions for each data set by the numbers of points in a quadrat, similar to the first two columns in Table 6.1.

4. What is λ in each data set? Given λ values, use the Poisson distribution to generate the expected frequency distributions for both data sets, similar to those columns in Table 6.3. Also, compute the cumulative proportions, and then compare the observed with the theoretical (Poisson) cumulative distributions to determine the K-S D statistic.

5. Use the Ch6 extension to perform Quadrat Analysis on both data sets. Verify if your answers from Question 4 are correct. Based upon the K-S D statistics for the two data sets, how would you characterize the spatial patterns of the two data sets? Do they differ?

6. Also, perform a Variance-Mean Ratio Test for each of the data sets. Do the results corroborate those of the K-S test?

7. In the previous questions, you used only one shape of quadrat (perhaps a square) and chose complete coverage in the quadrat generation. Select another quadrat shape and choose random placement of quadrats. Verify if using different quadrat shapes and an alternate sampling scheme will give consistent results or not.

8. Create a distance matrix for each of the two data sets using the Ch6 extension. Then bring the distance matrices back into ArcView. For each observation (point), identify the nearest neighbor and the corresponding nearest neighbor distance based upon the distance matrix for each data set. Then compute the average nearest neighbor distance.

9. Use the Ch6 extension to perform the Nearest Neighbor Analysis. Verify if the observed average nearest neighbor distances calculated in Question 8 are correct or not. What do the results tell us about

a. the spatial pattern of the facility distribution in each data set?

b. the differences in spatial patterns between the two types of facilities?

10. Rerun the Nearest Neighbor Analysis with Donnelly's correction. Do the results differ much from those not using the correction?

11. Examine the maps of the two types of facilities in detail. Do you see the need to use higher-order statistics? Why?

12. Regardless of your conclusion in Question 10, it is not a bad idea to explore if the spatial processes affecting the facility locations are homogeneous or not. Therefore, use the Ch6 extension to perform a *K*-function analysis for each type of facility. At what spatial lag(s) do the clusters exist? Do the two types of facilities cluster at different spatial lags?

13. All the questions asked so far concern only the locations of the facilities, but the data also include information on the amounts of pollutants released. Since the distance matrices for both data sets have been generated, the next logical step is to perform a spatial autocorrelation analysis. Compute Moran's I and Geary's Ratio for both data sets using the Ch6 extension function. Choose the fugitive and point source values as the attributes.

14. Which sampling assumption (randomization or normality) is more appropriate for this analysis? Why?

15. Interpret the results from Moran's I and Geary's Ratio. Do they support similar conclusions?

CHAPTER 7

LINE PATTERN ANALYZERS

In previous chapters, we have discussed how certain types of geographic features or events can be represented by point objects in an abstract manner in a GIS environment. In this chapter, we will shift our attention to the description and analysis of geographic features with linear extent that can be represented most appropriately by linear objects, such as line segments or arcs. We will describe two general types of linear features—vectors and networks—that can be represented in a GIS environment. Then, as in the previous chapters, we will discuss how geographic information can be extracted from the data to study these linear features. Most of these analyses are descriptive in nature.

7.1 THE NATURE OF LINEAR FEATURES: VECTORS AND NETWORKS

In a vector GIS database, linear features are best described as line objects. As discussed in earlier chapters, the representation of geographic features by geographic objects is scale dependent. For instance, on a small-scale map (1:1,000,000), a mountain range may be represented by a line showing its approximate location, geographic extent, and orientation. When a larger geographic scale is adopted (1:24,000) or more detailed information is shown, a line is insufficient to represent the detail of a mountain range with a significant spatial extent at that scale. Instead, a polygon object is more appropriate. In other words, a geographic feature with a significant spatial extent can be represented abstractly by a linear object at a smaller scale but by a polygon at a larger scale. This process is sometimes known as *cartographic abstraction.* Another example is the Mississippi River and its tributaries. They have significant widths when they are shown on large-scale maps as polygons, but they are represented by lines on small-scale maps, such as a map of the continental United States.

A line can be used to represent linear geographic features of various types. Most people, even GIS users, have a preconceived notion that linear features are for rivers and roads only, but actually they can represent many more types of geographic features. Features of the same type within the same system are generally connected to each others to form a network. For instance, segments of roads connected together typically form a road network, and segments of streams belonging to the same river system or the same drainage basin form a river network or a drainage network. Within these networks, individual line segments have their own properties. For example, in a river network, each river or tributary segment has its length, the beginning and ending points, the flow direction and volume. However, different segments in a network are related to each other in a topological manner. The end point of one segment is likely the beginning point of another segment unless the end point is a terminal point of a branch of the network. Therefore, these segments cannot be treated separately because of the topological relationship among them. Other examples of physical networks include networks of power utility lines, pipelines for gas transmission, sewage systems, and telephone or fiber-optic networks for Internet connection. There are also networks that are more abstract, such as airplane routes using the hub-and-spoke structure or even the social networks depicting the relationships among individuals in a community. The more abstract network without a clear geographic dimension can be depicted by a graph.

Some linear features, however, do not have to be connected to each other to form a network. Each of these linear segments can be interpreted alone. For instance, fault lines of a geologically active area may be noncontiguous. Other examples include spatially extensive features such as mountain ranges and touchdown paths of tornados. These spatially noncontiguous linear features can be analyzed as individual objects without referring to any topological relationships.

Line objects in a GIS environment are not limited to representing linear geographic features (either networked or nonnetworked). They can also be used to represent phenomena or events that have beginning locations (points) and ending locations (points). For instance, it is quite common to use lines with arrows to show wind directions and magnitudes, which can be indicated by the lengths of the lines. These are sometimes referred to as *trajectories* or *vectors.* Another example is tracking the movements of wild animals with Global Positioning System (GPS) receivers attached to them over a certain time period. In that case, the line objects represent where they started and stopped.

Figures 7.1a, 7.1b, and 7.1c are examples of these three types of linear objects in GIS. Figure 7.1a shows a set of fault lines in Loudoun County, Virginia. Some of these fault lines are joined together geologically. But top-

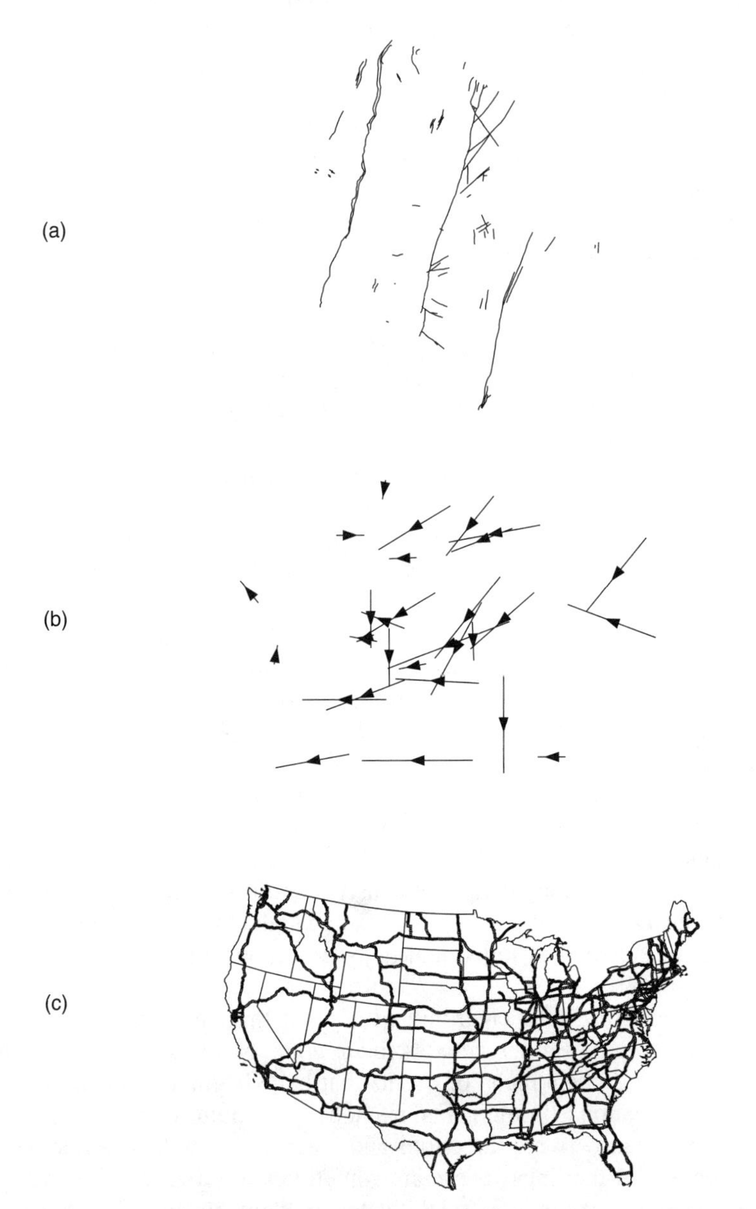

Figure 7.1 (a) Selected fault lines in Loudoun County, Virginia. (b) Trajectories of wind for selected San Francisco Bay locations. (c) Major road networks in the United States.

ologically, there is no reason why they have to be linked. In fact, Figure 7.1a shows many fault lines separately from each other. In contrast to fault lines, which are geographic or geologic features, the linear objects shown in Figure 7.1b represent events. These lines show the trajectories of wind direction and speed, which can also be thought of as magnitudes of these phenomena at given locations. There are no geographic features we can observe on the Earth's surface in this case, but there are geographic phenomena that can be represented by symbols. The line objects in Figure 7.1b reflect the two common attributes of linear geographic features or events: direction and length.

Different line segments can have similar orientations. For instance, airplane routes between a pair of locations exhibit similar orientations, running, for instance from north to south. However, the directions of some of these routes may be opposite to each other. That is, some run from north to south, but others run from south to north. To show the directions of line segments, an arrow can be added, for example, to each line to show the direction of the wind in Figure 7.1b. The length of the line can represent the spatial extent of the linear object if the line symbolizes a geographic feature. Alternatively, the length of the lines in Figure 7.1b is proportional to the strength or magnitude of the wind. The stronger the wind, the longer the line to reflect its magnitude. Lines can be connected to form a network. Figure 7.1c is a standard road network with line segments linking to each other. In addition to the characteristics of the other two types of linear features, a network of features shows how individual linear features are related topologically.

7.2 CHARACTERISTICS AND ATTRIBUTES OF LINEAR FEATURES

7.2.1 Geometric Characteristics of Linear Features

In most vector-based GIS packages, a linear object is defined either by a line segment or by a sequence of line segments sometimes referred to as an *arc* or a *chain.* If the object is relatively simple, such as the short, small fault lines shown in Figure 7.1a, then a simple line segment will be a sufficient representation. But if the object is one side of the curb of a winding street, then several line segments may be required to depict its curvilinear nature. Similarly a chain instead of a simple line will be effective and accurate in depicting an interstate highway that goes through various landscapes.

If a line is used to represent a simple linear geographic feature, two endpoints are needed to define it. The locations of these two points may be defined in the form of longitude-latitude, x-y, or another coordinate system. Using the fault lines as an example again, if we know that the fault is short and a simple line segment is adequate to represent it, then all we need are the locations of the two points at the ends of the line.

If a chain is required to describe a more complicated linear feature, in addition to the two endpoints, intermediate points depicting the sequence of line segments will be needed. The curb of a street is basically a collection of the edges of concrete blocks. Therefore, a line segment defined by two endpoints can represent the edge of a block, and a sequence of segments defined by two terminal points and a set of intermediate points can represent the curb. These chains may exist without linking to each other, like the fault lines in Figure 7.1a.

If chains are connected to each other, they form a network. In a network, linear objects are linked at the terminal points of the chains. A terminal point of a chain can be the terminal point of multiple chains, such as the center of a roundabout in a local street network or a major city such as St. Louis, where several interstate highways (I-70, I-55, I-64, and I-44) converge.

7.2.2 Spatial Attributes of Linear Features: Length

Linear features in GIS can have attributes just like other types of features. These attributes describe various characteristics of the linear features, including information on what those features are (roads or rivers) and other characteristics (types of roads, capacity, travel restriction, etc.). Here we focus on spatial attributes that can be derived from linear features. To simplify the discussion, we can treat a simple line and a chain in the same manner in the sense that both of them can generally be defined by two terminal locations. In fact, most GIS packages make the intermediate points transparent to users. Therefore, there is no need to differentiate a simple line segment and a chain in most cases. Given any linear feature, an obvious spatial attribute is its length.

When the locations of the two endpoints are available, the length of the linear feature can easily be calculated. After extracting the location of the two endpoints, we can apply the *Pythagorean Theorem* to calculate the distance between the points and thus the length of the linear feature. The Pythagorean Theorem states that for a right angle triangle (Figure 7.2), the sum of the squares of the two sides forming the right angle is equal to the square

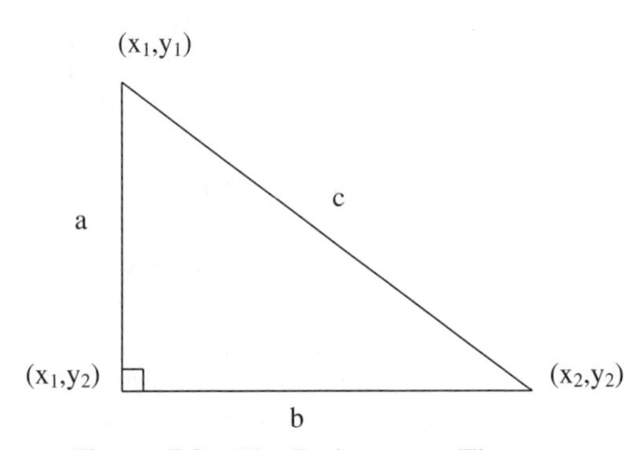

Figure 7.2 The Pythagorean Theorem.

of the longest side (hypotenuse). According to Figure 7.2, $a^2 + b^2 = c^2$. Therefore, if we know the x-y coordinates of the endpoints of c, the line segment whose length we want to find, then using the theorem, the length is

$$c = \sqrt{(x_1 - x_2)^2 + (y_1 - y_2)^2}. \tag{7.1}$$

The above discussion on the calculation of **linear distance** is applicable to a flat surface. If the coordinates of point locations are given in latitude-longitude readings, we can still apply the above method, but the result will be just an approximation, since the Earth's surface is not flat. To more accurately measure the distance between any two locations with longitude-latitude readings on the Earth's surface, we may want to calculate the **great circle distance.** Because the shape of the Earth can be approximated by a globe, the surface of the globe can be thought of as a collection of circles, or great circles circumscribing the Earth. In that case, the distance between any two locations on the globe is essentially the distance of the two points along a great circle, and thus it is called the great circle distance (DeMers, 2000). The relationship between the great circle distance and the longitude-latitude readings of any two given locations, A and B, are as follows:

$$\cos D = (\sin a \sin b) + (\cos a \cos b \cos(|\delta\lambda|)), \tag{7.2}$$

where

D is the great circle distance,
a and b are the latitude readings of locations A and B, and
$|\delta\lambda|$ is the absolute difference in longitude between A and B.

The great circle distance provides a more accurate distance measure than linear distance using the Pythagorean Theorem. However, the difference in results between these two methods is significant only if the area of concern is relatively extensive, such as at the continental scale. For studies at a local scale, such as within a city, using the great circle distance may not improve the results much, especially if the distances are for comparative rather than engineering purposes. The need for accuracy is dependent upon the nature of the application or analysis. If a high level of precision is required, such as for construction, then the more accurate distance calculation is preferable. For most socioeconomic analyses, precise distance calculation may not be necessary.

The above discussions on using the Pythagorean Theorem versus the great circle distance are appropriate for calculating the length of a simple line segment. If a set of line segments is linked together to form a chain, then the length of the chain can be defined as the sum of the lengths of individual line segments forming the chain. This is appropriate if we are interested in the aggregated length of the chain. But if we are interested only in the spatial extent of the linear geographic features, such as how far a river extends inland, then we can just calculate the straight-line length of the chain. This, of course, is just an approximation of the real distances of the meandering rivers. It can be accomplished by taking the terminal locations of the chain and deriving the length between them as if a straight line is formed using the two terminal locations.

7.2.3 Spatial Attributes of Linear Features: Orientation and Direction

Another obvious spatial attribute of a linear feature is its orientation. Orientation can be nondirectional or directional. For instance, features with an east-west orientation can be regarded as having an orientation different from that of features with a west-east orientation. Alternatively, orientation can be nondirectional when features with an east-west orientation are treated the same way as features with a west-east orientation.

Nondirectional orientation is appropriate when the linear feature does not have a directional characteristic. For instance, the fault lines shown in Figure 7.1a are nondirectional. There is no specific *from-to* characteristic of each of those fault lines even though we can describe them using "*from* location x *to* location y." However, if a fault line has an orientation but is nondirectional, then using "*from* location y *to* location x" to describe the fault line does not change its nature.

Other examples of nondirectional linear features include curbs, mountain ranges in small-scale displays, and sections of coastline. Usually the orientation of a set of linear features can be indicated by verbal descriptions, such as from north to south or from east to west, or vice versa. Another common approach is to use an angle measured counterclockwise in reference to east to describe the orientation precisely. Therefore, an orientation of 45 degrees means that the overall trend of the linear features is 45 degrees counterclockwise from the x-axis or east using the tradition in trigonometry. Sometimes, however, the orientation may refer to the north instead. In that case, the angle will be measured clockwise from the north. The referencing direction is situation dependent.

Stating that direction is not appropriate for some linear geographic features does not imply that GIS data used to represent these nondirectional features do not record direction information. In fact, most GIS packages capture the directions of linear features as the data are created even if the directions are not meaningful to the features. Depending on how the data are entered (or digitized) into GIS, quite often the beginning point and the endpoint of a chain during the digitizing process define the direction of the chain. Therefore, the directions of fault lines, for example, are stored in the GIS data when the data are created even though direction is not meaningful to describe fault lines.

Another attribute of linear features similar to orientation is their direction. Linear features have directional characteristics that are dependent on the beginning and ending locations. *From* location *x to* location *y* is not the same as *from* location *y to* location *x*. In fact, the directions of the two descriptions are exactly the reverse of each other. In addition, the two descriptions can refer to two different linear features. For instance, a two-way street can be described by two linear features with exactly the same geographic locations and extents but opposite directions. Linear objects representing events or spatial phenomena are often directional in nature. The wind trajectories described in Figure 7.1b are clearly of this type. Arrows are added to the lines to indicate the directions.

After the spatial attributes (direction/orientation and length) are extracted and added to the attribute table of the polyline theme in ArcView GIS, these attribute values can be analyzed further using classical statistical tools (Chapter 2), such as those included in the Field/Statistics menu item for ArcView. One may also use the statistical functions incorporated into the Ch2 extension discussed and demonstrated in Chapter 2.

7.2.4 ArcView Example: Linear Attributes

Given the importance of length and direction in analyzing linear features, an ArcView extension, Ch7.avx, was developed to assist users in calculating these attributes. Three groups of functions are included in Ch7.avx. A series of ArcView examples will follow each section in this chapter to demonstrate how these functions can be used.

ArcView Example 7.1 describes the steps used in calculating and adding the attributes of length and direction to a line (polyline) data theme. To calculate the length of a set of polylines, both the straight-line length and the true length of polylines are available. In terms of direction, arrows can be added as part of the linear data theme for visual evaluation of directional patterns in the data.

ArcView Example 7.1: Linear Attributes

Step 1 Data sources and load extension

As in previous ArcView examples, data for this example should be in C:\Temp\Data\ch7_data.

- Start ArcView with an empty View document.
- Use the **Add Theme** button to add `\winds.shp` and `\states.shp`.

The winds.shp is a linear data theme that contains wind direction data. The states.shp contains state boundaries.

- Use **Single Symbol** to display winds.shp.
- Change the display of states.shp to display only the outlines of the states.

- From the **File** menu, choose the **Extensions . . .** menu item.
- Check the box beside the `Ch7 extension` and click **OK.**

Now all three functions in Ch7.avx can be accessed from the Ch7 menu.

Step 2 Add attributes of lengths and directions
First, explore the winds.shp to examine what attributes are available.

- Click winds.shp in the Table of Contents area of the View document to make it the active theme.
- Use the **Open Theme Table** button to open the attribute table of winds.shp.

As can be seen, only two attributes are included: Shape and ID.

- Close the attribute table.

To calculate and add the attributes of lengths and directions:

- From the **Ch7** menu, choose **Add Length and Direction.**

In the Length and Direction dialog box:

- Click to select `winds.shp` as the line theme for analysis.
- Check both boxes for **True Length** and **Straight-Line Length.**
- Check the box to add directional information to the attribute table.
- For Direction or Orientation, click the `Direction` item to indicate that the directions are to be added. Note that when `Direction` is chosen, lines with directions from, for instance, east to west and from west to east are treated differently. If `Orientation` is chosen, the angle will not be larger than 180 degrees from either the north or the east.
- Click to choose `From North` as the direction.
- Also, check the box to add arrows to wind trajectories.
- Click the **Add** button to proceed.

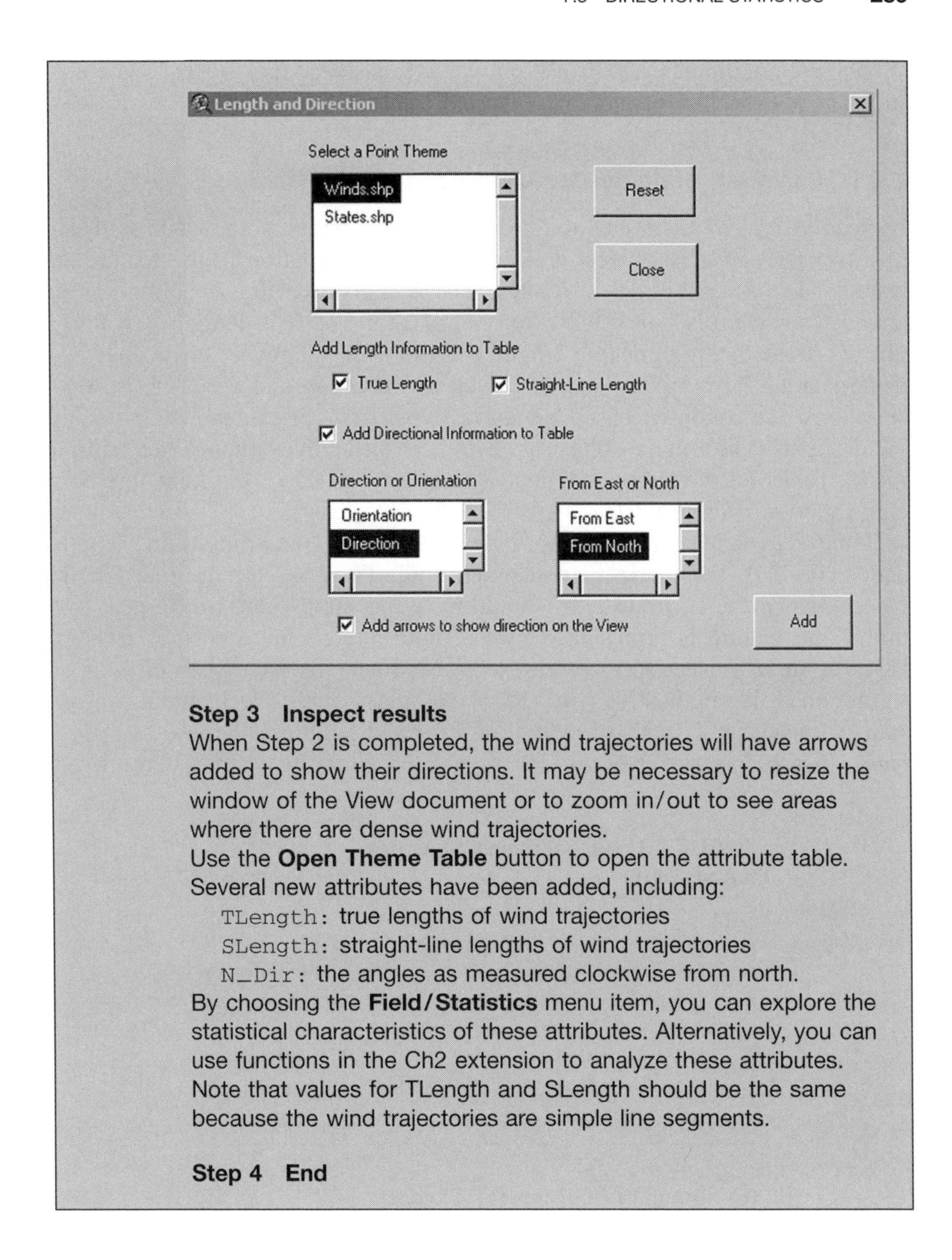

Step 3 Inspect results

When Step 2 is completed, the wind trajectories will have arrows added to show their directions. It may be necessary to resize the window of the View document or to zoom in/out to see areas where there are dense wind trajectories.

Use the **Open Theme Table** button to open the attribute table. Several new attributes have been added, including:

`TLength`: true lengths of wind trajectories
`SLength`: straight-line lengths of wind trajectories
`N_Dir`: the angles as measured clockwise from north.

By choosing the **Field/Statistics** menu item, you can explore the statistical characteristics of these attributes. Alternatively, you can use functions in the Ch2 extension to analyze these attributes. Note that values for TLength and SLength should be the same because the wind trajectories are simple line segments.

Step 4 End

7.3 DIRECTIONAL STATISTICS

To analyze linear geographic features in greater depth, we must rely on statistical techniques specifically designed for linear features. Unfortunately, not many statistical tools are available for analyzing these features. Most of them can be considered geostatistics developed and used mostly by geoscientists

(Swan and Sandilands, 1995). Before we discuss these techniques, some preliminary and exploratory analyses can be conducted.

7.3.1 Exploring Statistics for Liner Features

In Section 7.2, we discussed the process used to extract some basic statistics or attributes of linear objects and to store the information in the feature attribute table as additional attributes. These statistics offer opportunities to conduct some preliminary and exploratory analyses. The length of a linear object—both the straight-line length and the truth length—can be analyzed using standard descriptive statistics such as the mean and variance. In most situations, the analyses based on these two length measures will probably yield slightly different results. However, the differences should not be dramatic. Table 7.1 shows two statistical summary tables of the fault line coverage shown in Figure 7.1a.

Table 7.1a summarizes the attribute SLength for the straight-line length, and Table 7.1b summarizes another attribute, TLength, for the true length. Conceptually, the two attributes should have the same value for a linear feature if the feature is represented by a simple line segment but not a chain. When a chain is needed to represent a curvilinear feature, SLength will be shorter than TLength. The two sets of summary statistics, including mean, sum, variance, and standard deviation, reflect the nature of these two attributes. Please compare the two sets of statistics in the two tables carefully.

TABLE 7.1 (a) Summary Statistics for SLength

Sum:	932,332.65
Count:	422
Mean:	2,209.32
Max:	39,005.09
Min:	19.88
Range:	38,985.21
Variance:	16,315,314.64
Std Dev:	4,039.22

(b) Summary Statistics for TLength

Sum:	948,470.09
Count:	422
Mean:	2,247.56
Max:	41,313.70
Min:	19.88
Range:	41,293.83
Variance:	17,342,308.46
Std Dev:	4,164.41

Going one step further, based on the difference between the straight-line length and the full length of the chain, we can analyze the topological complexity of each linear feature in the data set. One simple method is to derive a ratio of the two length measures—**sinuosity** (DeMers, 2000). When the length of the entire chain is divided by the length of the straight-line distance, the ratio is 1 if the linear feature is simple enough to be represented by a simple line segment. The higher this ratio is, the more complex or curvilinear is the linear feature.

Table 7.2 shows selected fault lines with sinuosity ratios larger than 1.2 in descending order. As a result, 10 fault lines have ratios larger than 1.2. The first eight of them are shown in the top panel in Figure 7.3. All these faults are short but banned at one end, creating relatively high sinuosity ratios. The other two fault lines on the list are shown in the middle and bottom panels of Figure 7.3. The ninth fault line is crooked, and the tenth looks like a curve. The sinuosity ratio is very effective in identifying linear features that are relatively crooked or have a high degree of geometric complexity.

Another attribute of a linear feature is direction. A simple exploratory method used to study linear features with the directional attribute is to add arrows to those features in a display to provide visual recognition of the pattern, as in Figure 7.1b. Wind trajectories are shown quite effectively with the added arrows in Figure 7.1b. Based on the arrows' directions, we can develop an overall impression that in general the wind blows in the east-west or northeast-southwest direction.

Another example is presented in Figure 7.4, where information on wind directions over the United States was recorded with speeds represented by their lengths. Based on the displayed arrows in the figure alone, we can identify several circulation subsystems in the continental United States at that time. For instance, the New England region seemed to be influenced by one subsystem, while the Mid-Atlantic region throughout the South along the coast seemed to be affected by another subsystem.

TABLE 7.2 Length Measures and Sinuosity/Length Ratios of Selected Fault Lines

ID	Length	SLength	TLength	Sinuosity
1	596.841	386.532340	596.841079	1.544091
2	595.518	386.581012	595.517605	1.540473
3	698.879	483.387239	698.879055	1.445795
4	699.558	484.230785	699.558479	1.444680
5	651.754	469.902068	651.754268	1.387000
6	647.195	468.328944	647.194827	1.381924
7	603.488	441.679554	603.488095	1.366348
8	604.233	443.694169	604.232631	1.361822
9	3,400.414	2,779.877117	3,400.413628	1.223224
10	217.739	180.057803	217.738803	1.209272

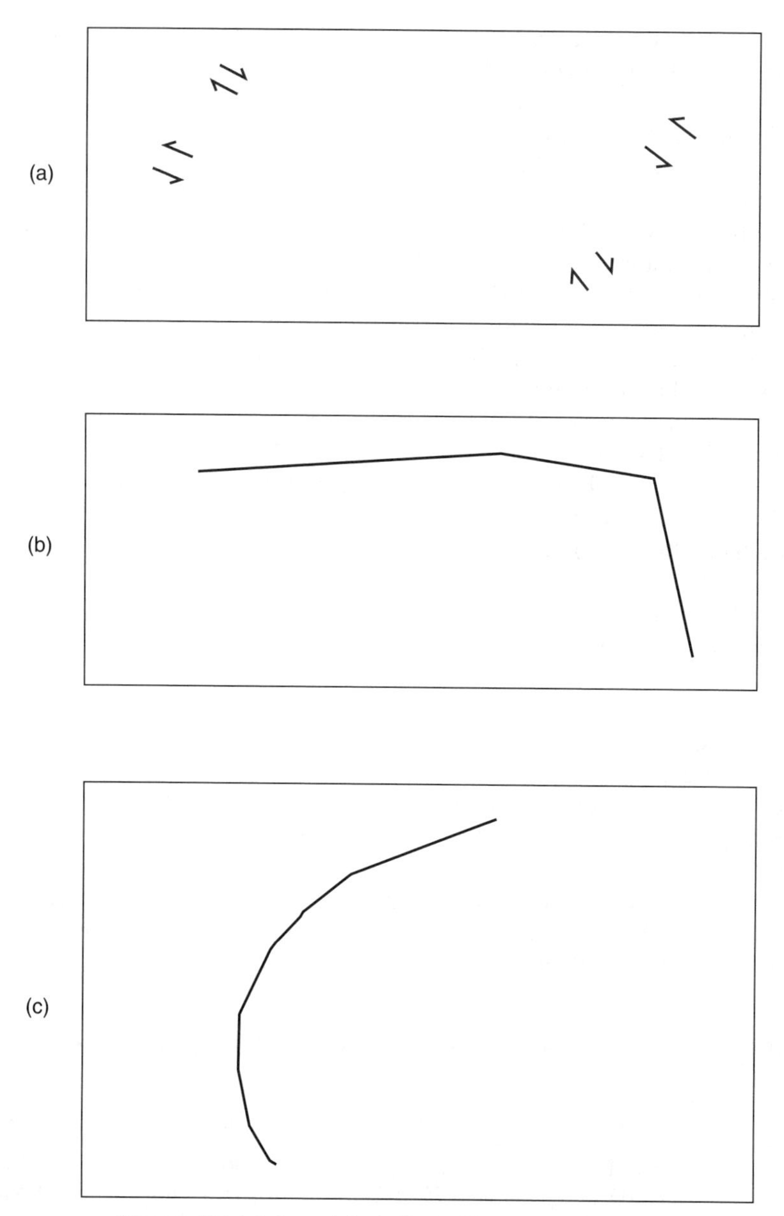

Figure 7.3 Selected fault lines with high sinuosity.

While it is relatively easy and straightforward to visually analyze the length and directional attributes of linear features, it is also possible to use simple statistical methods to quantitatively analyze the directional or orientation aspects of linear features. With the orientation and/or directional information of linear features extracted and added to the attribute data, we would naturally calculate descriptive statistics based on the angles (orientation or direction)

Figure 7.4 Trajectories of wind for selected U.S. locations.

of linear features. Unfortunately, descriptive classical statistics such as mean and variance are in general not appropriate to analyze angles of linear features. Figure 7.5 provides a simple example to illustrate this problem.

Figure 7.5 shows two vectors, A and B, with 45 degrees and 315 degrees, respectively, clockwise from the north. If we use the concept of the mean in classical statistics to indicate the average direction of these two vectors, the mean of the two angles will be 180 degrees, that is, a vector pointing south,

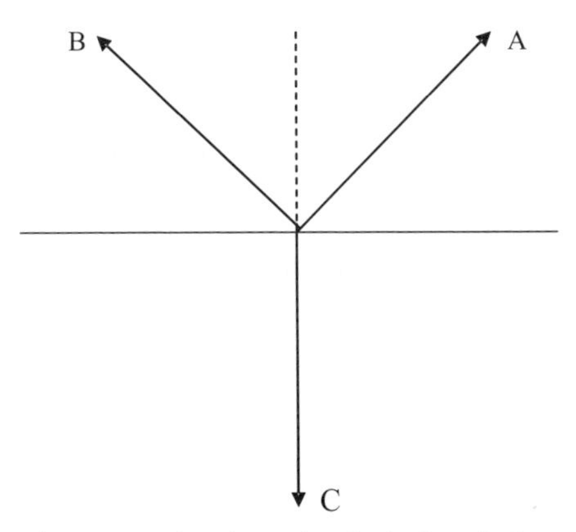

Figure 7.5 Inappropriateness of using classical descriptive statistics for angular measures.

as shown by vector C. But graphically, given the directions of vectors A and B, the average direction should be 0 degree, that is, a vector pointing north. Therefore, using classical statistical measures is inappropriate in this situation. Because the concept of the arithmetic mean of the two angles cannot reflect the average direction, other measures such as variance cannot be defined meaningfully. To analyze angular information, we have to rely on **directional statistics** specifically designed to analyze vectors.

Note that the attributes of length, orientation, and direction associated with linear features as discussed here are applicable to both linear features that are not connected and linear features that are topologically connected, forming a network. Linear features in a network may have additional attributes. They will be discussed in later sections. In the next section, we will discuss how different attributes of spatially noncontiguous linear features can be used to support various analytical tools specifically designed for these features.

7.3.2 Directional Mean

The concept of a **directional mean** is similar to the concept of an average in classical statistics. The directional mean should be able to show the general direction of a set of vectors. Because a directional mean is concerned with the direction but not the length of vectors, vectors can be simplified to 1 unit in length (unit vectors). Figure 7.6a shows three vectors of unit length originating from O. Each vector shows a direction in reference to the origin. According to coordinate geometry, a vector is the distance between the origin and a point shown by the horizontal (along the x-axis) and vertical (also the y-axis) movements. Therefore, if the origin has a coordinate of (0, 0), a vector can be (1, 2), indicating that the point is located 1 unit east and 2 units north of the origin.

Also, according to coordinate geometry, the directional mean of a set of vectors is defined as the direction of the resultant vector, while the resultant vector is derived by "adding" all the vectors together. For instance, we have two vectors: (1, 2) and (3, 4). By adding the corresponding x and y "movements" of these two vectors, such as (1 + 3, 2 + 4), we can obtain the resultant vector (4, 6). Graphically, adding any two vectors together means appending the beginning point of the second vector to the end point of the first vector. Figure 7.6b shows how the three unit vectors in Figure 7.6a are added together graphically. The result of the vector addition is the resultant vector, OR. The directional mean of the three vectors, in this case θ_R, is the direction of the resultant vector OR. The direction of the resultant vector, θ_R, can be derived from the following trigonometric relation:

$$Tan\,\theta_R = \frac{o_y}{o_x}, \tag{7.3}$$

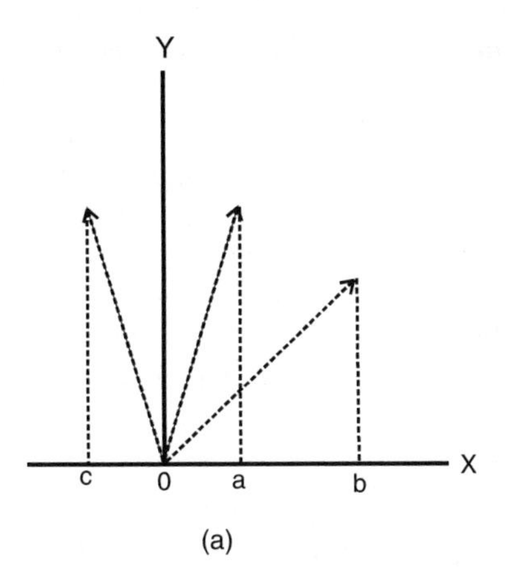

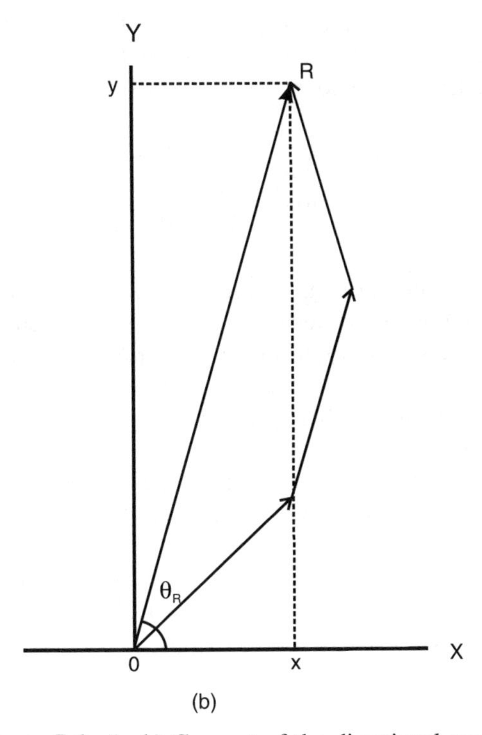

Figure 7.6 (a, b) Concept of the directional mean.

where

o_y is the sum of the heights of the three vectors and
o_x is the total horizontal extent of the three vectors.

Because all three vectors are unit vectors, the height of a vector (in the y-axis) is basically the sine function of the angle of the vector, and the horizontal extent (the x-axis, as shown in Figure 7.6a) of a vector is the cosine function of the angle of the vector. Therefore, the sum of the sine of all angles will give us the height of the resultant vector and the sum of the cosine of all angles will give us the horizontal extent of the resultant vector. If the three vectors are identified as a, b, and c, and their corresponding angles are θ_a, θ_b, and θ_c, then

$$\tan\theta_R = \frac{\sin\theta_a + \sin\theta_b + \sin\theta_c}{\cos\theta_a + \cos\theta_b + \cos\theta_c}. \tag{7.4}$$

To generalize, assuming that there are n vectors v, and the angle of the vector v from the x-axis is θ_V, the resultant vector, OR, forms an angle, θ_R, counter-clockwise from the x-axis. Because each vector is of unit length, we have

$$\tan\theta_R = \frac{\Sigma \sin\theta_v}{\Sigma \cos\theta_v}, \tag{7.5}$$

where tan θ_R is the tangent of the resultant vector. In order to find the directional mean (the angle of the resultant vector θ_R), the inverse of $\tan\theta_R$ (or $\tan^{-1}\theta_R$, the arc tangent) has to be taken from Equation 7.5.

The directional mean is the average direction of a set of vectors. The idea of vector addition, as shown in Figures 7.6a and 7.6b, utilizes the fact that the resultant vector from vector addition shows the general direction of the set of vectors. If all the vectors have similar directions, after all vectors are appended to each other, the resultant vector will be pointing at a direction somewhere among this group of vectors.

If two vectors have different directions, such as a vector of 45 degrees and a vector of 135 degrees, the resultant vector will be 90 degrees. As discussed before, just taking the average angle of those vectors may not be appropriate. If *all* vectors under consideration are smaller or larger than 180 degrees, then the angle based upon their arithmetic *average* will yield the correct answer. However, if some angles are smaller than 180 degrees and others are larger, then the averaging method will be incorrect.

The result derived from the aforementioned equation for the directional mean must be adjusted to accommodate specific situations in different quadrants according to Table 7.3. The table shows the trigonometric results for angles in each of the four quadrants. Because of these specific situations, the results from directional mean calculation should be adjusted accordingly.

1. If the numerator and the denominator are both positive in Equation 7.5, no adjustment of the resultant angle is needed (the resultant angle lies in the first quadrant).
2. If the numerator (sine) is positive and the denominator (cosine) is negative (second quadrant), then the directional mean should be $180 - |\theta_R|$. This is because the arc tangent from Equation 7.5 will be a negative angle, implying an angle clockwise from 180 degrees. But we would like the directional mean measured counterclockwise from 0 degree.
3. If both the numerator (sine) and the denominator (cosine) are negative (third quadrant), then the directional mean should be $180 + \theta_R$. The positive angle from Equation 7.5 is measured counterclockwise from 180 degrees. Therefore, we must add the angle to 180 in order to obtain the directional mean measured counterclockwise from 0 degree.
4. If the numerator (sine) is negative and the denominator (cosine) is positive (fourth quadrant), then the directional mean should be $360 - |\theta_R|$. This is because the arc tangent from Equation 7.5 will be a negative angle, implying an angle clockwise from 0 degree. But we would like the directional mean measured counterclockwise from 0 degree.

The directional mean can easily be computed if the angles of the linear feature are available. The first step is to compute the sine of each angle; the second step is to compute the cosine of each angle. Treating the sine and cosine results as ordinary numbers, we can calculate their sums to form the ratio, as in Equation 7.5, in order to derive the tangent of the resultant vector. Taking the inverse of the tangent (arc tangent) on the ratio gives us the directional mean. Table 7.4 shows the major steps in deriving the directional mean of all 422 fault lines described in Figure 7.1a, but only a subset of the

TABLE 7.3 Adjusting the Directional Mean

2nd		1st	
sin	+ve	sin	+ve
cos	−ve	cos	+ve
sin	−ve	sin	−ve
cos	−ve	cos	+ve
3rd		4th	

TABLE 7.4 Directional Mean Calculation of Fault Lines

Number	LENGTH	E_DIR	sin_angle	cos_angle
1	1284.751	68.56	0.930801	0.3655267
2	118.453	74.58	0.964003	0.2658926
3	255.001	94.09	0.997453	−0.071323
4	6034.442	49.34	0.758589	0.651569
5	22.963	42.24	0.672238	0.7403355
6	130.147	43.6	0.68962	0.7241719
7	482.716	95.06	0.996118	−0.088025
8	1297.714	89.74	0.99999	0.0045378
9	97.953	90.62	0.999941	−0.010821
10	3400.414	57.28	0.841322	0.540534
.	.	.	.	.
.	.	.	.	.
.	.	.	.	.
414	2976.265	58.1	0.848972	0.5284383
415	1936.928	74.49	0.963584	0.2674066
416	10528	74.19	0.96217	0.2724482
417	2569.188	50.47	0.771291	0.6364822
418	167.585	70.12	0.940407	0.3400513
419	365.049	75.77	0.969317	0.245815
420	662.586	75.61	0.968627	0.2485208
421	274.119	77.3	0.975535	0.2198462
422	107.418	76.54	0.972533	0.2327665
		Sum =	301.914	172.76549

records is shown in the table for illustrative purpose. First, the angles of the fault lines were extracted. Note that directions of fault lines are not really meaningful, thus orientations from the east should be used. Therefore, all angles are no larger than 180 degrees, and the directional mean is based upon the orientations rather than directions of the fault lines from the east.

The angles shown in Table 7.4 are orientations of fault lines from east. Orientation in reference to east instead of north was used to conform to the trigonometry tradition that angles are measured counterclockwise from the x-axis. Regarding the procedure used to calculate the directional mean in Table 7.4, the sines of the angles were derived first, followed by their cosines. In some spreadsheet packages, especially in Microsoft Excel, trigonometric functions (sine, cosine, tangent, and their inverses) expect the input angles to be expressed in radians instead of degrees. Therefore, the degrees of the angles need to be converted into radians before the sine and cosine functions are used. There are 422 fault lines in the data set. The sine values for all angles were summed, and so were the cosine values. The ratio of the two sums is 1.7536, which is the tangent of the directional mean. Thus, taking the inverse tangent of 1.7536 gives us 60.31 degrees counterclockwise from the east. This is the angle of the resultant vector or the directional mean.

7.3.3 Circular Variance

Analogous to classical descriptive statistics, directional mean reflects the "central tendency" of a set of directions. As in many cases, the central tendency, however, does not reflect the distributional characteristics of the observations. For instance, the directional mean of the two vectors, 45 degrees and 135 degrees, is 90 degrees. Apparently, the directional mean is not efficient in representing two vectors pointing in very different directions. A more extreme case is one in which two vectors point in opposite directions. In that case, the directional mean will point at the direction in between, but it fails to reflect the opposite directions of the vectors and provides no information on the effectiveness of the directional mean in representing all observations or vectors. A measure showing the variation or dispersion among the observations is therefore necessary to supplement the directional mean statistic. In directional statistics, this measure is known as the **circular variance.** It shows the angular variability of the set of vectors.

If vectors with very similar directions are added (appended) together, the straight-line distance of all of these unit vectors or the length of the resultant vector will be relatively long, and its length will be close to n if there are n unit vectors. On the other hand, if the vectors are in opposite or very different directions, the resultant vector will be relatively short compared to n for n vectors. This is because the resultant vector will be the straight line connecting the first point and the last point of the set of zigzag lines or even lines with opposite directions. In Figure 7.6, if all three vectors have very similar directions or the same direction forming a linear trend, the resultant vector (OR) should be on the top of the three vectors when these vectors are appended to each other. But in our example, the three vectors do vary in direction. Therefore, the graphic addition of the three vectors deviates from the straight-line resultant vector, OR, substantially. Therefore, OR is shorter than the actual length of the three appended vectors (which is 3). Then the length of the resultant vector can be used as a statistic to reflect the variability of the set of vectors. Using the same notations as above, the length of the resultant vector is

$$OR = \sqrt{\left(\sum \sin\theta_v\right)^2 + \left(\sum \cos\theta_v\right)^2}. \tag{7.6}$$

Circular variance, S_v, is

$$S_v = 1 - \left(\frac{OR}{n}\right), \tag{7.7}$$

where n is the number of vectors.

S_v ranges from 0 to 1. When $S_v = 0$, this means that OR should equal n; therefore, all vectors have the same direction or no circular variability. When $S_v = 1$, OR is of length 0 when all vectors are in opposite directions, and the resultant vector is a point at the origin. Note that the concept of circular variance is the same as the concept used when we compared the straight-line length of a chain with the length of the entire chain—**sinuosity.**

Using the fault lines as an example again,

$$\sum \sin\theta_v = 301.914$$

$$\sum \cos\theta_v = 172.765$$

$$\left(\sum \sin\theta_v\right)^2 = 91{,}152.063$$

$$\left(\sum \cos\theta_v\right)^2 = 29{,}847.745$$

$$OR = \sqrt{(91{,}152.063) + (29{,}847.745)} = 347.85$$

$$S_y = 1 - \left(\frac{347.85}{422}\right) = 0.1757.$$

Because circular variance ranges from 0 to 1, this magnitude of circular variance is rather small, indicating that most of the fault lines have similar directions.

In Figure 7.7, two sets of wind vectors represent the wind data captured on a day in July and a day in December at selected stations in the San Francisco Bay area. From the arrows added to the vectors, it is apparent that the wind had a general northeast or northwest direction in July and a general westward direction in December. However, it is difficult to tell which month has a more consistent wind direction than the other month. Therefore, we can compute the directional mean for each set of vectors to show the general direction. In addition, we can calculate the circular variance to show the consistency of the wind direction. Table 7.5 shows the results.

The results confirm the visual estimation that the general wind direction is northeast in July and west in December. More precisely, the circular variance indicates that the December wind directions from selected stations are more consistent than those in July.

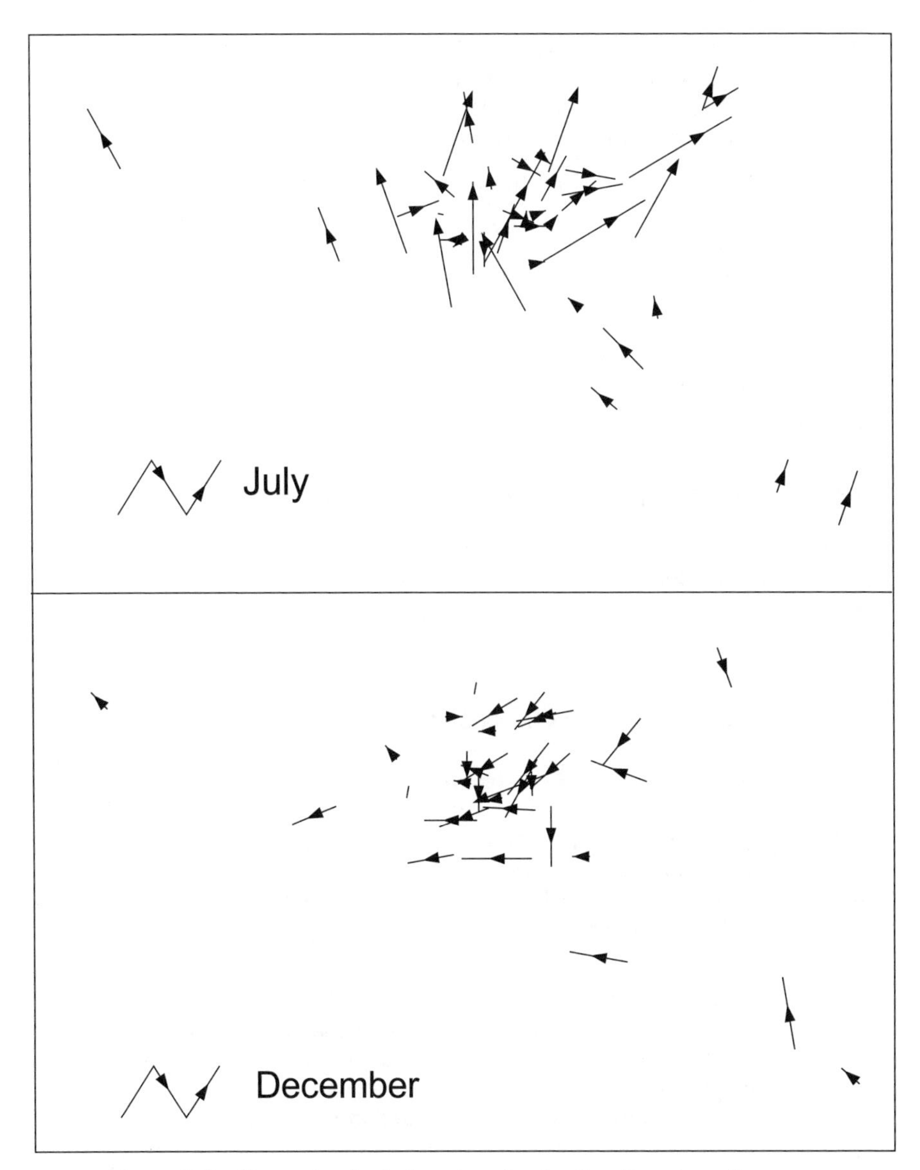

Figure 7.7 Two sets of wind vectors for the San Francisco Bay area.

Directional statistics have been used in geological and earth science applications (e.g., Swan and Sandilands, 1995). For instance, variations of directional statistics discussed in this chapter have been used to study surface wind directions (Klink, 1998) and other phenomena with directional or orientation characteristics (Jelinski et al., 2002). The analyses of length and directional information discussed above are applicable to linear features that are not necessarily spatially contiguous (vectors and line segments) and that

TABLE 7.5 Directional Mean and Circular Variance of Wind Directions in July and December in the San Francisco Bay Area.

	Directional Mean	Circular Variance
July	69.56	0.45
December	199.40	0.34

are connected—a network. After presenting an example of using directional statistics in the next section, we will discuss techniques specifically used in analyzing a network.

7.3.4 ArcView Example: Directional Statistics

In this example, the functions for calculating directional means and circular variance will be demonstrated. Only a simple step is needed to run these functions already incorporated into the ArcView Ch7.avx extension.

ArcView Example 7.2: Directional Statistics

Step 1 Data and extension

If not already opened and loaded, as in ArcView Example 7.1, refer to Step 1 of the ArcView example to load the Ch7.avx extension and to add the two data themes: `winds.shp` and `states.shp`. Similarly, Ch7.avx can be loaded from the **File/Extensions** menu item.

Step 2 Calculate directional statistics

Directional statistics include directional mean and circular variance. For winds.shp, we will use the Ch7 extension to calculate these statistics.

From the **Ch7** menu, choose **Directional Statistics.**

In the Directional Statistics dialog box:

- Select `winds.shp` as the line theme for analysis.
- Make sure that the check box beside **Use Selected Records Only** is left unchecked (since we have not selected any records prior to this step).
- Click the **Yes** option to indicate that the flow directions of lines are important.
- Click the **Calculate Directional Mean and Circular Variance** button to proceed.

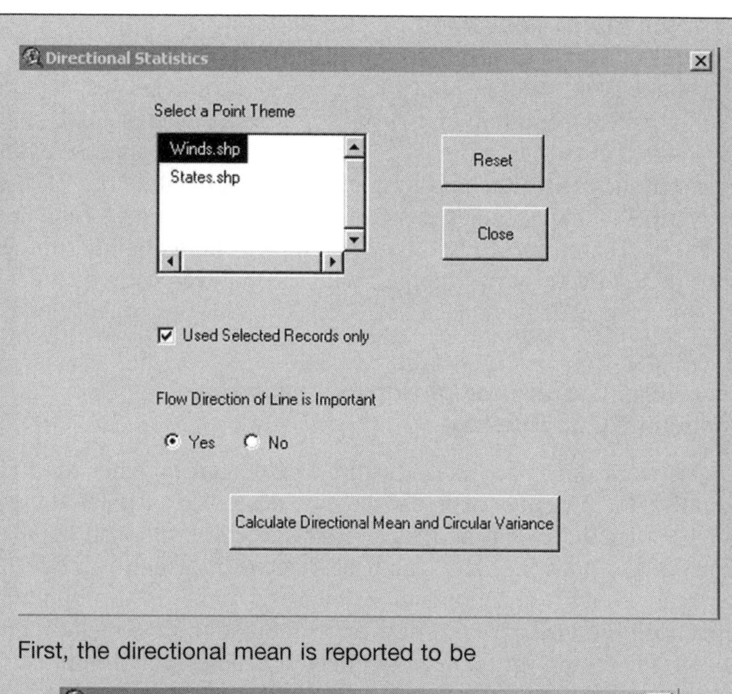

First, the directional mean is reported to be

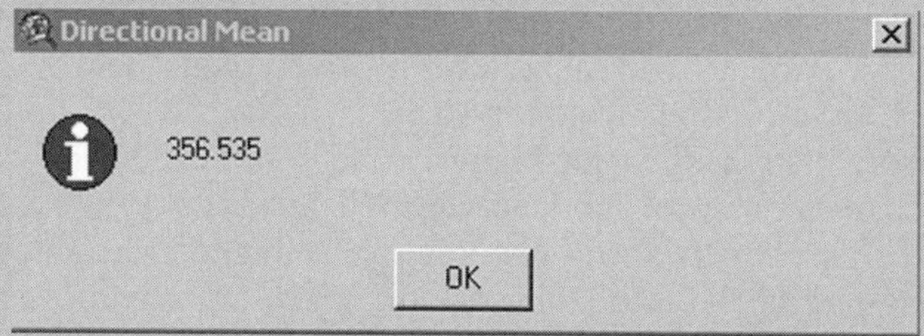

Then circular variance is reported as

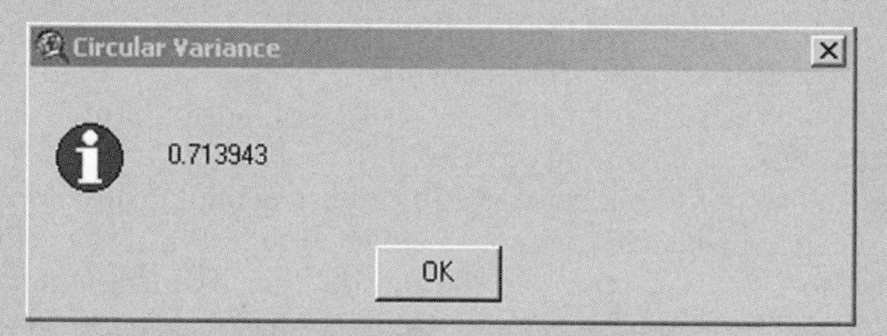

Now we know that the wind trajectories have an average direction of 356.535, almost a northerly direction, with a circular variance of 0.7139 degree, a relatively large degree of variability. In other words, the average direction of wind trajectories as reflected by the directional mean of 356.535 degrees is not a very representative direction for all the wind trajectories because the wind trajectories vary substantially.

Step 3 End

7.4 NETWORK ANALYSIS

To execute a full suite of network analytical techniques, a special computation environment is required. ArcView 3.X adopted in this book cannot support this kind of environment without extensive programming development effort. However, the Environmental Systems Research Institute (ESRI) has developed the Network Analyst, an extension for ArcView 3.X and for the newer ArcGIS packages. Therefore, we will focus on topics and techniques in network analysis not significantly overlapping with those covered by commercially available software such as the Network Analyst extensions of ArcView and ArcGIS.

7.4.1 Spatial Attribute of Network Features: Connectivity or Topology

In a network database, linear features are linked together topologically. Apparently, the attributes of linear features described in previous sections, such as length/spatial extent and orientation/direction, are also applicable to segments of the network. The length of a network, which can be defined as the aggregated length of individual segments or links, is an important feature in supporting the analysis of a network. Orientation or direction, depending on the specific nature of the network, may also be essential in understanding the geographic setting of the network. For instance, the flow directions of tributaries of a river network should be relatively consistent if the watershed is not very large or is elongated in shape.

The orientation of major highways, to some extent, may partially reflect some characteristics of the landscape. Clearly, at a local scale, the direction of a local street network is important in planning the traffic pattern. All the concepts and analyses of linear features that we have discussed so far are applicable in analyzing segments or links in a network. An additional aspect of a network, however, is how different segments are linked together and how these segments are related to each other. This is part of the general topic of the topological structure of a network.

The most essential topological aspect or attribute of a network is how different links or edges are connected to each other. This is sometimes known as the *connectivity* of a network. To capture quantitatively how different links are joined together, a common method is to use a matrix, the *connectivity matrix,* to store and represent the information. We can assume that the network has n links (or edges) and each link has a unique identifier (ID). In a conventional connectivity matrix, the labels of the columns are the IDs of the links in the network. The labels of the rows in the connectivity matrix are also the IDs of the links in the network. This matrix is a square matrix. That is, the number of rows equals the number of columns. A cell in the matrix captures the topological relationship between the two links denoted by the IDs in the corresponding row and column labels. If the two links are directly joined to each other, the cell will have a value of 1. Otherwise, the value will be 0. To a large degree, the connectivity matrix for a network is the same as the distance matrix discussed in Chapter 6 in point pattern analysis and the spatial weights matrix to be discussed in Chapter 8 for measuring spatial autocorrelation.

In most situations, we assume that a link or an edge is not connected to itself, and therefore all diagonal elements in the matrix are 0's. Because each cell carries a value of either 0 or 1, this type of matrix is also called a *binary matrix.* Because the relationship between any pair of edges is symmetrical, the connectivity matrix is also symmetrical. That is, if Link A is connected to Link B, then Link B is also connected to Link A. In other words, the upper right triangle of a matrix separated by the major diagonal is a mirror image of the lower left triangle.

Figure 7.8 shows the railroads centering at Washington, D.C., covering part of the suburban counties. Each railroad segment is labeled with a unique

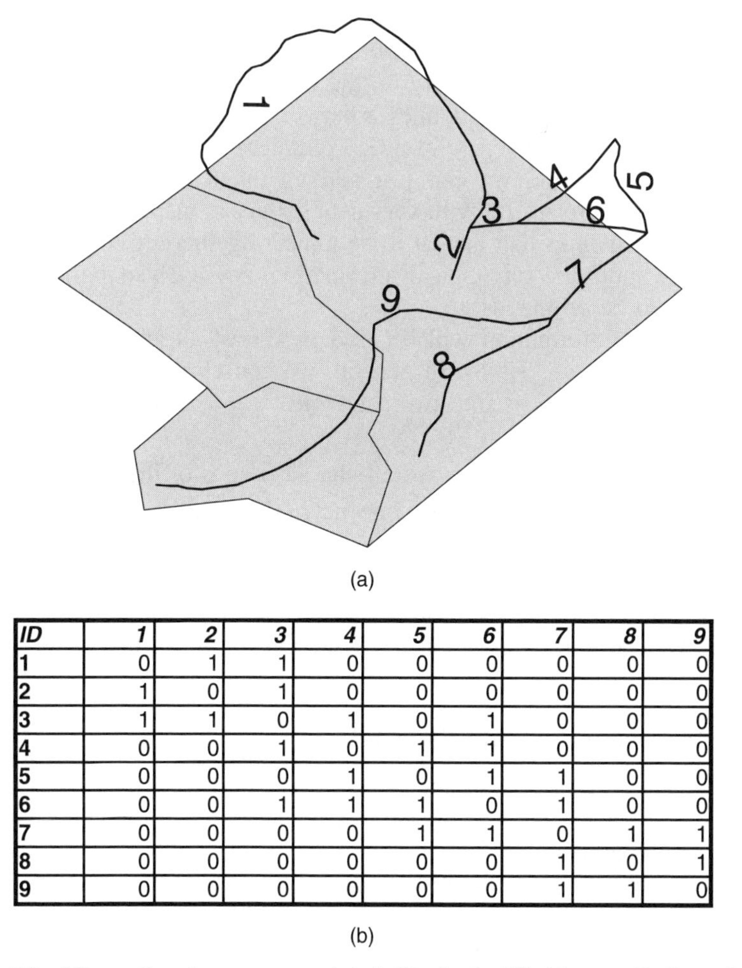

(a)

ID	1	2	3	4	5	6	7	8	9
1	0	1	1	0	0	0	0	0	0
2	1	0	1	0	0	0	0	0	0
3	1	1	0	1	0	1	0	0	0
4	0	0	1	0	1	1	0	0	0
5	0	0	0	1	0	1	1	0	0
6	0	0	1	1	1	0	1	0	0
7	0	0	0	0	1	1	0	1	1
8	0	0	0	0	0	0	1	0	1
9	0	0	0	0	0	0	1	1	0

(b)

Figure 7.8 Nine railroad segments and their IDs in the Washington, D.C., region (a) and the associated connectivity matrix (b).

ID. There are nine railroad segments in this system. To capture the topological relationship among them, a connectivity matrix was created. Figure 7.8 also includes the connectivity matrix of the railroad network. The row and column labels refer to the IDs of the links or segments in the network. In the matrix, all diagonal elements are 0's, indicating that each network segment is not connected to itself. This matrix is also known as the *C matrix* in the spatial analysis literature.

In Figure 7.8, railroad segment 7 is in the middle section of the network. It is connected to segments 5 and 6 to the northeast and to segments 8 and 9 to the southwest. Therefore, in the seventh row of the matrix, which indicates which railroad segments segment 7 is connected to, 1's are found in the cells for segments 5, 6, 8, and 9. The same principle is applied to other rows in the matrix. But if we focus on the seventh column, the information captured there is the same as that in the seventh row. Therefore, this is a symmetrical matrix.

Connectivity of links is the most fundamental attribute of a network. Any analysis of the network must rely on the connectivity attribute. In fact, one may argue that connectivity defines a network. For instance, a set of linear features is shown on the screen of a GIS package. If we want to find the total length of that network, we can just add up the lengths of the individual segments on the screen. But with very detailed visual inspection, we may find that two of those links that appear to be joined together are in fact separated. Therefore, without the topological information, any analysis performed on a network could be erroneous.

Connectivity information will be used in almost all analyses involving a network (Taaffe et al., 1996). In general, we can classify different types of tools for network analysis into two categories:

- The type that assesses the overall characteristics of the entire network.
- The type that describes how one network segment is related to other segments or the entire network system.

The first type will be discussed in the next section on assessing connectivity. The second type will be covered in the section following that one.

7.4.2 Assessing Connectivity Level

Before starting, we have to define several conventional terms used in network analysis. In network analysis, a segment of linear feature is called a *link* or an *edge*. An edge is defined by the two vertices or nodes at either end of the edge in the network. The number of edges and vertices in a network is often used to derive statistics reflecting the structural characteristics of the network.

So far, we have discussed only how the connectivity information is captured and represented in the connectivity matrix. We have not analyzed the level of connectivity of a network. For a given set of vertices, there are various ways that the vertices can be connected by edges to form different networks. When the number of vertices is fixed, networks with more edges are better connected. That is, by adding more edges to the network, we increase the level of connectivity in the network. On the other hand, there is a minimum number of edges required to connect all the vertices to form a network. If the number of edges is below this minimum number, not all the vertices will be connected together to form a network.

If v denotes the number of vertices and e denotes the number of edges in the network, then the minimum number of edges, e_{min}, that is required to link all these vertices to form a network is

$$e_{min} = v - 1. \tag{7.8}$$

If a network has a number of edges equal to e_{min}, then the network is a *minimally connected network.* In a minimally connected network, if any one edge is removed from the system, the network will be broken up into two unconnected subnetworks. In the simple network shown in Figure 7.8, there are nine vertices and nine edges. Since the number of vertices is nine, a minimally connected network should have eight (9 − 1) edges. However, there are nine edges. Therefore, the network in Figure 7.8 is not a minimally connected network. If edge 6 is removed, then this is a minimally connected network. If either edge 3 or edge 7 is removed, the network will be broken up into two systems. If the network in Figure 7.8 represents a critical infrastructure, the above analysis can shed light on the protection of the infrastructure.

Similarly, given a fixed number of vertices, there is a maximum number of edges one can construct to link all the vertices together. However, edges in this *maximally connected network* do not cross or intersect with each other—a requirement of the planar graph topology. If any two edges cross each other, a vertex has to be formed. The maximum possible number of edges, e_{max}, in the network is

$$e_{max} = 3(v - 2). \tag{7.9}$$

Therefore, the network in Figure 7.8 can have up to 21 edges (3 × (9 − 2)) without crossing each other for the given number of vertices. But the actual network has only nine edges. Thus, we can form a **Gamma Index** (γ), which is defined as the ratio of the actual number of edges to the maximum possible number of edges in the network. Formally, the index can be computed as

$$\gamma = \frac{e}{e_{max}}. \tag{7.10}$$

The Gamma Index for the network in Figure 7.8 is $\gamma = 9/21 = 0.4286$. The Gamma Index is most useful in comparing different networks to differentiate their levels of connectivity. To illustrate this aspect of the index, another network selected from the railroad network in the Cleveland, Ohio, region is shown in Figure 7.9. It is not easy to tell by visual inspection which of the two railroad networks is better connected than the other. Intuitively, it seems that the Cleveland railroad system is better connected than the D.C. system. To objectively assess their levels of connectivity, we can also calculate the Gamma Index for the Cleveland railroad network and use this index as the basis of comparison. In the Cleveland network, there are 8 vertices and 12 edges. Therefore, the Gamma Index is 0.6667, clearly higher than 0.4286, the Gamma Index for the D.C. network. Thus, the objective evaluations of these two networks based upon the Gamma Index values do support the initial conclusion from visual inspection that the Cleveland railroad network has a higher level of connectivity than the D.C. railroad network.

Another characteristic of connectivity is the number of circuits that a network can support. A *circuit* is defined as a closed loop along a network. In a circuit, the beginning node of the loop is also the ending node. The existence of a circuit in a network implies that there are multiple paths to connect any two vertices in the network. If the network is a road network, the presence of a circuit means that travelers can use alternate routes to commute between any two locations in the network. A minimally connected network, as discussed above, barely links all vertices together, and does not contain any circuit. In the context of critical infrastructure protection, a minimally connected network is vulnerable, but the presence of a circuit will increase reliability. If an additional edge is added to a minimally connected network, a circuit may emerge. Therefore, the number of circuits can be obtained by subtracting the number of edges required for a minimally connected network from the actual number of edges. That is, $e - (v - 1)$ or $e - v + 1$.

In the DC railroad network shown in Figure 7.8, with $e = 9$ and $v = 9$, the number of circuit = 1. This is formed by edges 4 and 5. For a given number of vertices, the maximum possible number of circuit is $2v - 5$. Therefore, with these two measures of the circuit, we can derive a ratio of the

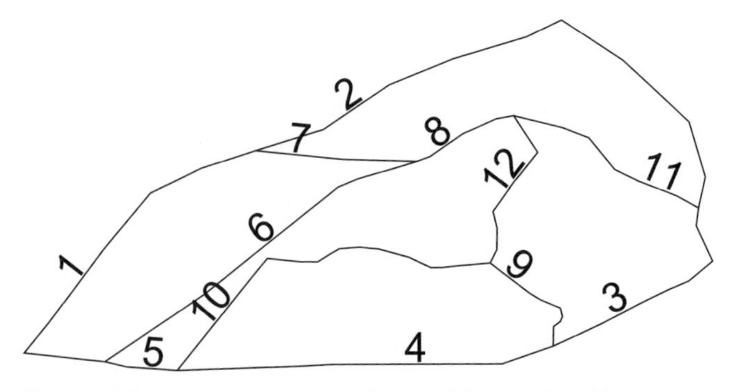

Figure 7.9 Railroad network in the Cleveland, Ohio, region.

number of actual circuits to the number of maximum possible circuits. This ratio, which is sometimes known as the **Alpha Index,** is defined as

$$\alpha = \frac{e - v + 1}{2v - 5}. \tag{7.11}$$

Using the Alpha Index, we can compare the two networks in Figures 7.8 and 7.9. In the D.C. railroad network, if not for the circuit formed by edges 4 and 5, this would be a minimally connected network. Thus the Alpha Index of this network is only 0.077, a relatively low value. The Cleveland network is better connected, with several circuits identifiable through visual inspection. The Alpha Index turns out to be 0.455, much higher than that of the D.C. network.

7.4.3 Evaluating Accessibility

The two indices discussed above assess the characteristics of the entire network. Individual elements in the network, either the vertices or edges, have different characteristics or relationships in regard to other elements in the network. Therefore, it is necessary to analyze the characteristics of each of these elements. In general, the procedure is to evaluate the **accessibility** of individual elements in regard to the entire network or to different elements of the network.

Conventionally, network analysis literature uses vertices or nodes as the basic elements of analysis (Taaffe et al., 1996). Using links or edges as the basic elements is also appropriate because the accessibility results are now applied to the edges but not the vertices. In the rest of this chapter, our discussion will focus on the accessibility of individual edges.

A simple analysis of edge accessibility in a network consists of identifying how many edges a given edge is directly connected to. If an edge is well connected, then many edges will be linked to it directly. For instance, in the D.C. railroad network, Link 3 is well connected, because it has four *direct links* to other edges, and so is Link 7. Therefore, Links 3 and 7 have the same level of accessibility. On the other hand, Links 1, 2, 8, and 9 share the same level of accessibility based on their number of direct links.

To a large degree, this information is already captured in the connectivity matrix C such as the one shown in Figure 7.8. The element of the binary connectivity matrix is 1 if the corresponding edges are directly linked and 0 otherwise. Therefore, in order to find out how many direct links are established for an edge, we just need to add all the 1's across the column for the corresponding rows. Table 7.6 shows the connectivity matrix of the D.C. railroad network with the sum of direct links for each edge. The results confirm our conclusions from earlier visual inspection based upon the map in Figure 7.8 that Links 3 and 7 have the highest accessibility, as both have four

TABLE 7.6 Connectivity of the DC Railroad Network and the Number of Direct Links

ID	1	2	3	4	5	6	7	8	9	Sum Links
1	0	1	1	0	0	0	0	0	0	2
2	1	0	1	0	0	0	0	0	0	2
3	1	1	0	1	0	1	0	0	0	4
4	0	0	1	0	1	1	0	0	0	3
5	0	0	0	1	0	1	1	0	0	3
6	0	0	1	1	1	0	1	0	0	4
7	0	0	0	0	1	1	0	1	1	4
8	0	0	0	0	0	0	1	0	1	2
9	0	0	0	0	0	0	1	1	0	2

direct links. However, our visual inspection missed another link, Link 6, which has the same level of accessibility and also four direct links. Therefore, even for simple networks like the D.C. railroad network, the quantitative approach based upon the connectivity matrix is much more reliable than the visual inspection.

Note that the number of direct links to an edge reflects only one aspect of accessibility. An edge with a large number of direct links may not be very accessible if that edge is located at one end of the network. On the other hand, an edge with a small number of direct links may still be quite accessible, depending on its relative (topological) location in the network. For instance, an edge may have a small number of direct links but may be centrally located in the network, so that it is still well connected to other edges. In our network examples, we do not have such a case to illustrate this situation. But if there was an edge connected to the middle section of Link 7 in Figure 7.8 in the D.C. railroads case, it would have only two direct links (to either side of Link 7), but it is centrally located. Thus, it is easy to travel from this hypothetical edge to all other edges in the network as compared to Link 1, which is located at the end of the network. To capture the relative location of a given link in the network, we need to find out how many links are required to reach the farthest part of the network from that given link.

In a simple network like the one in D.C., it is quite easy to visually derive the number of links or steps required to reach the *farthest edges*. Using the edges identified for Link 3, one of the most accessible edges according to the number of direct links, three links or steps are required to reach the farthest edge (Link 8 or 9). Link 9, one of those with the smallest number of direct links requires four steps to reach the farthest edges (Link 1 or 2).

Quantitatively, we can rely on the C matrix to derive all this information. Recall that the element of the matrix $c_{ij} = 1$ indicates that links i and j are connected. We can perform matrix algebra to obtain C^2, squaring the matrix. In this matrix, the cell values indicate the number of two-step links between the corresponding i-j links. Similarly, C^3 gives us the number of three-step links between the corresponding *i-j* links. By combining information from these matrices, we can obtain the total number of direct and indirect links.

Table 7.7 shows the number of direct links and the number of steps or links required to reach the farthest part of the network for each link. The two measures do not necessarily yield the same conclusion because they evaluate different aspects of the edges with regard to other edges. Note that 1 plus the highest number of links or steps required to reach the farthest edge of the entire network is also known as the *diameter* of the network. In Figure 7.8, the largest number of steps required to reach the farthest link of the network is four, which is defined by Links 1, 2, 8, and 9. Then the diameter of the network is 5, that is, five edges are required to link the farthest parts of the network together.

An analysis based solely on the number of direct links is not the most reliable. An edge may not have many direct links, but because of its location, it can be reached from other edges indirectly and thus may be reasonably accessible. Therefore, in addition to direct links, our analysis should consider indirect links. But obviously, a direct link is better than an indirect link; therefore, the two types of links should be treated differently. Indirect links also have different degrees of connectivity. For instance, in the DC railroad network in Figure 7.8, Links 1 and 6 are indirectly linked. Because Link 3 is between them, this indirect link is inferior to a direct link. Links 1 and 7 are also indirectly linked. But between them are Links 3 and 6. Therefore, two steps are required to join Links 1 and 7, one step more than the indirect link between Links 1 and 6. Thus, using the number of steps between links can indicate the quality of the link. If more steps are required, the indirect link is less desirable.

Based upon this idea, we can derive the number of direct and indirect links that will require all edges to be joined together for each given edge, but the indirect links will be weighted by the number of steps or the degree of in-

TABLE 7.7 Number of Direct Links, Steps Required to Reach the Farthest Part of the Network, and Total Number of Direct and Indirect Links for Each Edge

ID	Direct Links	Steps	All Links
1	2	4	20
2	2	4	20
3	4	3	14
4	3	3	15
5	3	3	15
6	4	2	12
7	4	3	14
8	2	4	20
9	2	4	20

directness. Apparently, the larger the number of these total links, the less accessible is an edge. A small number of total links implies that the edge is well connected directly and indirectly to other edges. Table 7.7 also includes a column of all links (direct and indirect weighted by steps) for each edge. Links 1, 2, 8, and 9 require the highest number (20) of direct and indirect links in order to connect the entire network; thus, they may be regarded as the most inaccessible. This conclusion confirms the result of counting the number of direct links and the maximum number of steps required to reach the farthest part of the network. However, the total number of direct and indirect links is better in differentiating Link 3 from Link 6. The two edges have the same number of direct links, but in term of the total number of links, Link 6 is more efficient than Link 3.

7.4.4 ArcView Example: Network Analysis

For network analysis, the winds.shp data used in previous two ArcView examples will not be appropriate because the wind trajectories do not connect to form networks. For this example, we will use the road network data in roads.shp.

ArcView Exmaple 7.3: Network Analysis

Step 1 Data and Ch7 extension

As in previous examples, the Ch7 extension can be loaded by using the **File/Extensions** menu item.

For the data theme, this example will use C:\Temp\Data\ch7_data\roads.shp themes.

- Once roads.shp is added with the **Add Theme** button, use the **Open Theme Table** button to open the attribute table of roads.shp to explore its attribute data.
- Among other attribute fields, there is an ID field, which will be useful in the next steps.

Step 2 Network analysis

As discussed in Section 7.4, a connectivity matrix is useful in assessing how each edge in a network can be accessed through other edges. For statistics, both Alpha and Gamma indices are useful in providing detailed information on the accessibility of the network. Using the Ch7 extension, the number of links and the number of steps in links can be added to the attribute table.

From the **Ch7** menu, select the **Network Analysis** menu item.

In the Network Analysis dialog box, click to select `roads.shp` as the line theme for analysis.

- Check the box beside **Construct a Connectivity Matrix.**
- For IDs, click to select the `ID` field.
- Check the box beside **Alpha and Gamma** Indices.
- Check the box beside **Add # of Links to Table.**
- Check the box besides **Add # of Steps and All Links to Table.**
- Click the **Proceed** button to continue.

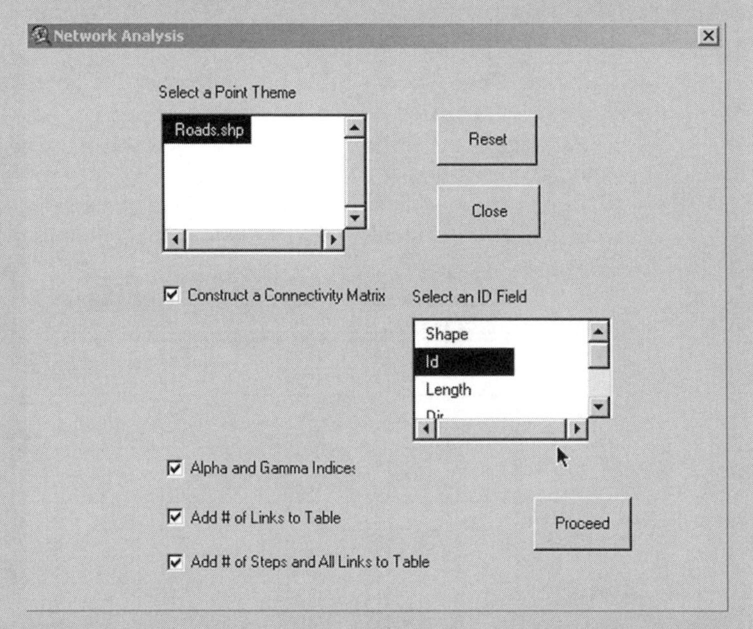

When prompted, navigate to a folder and give a name to store the connectivity matrix.

- Click **OK** to acknowledge completion of the calculation of the connectivity matrix.

For the network in roads.shp, we have the following results:

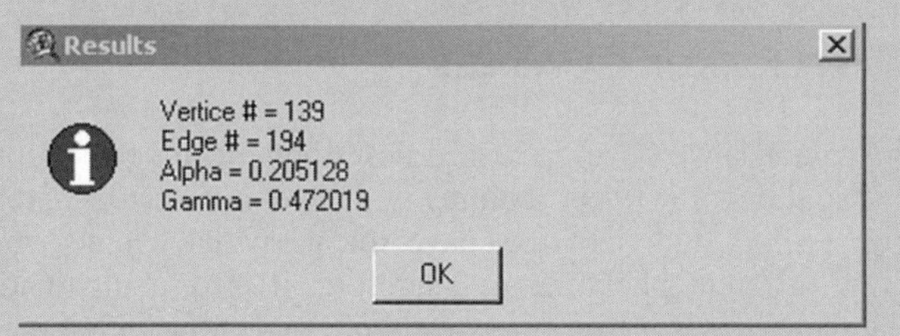

At this point, the program continues to perform the tasks of adding links and steps to the attribute table. It normally takes some time for this task to be completed, as it involves many iterations in the computation processes. Users need to be patient, especially when adding the number of links and steps in a dense network.

Users also need to be aware that the functions of adding total step numbers and link numbers to the attribute table will *not* perform correctly if some line segments in the polyline theme are not connected. That is, the data set consists of more than one network.

When the task is completed:

- Use the **Open Theme Table** button to open the attribute table of roads.shp.
- At the right end of the table, several new fields have been added, including `Link_`, `Step_`, `Alllinks`, `Link#`, and `Step#`. By stretching the Table document on its side, users will be able to see the full labels of the added columns.

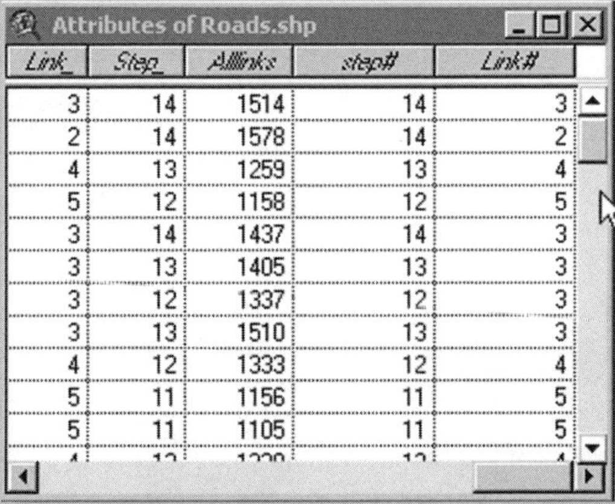

Attributes of Roads.shp

Link_	Step_	Alllinks	step#	Link#
3	14	1514	14	3
2	14	1578	14	2
4	13	1259	13	4
5	12	1158	12	5
3	14	1437	14	3
3	13	1405	13	3
3	12	1337	12	3
3	13	1510	13	3
4	12	1333	12	4
5	11	1156	11	5
5	11	1105	11	5

Step 3 End

7.5 APPLICATION EXAMPLES

In this chapter, we have discussed descriptive techniques appropriate to analyze spatial data for linear features in the form of vectors and networks. Even though vectors and networks have linear spatial extents, they are used to describe different phenomena. Therefore, it is difficult to identify a single spatial dataset as an appropriate example to demonstrate the analyses of both vector and network data. As a result, we will use several spatial datasets accessible through the National Atlas of the United States website (*http://www.nationalatlas.gov/*). For detailed information on this effort, review the website. In short, this is an effort led by several U.S. federal government agencies in providing national coverage of different themes. Note that all data sets include metadata providing quality data users. Through the website, users can create interactive maps or download map data from the data warehouse.

7.5.1 Length Attribute Analysis of Linear Features

The first spatial dataset for the application example in this section is the Breeding Bird Survey Routes of North America from the National Atlas (*http://nationalatlas.gov/birdm.html*). The dataset includes routes for the annual bird survey in the continental United States. Routes for the survey are represented as polyline segments. Refer to the website for detailed information on the dataset. As a reference, we also use the data describing the Continental Divide—the Rocky Mountains from the National Atlas (*http://nationalatlas.gov/condivm.html*). Figure 7.10 shows the Continental Divide with all the routes for the breeding bird survey.

In general, the topography along the Continental Divide and its vicinity is quite rugged. Therefore, it may be logical to assume that those survey routes closer to the Continental Divide are likely more meandering or geometrically more complex than routes farther away so that we can get around the rugged terrain. To test this hypothesis, we selected those routes within 100 miles of the Divide and those routes more than 100 miles but less than 200 miles from

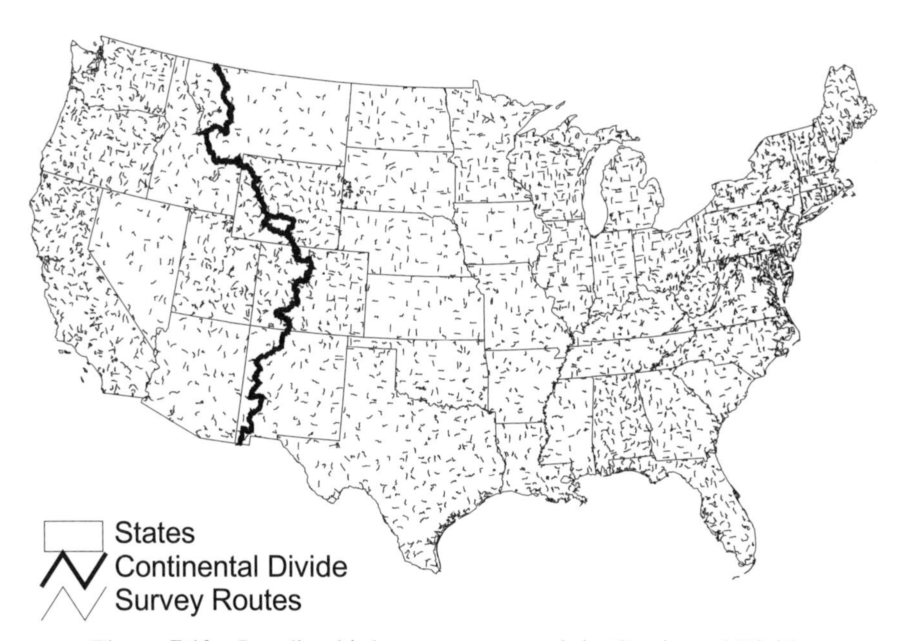

Figure 7.10 Breeding bird survey routes and the Continental Divide.

the Divide. To create subsets of these datasets, we applied a series of buffering procedures and spatial selection procedures to derive routes subsets, as depicted in Figure 7.11.

The two sets of survey routes were projected to UTM coordinates so that the distances calculated for the routes were less distorted than those calculated using the longitude-latitude coordinate system. Then, using the ArcView Ch7.avx extension, we added the true length and the straight-line length of all routes to the attribute tables. After the two types of lengths were added, we created a new field—sinuosity—in the attribute table. Using the Field/ Calculate function under Table, we calculated the sinuosity of each route by dividing the true length (TLength) by the straight-line length (SLength) of each route. Part of the results for the survey routes within the 100-mile buffer are shown in Table 7.8.

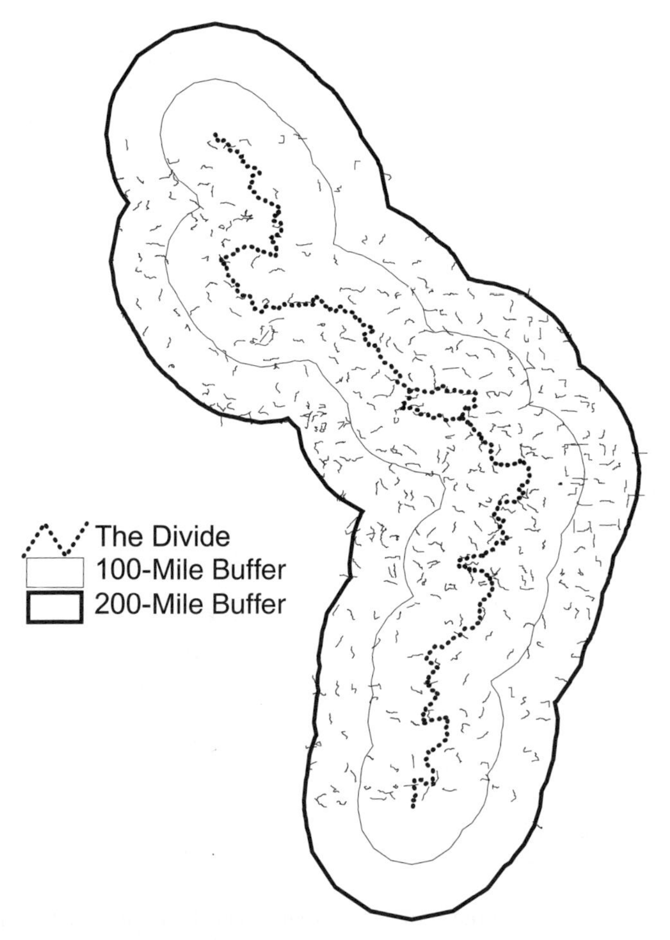

Figure 7.11 Breeding bird survey routes at 100 miles and between 100 and 200 miles from the Continental Divide.

TABLE 7.8 Part of the Attribute Table with True Length and Straight-Line Length with Sinuosity for Routes within 100 Miles of the Continental Divide

TLength	SLength	Sinuosity
34730.986166	23871.052004	1.4549
40784.159021	36206.173056	1.1264
40098.912908	34025.639116	1.1785
51705.503644	33294.594127	1.5530
41477.113442	25909.845919	1.6008
38804.011837	27612.350305	1.4053
39210.498371	26080.856814	1.5034
35548.165992	17125.907446	2.0757
41030.783075	26039.971732	1.5757
15613.064663	7373.045642	2.1176
20682.181283	11297.242020	1.8307
1956.020993	1817.758664	1.0761
38216.759167	16582.132531	2.3047
40370.324372	32827.255711	1.2298
38536.584150	22191.949259	1.7365
51211.101276	40850.888992	1.2536
40314.563984	38382.428655	1.0503
39198.858948	32278.656833	1.2144
44615.092065	25242.634661	1.7674
16526.136622	12253.899850	1.3486

In Table 7.9, the summary statistics of the new attribute sinuosity for both sets of routes are reported.

From these descriptive statistics of sinuosity, it is quite obvious that, based upon the means, the routes closer to the Continental Divide have a slightly higher degree of geometric complexity than those farther away. Still, we would like to confirm if the difference in the mean is due to sampling error (by chance) or to some systematic processes by performing the difference-of-means test described in Chapter 4. We will leave this task to the readers.

7.5.2 Application Example for Directional Statistics

For an application example of this subsection, we will use another dataset from the National Atlas. This dataset is the Historical North Atlantic Hurri-

TABLE 7.9 Descriptive Statistics of Sinuosity for Survey Routes within the 100-Mile and 100–200-Mile Buffers

Subsets	Count	Mean	Max	Min	Range	Std Dev
100 miles	278	1.5275	4.9232	1.0081	3.9151	0.4485
100–200 miles	211	1.4775	3.3766	1.0011	2.3755	0.3852

cane Tracks—Major Storms with Landfall in the United States (*http://nationalatlas.gov/hurmjrm.html*). Refer to the website for detailed information on this dataset and metadata.

The dataset includes tracks of hurricanes all the way back to 1851. Hurricanes were not named until 1950. Therefore, we selected only hurricanes with names for this analysis. Also, the data depict the track of each hurricane with a set of arcs or segments. To facilitate the analysis in this example, we merged all segments belonging to the same hurricane into one segment. Eventually, the dataset had 28 named hurricanes from 1950 to 1999 that resulted in landfalls in the United States. These tracks are shown in Figure 7.12.

Though the data on hurricane tracks are most appropriate for analysis using directional statistics, we can add arrows to the tracks for a better visual display of the directions. In addition, we can analyze the curvature of the tracks. Thus, the true lengths and straight-line lengths were added to the attribute table, and sinuosity was calculated for each hurricane track. Table 7.10 shows the results sorted according to sinuosity.

Based on the sinuosity displayed in this table, we can tell which hurricanes hit the United States via convoluted routes and which ones followed more direct routes. We can also examine the angles from which these hurricanes

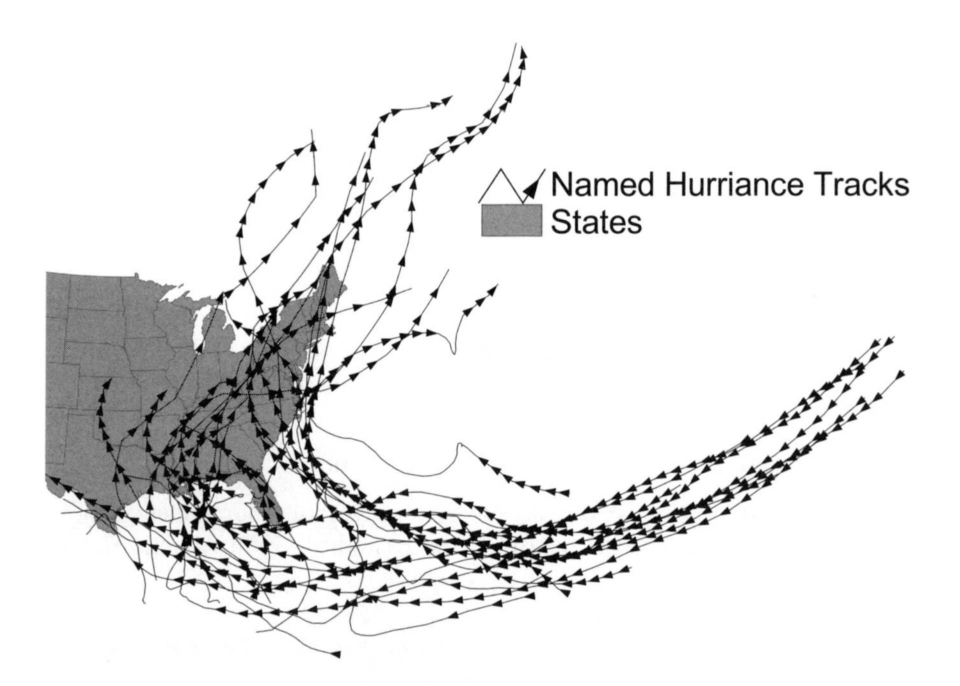

Figure 7.12 Tracks of 28 named hurricanes from 1950 to 1999 that resulted in landfalls in the United States.

TABLE 7.10 Hurricane Tracks Sorted by Sinuosity

Name	Year	TLength	SLength	Sinuosity
Gracie	1959	55.791764	23.307837	2.39369119
Emily	1993	61.887787	26.338302	2.34972577
Gloria	1985	117.812966	55.626752	2.1179192
Hilda	1964	26.509872	12.971512	2.04369945
Ione	1955	98.418947	52.425949	1.87729452
Edna	1954	78.168721	47.369190	1.65020177
Donna	1960	102.595092	62.773866	1.63435994
Hazel	1954	73.469371	46.773603	1.57074431
Carla	1961	75.135432	48.992346	1.53361572
Hugo	1989	101.957232	66.778118	1.52680601
Camille	1969	56.764118	37.354700	1.51959775
Elena	1985	37.512745	24.821384	1.51130755
Eloise	1975	54.010157	36.360034	1.48542647
Frederic	1979	98.265519	66.661020	1.47410764
Easy	1950	26.218546	17.873267	1.46691402
Betsy	1965	68.379088	46.692902	1.46444288
Bret	1999	14.736625	10.720383	1.37463606
Opal	1995	36.534413	27.600878	1.32366851
King	1950	21.372573	16.211868	1.31832883
Fran	1996	85.443867	65.422234	1.30603713
Alicia	1983	18.007059	14.317681	1.25767986
Andrew	1992	71.528196	59.421800	1.20373661
Carmen	1974	51.315354	43.394541	1.18253017
Connie	1955	65.248416	56.637078	1.15204418
Audrey	1957	30.475491	27.401478	1.1121842
Beulah	1967	49.865280	44.879458	1.11109363
Celia	1970	29.115292	28.119740	1.03540403
Allen	1980	76.368912	75.065913	1.01735807

hit. Using the ArcView Ch7.avx extension, we computed the directional mean for all hurricane tracks. It was 63.3122 degrees with a circular variance of 0.6901. Though the average direction of all these hurricanes was 63 degrees, their directions were not very consistent, as the circular variance was relatively high. Note that the calculation of the directional mean is based on the two terminal locations of the tracks. It does not account for the curvilinear nature of most tracks. Therefore, the directional mean information must be interpreted together with the sinuosity information.

7.5.3 Application Example for Network Analysis

For the application example for network analysis, we will use a dataset modified from the shapefile of major U.S. interstate highways included in the dissemination of ArcView GIS by ESRI. The data theme is Roads_rt.shp with 147 line segments. These line segments represent major interstate highways and some state highways. For the network analysis tools we discussed in the chapter, the data must conform to the properties of a planar graph, which stipulates that when two lines cross each other, a vertex will be created. Unfortunately, the highway data, like many transportation data, do not meet the planar graphic requirement. Also, the essence of network analysis is to analyze topology.

The data for network analysis must maintain a high level of topological integrity. Many polyline data, including Roads_rt.shp, do not have arcs or segments topologically built in. Vertices were not snapped together, and the network was not formed. Thus, heavy manual editing operations were needed to modify the data structure and ensure that the segments were topologically connected in order to perform network analysis.

One of the major problems in the Appalachian region of the eastern United States is difficulty in promoting economic development. Some studies claim that the relatively low level of economic development there is due to the low level of accessibility partly created by the rugged terrain. The two Appalachian states partly affected by this problem are Kentucky and Tennessee. In this example, we will evaluate the highway accessibility in these two states using network analysis tools described in this chapter. Figure 7.13 shows the major interstate highway networks of the two neighboring states.

The purpose of this example is not to evaluate the accessibility of individual locations, but to develop an overall assessment of the entire state; therefore, only the Alpha and Gamma indices were calculated. Table 7.11 summarizes the values of the two indices. The Gamma Index shows how the observed network approaches a fully connected network given the number of vertices. The Gamma Index values of both states are relatively low, much lower than that of the railroad network in the Cleveland area discussed earlier, though Kentucky has a value slightly higher than that for Tennessee. The Alpha Index values of both states are also very low. The index reflects how well the area is connected through network circuits. Visually, we can see that the highway network in Tennessee developed in the linear fashion across the state so that circuits are only found in large cities. On the other hand, several highways in Kentucky form at least two circuits. Still, the interstate highways layout in Tennessee is slightly more efficient, mainly due to the circuits in the two large cities (Alpha index).

One can see that there is a highway at the western end of Kentucky not connected to the highway network of the rest of the state. A similar situation is found in the northwest corner of Tennessee. When these two highway segments were removed from the analysis, the results were slightly better, as reflected by the index values in parentheses in Table 7.11.

We can also use the functions in the Ch7.avx extension to evaluate the accessibility of each network segment in regard to the entire network. How-

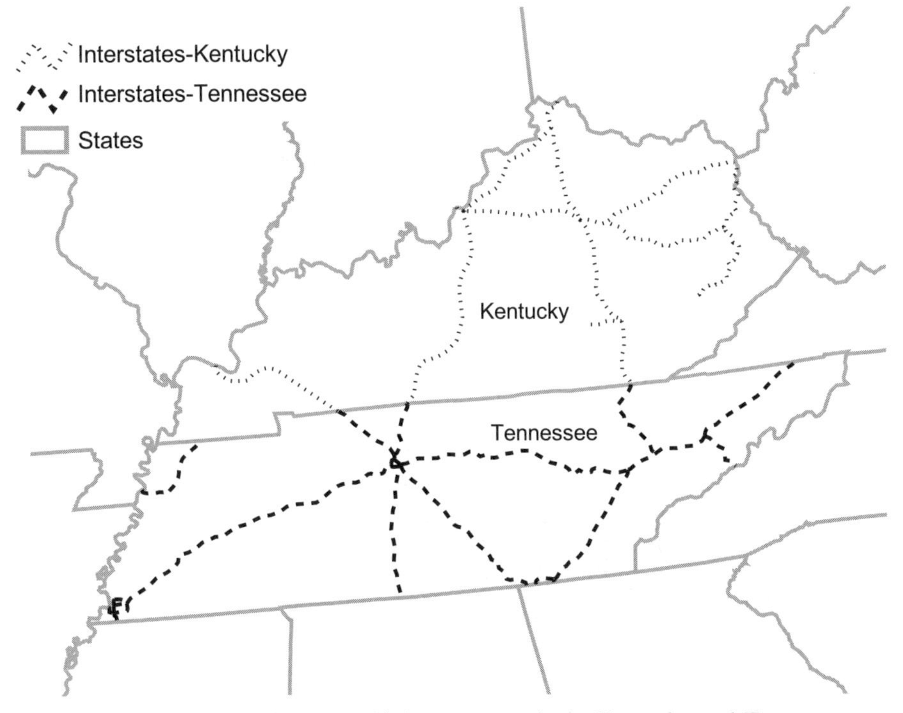

Figure 7.13 Major interstate highway networks in Kentucky and Tennessee.

ever, as mentioned above, there is at least one highway segment not connected to the rest in both states. In other words, there is more than one interstate highway network in each state. The functions of adding numbers of steps and total links will not work correctly if there is more than one network. Therefore, the isolated highway segment in western Kentucky and the one in northwestern Tennessee were first removed from the data. Then, under the Network Analysis menu item for the Ch7 menu, we checked the boxes for "Add # of Links to Table" and "Add # of Steps and All Links to Table." The results are shown in Table 7.12.

In general, the smaller the number of steps, the more centrally located is the segment in the network. Similarly, the smaller the numbers of links, the

TABLE 7.11 Alpha and Gamma Index Values for the Major Interstate Highway Systems in Kentucky and Tennessee

States/Indices	No. of Vertices	No. of Edges	Alpha	Gamma
Kentucky	17	17	0.0345 (0.08)	0.3778 (0.4103)
Tennessee	29	30	0.0377 (0.0638)	0.3704 (0.3889)

TABLE 7.12 Numbers of Steps and Total Links for the Selected Interstate Highways in Kentucky and Tennessee

Kentucky Routes	LINK_	STEP_	ALLLINKS	Tennessee Routes	LINK_	STEP_	ALLLINKS
Interstate 64	5	4	29	Interstate 24 (1)	2	7	112
Interstate 65	2	6	50	Interstate 24	2	8	130
Interstate 71	4	5	37	Interstate 40	5	7	88
Interstate 75	5	4	31	Interstate 55	3	9	128
Kentucky State Hwy 402	4	4	35	Interstate 65	2	7	112
U.S. Hwy 23	3	5	46	Interstate 75	2	8	126
Kentucky State Hwy 80	1	6	60	Interstate 81	2	9	149
Kentucky State Hwy 80	2	5	44	Interstate 240	3	9	128
U.S. Hwy 23 (Ext)	3	5	46	Interstate 265	5	6	79
Interstate 75 (3)	2	5	44	Interstate 240 (1)	5	8	106
Interstate 75 (2)	5	4	31	Interstate 240 (2)	6	8	105
Interstate 75 (1)	2	5	41	Interstate 240 (3)	6	9	125
Interstate 64 (1)	2	6	50	Interstate 55 (1)	3	9	128
Interstate 64 (2)	4	5	37	Interstate 40 (2)	5	8	106
Interstate 64 (3)	5	3	28	Interstate 40 (6)	4	6	88
Interstate 64 (4)	3	4	37	Interstate 40 (7)	4	7	103
				Interstate 40 (8)	4	8	124
				Interstate 40 (9)	2	9	149
				Interstate 24 (2)	3	6	89
				Interstate 24 (3)	5	5	76
				Interstate 24 (4)	4	5	79
				Interstate 24 (5)	2	6	96
				Interstate 65 (1)	2	6	95
				Interstate 65 (2)	5	6	87
				Interstate 75 (1)	2	8	130
				Interstate 75 (2)	4	7	105
				Interstate 40 (4)	4	5	79
				Interstate 65 (3)	4	6	82

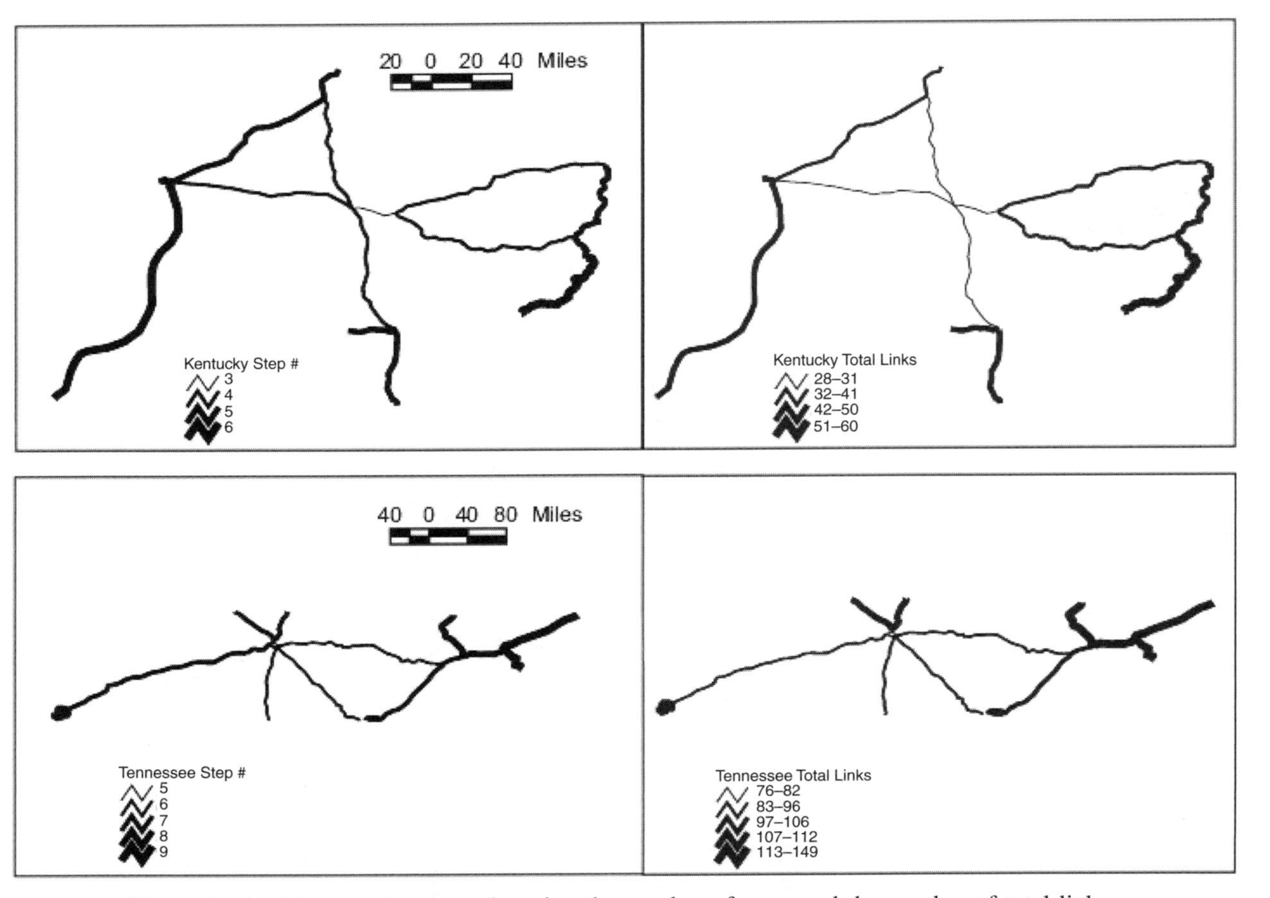

Figure 7.14 Mapping the networks using the number of steps and the number of total links.

more accessible is the network segment, though the two measures do not capture the same network characteristics. As the results of these two measures are recorded in the attribute data, it is logical to examine the results using maps. By using these measures as variables, we can create maps (Figure 7.14) to display the level of accessibility of every network segment.

Besides technical details in implementing the spatial analytical techniques, one must be aware of the quality of the data. Many network data are good for cartographic display and general map making, but the structure may not be appropriate for spatial analysis. The planar graph properties are important in implementing network analysis. Data not meeting these requirements may need to go through certain data management processes before being used in the analysis. Users must examine the data carefully, especially to ensure that vertices are snapped together.

REFERENCES

DeMers, M. N. 2000. *Fundamentals of Geographic Information Systems.* New York: John Wiley & Sons.

Jelinski, D. E., C. C. Krueger and D. A. Duffus. 2002. Geostatistical analyses of interactions between killer whales (*Orcinus orca*) and recreational whale-watching boats. *Applied Geography* 22: 393–411.

Klink, K. 1998. Complementary use of scalar, directional, and vector statistics with an application to surface winds. *The Professional Geographer* 50(1): 3–13.

Swan, A. R. H. and M. Sandilands. 1995. *Introduction to Geological Data Analysis.* Oxford: Blackwell.

Taaffe, E. J., H. L. Gauthier and M. E. O'Kelly. 1996. *Geography of Transportation.* Upper Saddle River, NJ: Prentice-Hall.

EXERCISES

We will use several datasets for the exercise in this chapter. The first dataset is the simple one depicting the Continental Divide available from the National Atlas (*http://nationalatlas.gov/condivm.html*). In this data set, there are 11 line segments. The Ch7.avx extension was used to add the true length and straight-line length to the attribute data. The shapefile was originally in longitude-latitude. But because the analysis involves distance calculation, it was desirable to project the shapefile into coordinate systems with less distortion in distance. Therefore, the Continental Divide shapefile was reprojected into UTM coordinates using zone 13 north. Because the data are now in UTM, the lengths are meters. The direction of each segment from the north was also added. The data are shown in the following table.

Segment ID	TLength	SLength	N_DIR
1	535251.55	355360.03	345.88
2	471872.54	193577.13	355.53
3	372506.49	213475.41	111.42
4	271133.61	191752.76	337.68
5	319488.11	148590.97	95.64
6	247941.05	141011.70	108.04
7	414733.34	242922.93	329.20
8	357103.66	211202.86	23.52
9	509185.33	281568.68	19.82
10	697742.67	400913.42	3.31
11	169098.67	86397.92	7.06

1. Use descriptive statistics to describe the statistical characteristics of the true lengths and straight-line lengths of all segments.

2. Calculate the sinuosity of each segment of the Continental Divide. Which segment has the most complicated geometric structure and why?

3. The direction of each segment from the north is also provided. What is the average direction of all the segments forming the Divide? To what extent do the segments vary in direction?

In Figure 7.15, a set of cities (vertices) is connected by interstate highways, which are numbered.

4. Create a connectivity matrix of the network depicted in Figure 7.15.

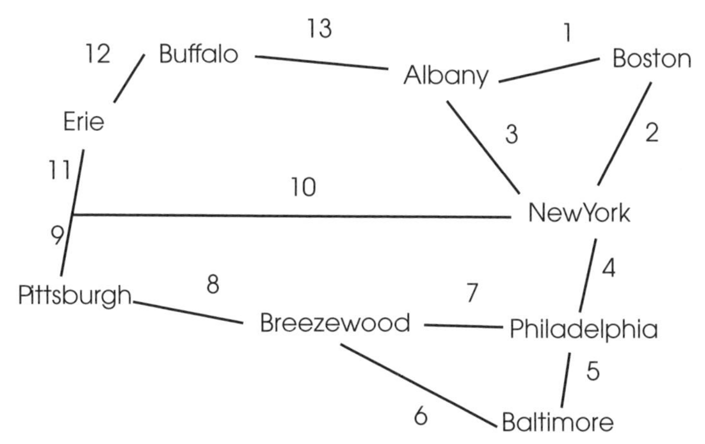

Figure 7.15 Schematic representation of several major interstate highways in several northeastern states.

5. What are the numbers of vertices and edges in the network?
6. What is the minimally connected network in this highway system? What is the diameter of the network?
7. Access the overall connectivity of the highway network in Figure 7.15 in terms of the presence of a circuit and the level of connectivity using the Alpha Index and Gamma Index.
8. If segments 3 and 10 were removed from the system, how will that affect the overall connectivity of the system?
9. How could the connectivity of the system be improved if one was allowed to add edges or links to it? Which addition would be most effective and least costly?
10. Which segment has
    - **a.** the largest number of direct links?
    - **b.** the smallest number of direct links?
11. Enter the connectivity matrix into an Excel spreadsheet. Using the matrix multiplication function in Excel, calculate the square and cube of the connectivity matrix. What do these matrices indicate?
12. Which segment in the entire network is most accessible? Which one is least accessible? Can you use the power (square, cube) of the connectivity matrix to help determine the overall accessibility of each network segment?

CHAPTER 8

POLYGON PATTERN ANALYZERS

8.1 INTRODUCTION

Human settlements occur mostly in places where resources are available to support the population and where the climate is reasonably decent. Changes of animal habitats often happen when certain events occur to alter the environments. Geographic objects form various spatial patterns according to their characteristics or human behaviors. Changes in spatial patterns over time reflect the dynamic spatial processes dictated by the underlying environmental or cultural-human factors.

The spatial patterns of geographic objects and phenomena are often the result of physical or cultural-human processes taking place on the surface of the Earth. *Spatial pattern* is a static concept since a pattern only shows how geographic objects distribute at one given time. *Spatial process* is a dynamic concept because it depicts and explains how the distribution of geographic objects comes to exist and may change over time. For any given geographic phenomenon, it is important to study both its spatial patterns and its associated spatial processes.

Describing the spatial patterns comprehensively allows us to understand how the geographic phenomenon distributes and how its spatial patterns can be compared with those of other phenomena. The ability to identify and describe spatial processes enables us to determine the underlying environmental and/or cultural-human factors that influence the patterns. If the changes are desirable, we should find ways to promote them in the future. If they were undesirable, we should find ways to correct the problems.

Spatial statistics provide useful tools for describing and analyzing how various geographic objects (or events) occur or change across the study area and over time. These statistics are formulated specifically to take into account the location attributes of the geographic objects. We can use spatial statistics to describe the spatial patterns formed by geographic objects in one study area so that we can compare them with patterns found in other study areas. For spatial processes associated with these patterns, we can use spatial statistics to describe their forms, to detect changes, and to analyze how spatial patterns change over time.

In earlier chapters, we demonstrated the use of descriptive spatial statistics or centrographic measures to measure central tendency and dispersion among point-based geographic objects. Such objects are appropriate to represent object features with no or little spatial extent. However, many spatial features, such as land parcels, census units, and wetlands, have significant spatial extents. They are represented most appropriately by polygons in spatial models. In this chapter, we will discuss the use of spatial statistics to describe and measure spatial patterns formed by geographic objects that are associated with *areas,* or *polygons.* Some of the spatial statistics, such as Moran's I Index, have been discussed and applied in Chapter 6 for analyzing point features. In this chapter, we will use spatial statistics to describe spatial patterns manifested in a set of polygons according to their characteristics or attributes. In addition, we will examine how the spatial patterns can be compared to other patterns by using statistical tests.

8.2 SPATIAL RELATIONSHIPS

When studying a spatial pattern, we may want to compare the observed pattern with a theoretical pattern, a pattern whose statistical properties and associated spatial process or processes are known. By comparing the observed pattern with a theoretical pattern, we will know if the theory or process describing the theoretical pattern holds for the observed pattern. Alternatively, we may want to classify spatial patterns into existing categories of known patterns. If the spatial patterns we study correspond to a particular theoretical pattern, we will be able to apply known properties of the theoretical pattern to interpret the observed spatial pattern. Moreover, if the spatial pattern closely resembles a known pattern, we will be able to borrow from our experience with and knowledge of the pattern for further study. In either case, it is necessary to establish a category of spatial patterns.

A spatial pattern can generally be categorized as *clustered, dispersed,* or *random.* In Figure 8.1, these three categories are shown in hypothetical patterns in the seven counties of northeast Ohio (Cuyahoga, Summit, Portage, Lake, Geauga, Trumbull, and Ashtabula). In Case 1, darker shades representing a certain characteristic associated with the counties appear to cluster on the western side of the seven-county area. On the eastern side, the lighter shade (the absence of that specific characteristic) prevails among the remaining counties. Perhaps the darker shade indicates the growth of the urban

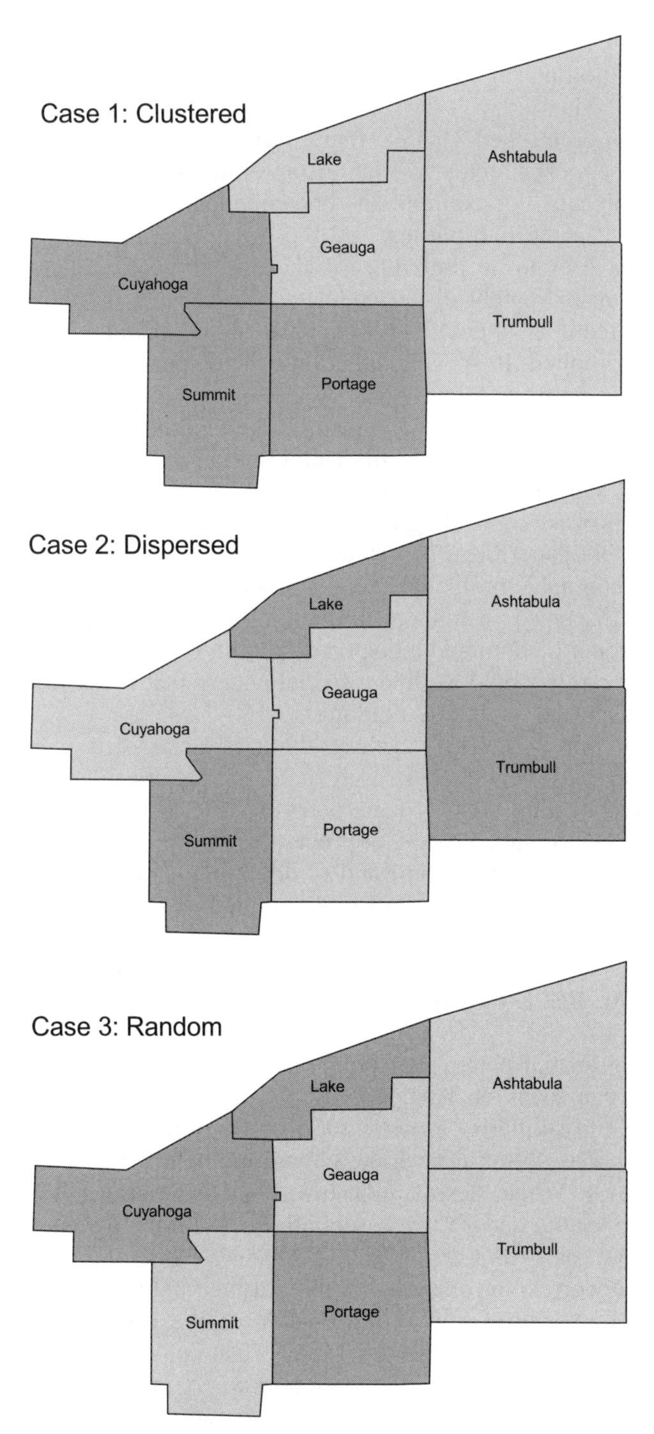

Figure 8.1 Types of patterns: clustered, dispersed, and random.

population in recent years, whereas the lighter shade indicates the loss of urban population during the same period.

In Case 2, counties with darker shades appear to be spaced evenly and to be apart from each other. This is often referred to as a dispersed pattern. The pattern may suggest a repulsive relationship of the phenomenon among nearby geographic objects. For example, the phenomenon could be the political preferences between the two parties, and these neighboring Ohio counties may happen to have opposite preferences. The checkerboard pattern is a typical example of an extremely dispersed pattern because it has black and white cells distributed in an alternate fashion. A repulsive spatial relationship among neighbors is implied. In specific situations, the dispersed pattern may also be perceived as a uniform pattern.

The spatial pattern in Case 3 appears to be neither clustered or dispersed. It may be close to what we typically call a random pattern. If a spatial pattern is random, it suggests that there may be no particular systematic structure or mechanism/process controlling the way these polygons are distributed. Thus, no recognizable pattern can be detected.

In most geographic problems, there is no definite indication that the pattern is clustered, dispersed, or random. Rather, most real-world patterns are somewhere between a random and a dispersed pattern or a random and a clustered pattern. Very rarely would we find a spatial pattern that is extremely clustered, extremely dispersed, or purely random.

Since real-world patterns can rarely be easily classified as random, clustered, or dispersed, how close is a given spatial pattern to any of these three categories of patterns? Also, if the pattern is close, is the resemblance due to chance or to a systematic process or processes? In this chapter, we will discuss how to measure a spatial pattern and to determine its closeness to a random pattern so that the pattern can be further studied.

8.3 SPATIAL DEPENDENCY

In classifying spatial patterns of polygons as either clustered, dispersed, or random, we can focus on how various polygons are arranged spatially. We can measure the similarity or dissimilarity of any pair of neighboring polygons, or polygons within a given neighborhood, which may be defined by a certain distance. When these similarities and dissimilarities are summarized for the entire spatial pattern, we essentially measure the magnitude of spatial autocorrelation, or spatial dependency (Odland, 1988).

Spatial autocorrelation means that the attribute values are correlated over space or that the correlation is attributable to the geographic ordering or locations of the objects or features. Many situations exhibit some degree of spatial autocorrelation. When the agricultural production levels of all the farms in a region are evaluated, not all farms are found to produce at the exact same level. Even though the local climate may be the same for all of these farms, the soil conditions or water supplies within the region may vary.

Still, neighboring farms within the region do share similar soil and moisture conditions. Thus, these farms probably have similar levels of production.

In regard to population density at the county level, it is likely that those counties located in or close to a large central city tend to have higher population densities than those located far away from the central city. In general, locations closer to cities are more desirable for many economic activities and therefore attract more people, creating areas of higher population density. While the most desirable locations in the cities are crowded, neighboring locations are still preferable—the so-called spillover effect. Figure 8.2 is a three-dimensional map of population densities by county for the northeastern United States from Virginia to Maine. Apparently, high-density counties cluster around several urban centers, while most of the outlying areas are of low density.

The basic property of spatially autocorrelated data is that the values are not randomly related. Instead, they are spatially related to each other, even though they may be related in different ways. Referring back to the three

Figure 8.2 Population density by county in the northeast United States, 2000 Census.

cases in Figure 8.1, Case 1 has a positive spatial autocorrelation, with adjacent or nearby polygons having similar shades (values). Case 2 has a dispersed pattern that has a negative spatial autocorrelation, with changes of shade often occurring in adjacent polygons. Case 3 appears to be close to a random pattern in which very little or no spatial autocorrelation can be recognized.

In addition to its type or nature, spatial autocorrelation can be measured by its strength. Strong spatial autocorrelation means that the attribute values of adjacent geographic objects are strongly related (either positively or negatively). If attribute values of adjacent geographic objects do not appear to have a clear order or a relationship, the distribution is said to have a weak spatial autocorrelation or a random pattern. Figure 8.3 shows five different structures with darker and lighter shades across census tracts in Washington, D.C. The darker shade in a census can be interpreted as the presence of a certain characteristic. Patterns toward the left end are examples of positive spatial autocorrelation, while patterns toward the right end are examples of negative spatial autocorrelation. In the middle is a pattern of weak spatial autocorrelation that is close to a random pattern.

Spatial autocorrelation can be a valuable tool in exploring how spatial patterns in a set of points or polygons change over time. Results of this type of analysis often lead to a better understanding of how spatial patterns have changed from the past to the present. Once the change is described, explained and understood, it is possible to estimate how the spatial pattern will evolve from the present to the future. Therefore, useful conclusions can be drawn to advance our understanding of the underlying factors that drive the changes in spatial patterns.

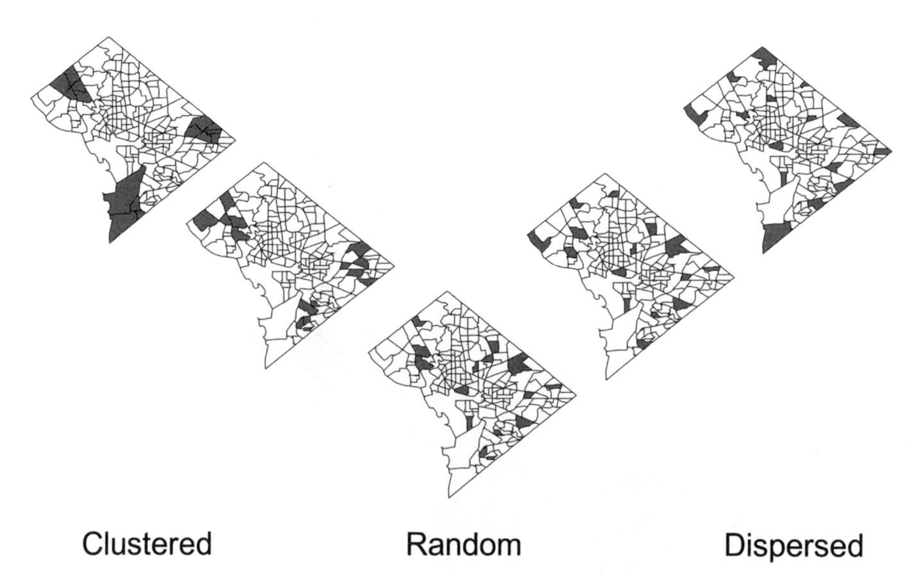

Figure 8.3 Spectrum of spatial autocorrelation.

In addition, spatial autocorrelation has significant implications for the use of statistical techniques in analyzing spatial data. For most classical statistics, including regression models, a fundamental assumption is that the sample observations are randomly selected and therefore independent of each other. That is, sample values are not affected by each other. But when spatial data are analyzed using these classical statistical methods, this assumption of independence is often violated because most spatial data have a certain degree of spatial autocorrelation (Anselin and Griffith, 1988). Recalling the *First Law of Geography* from an earlier discussion, everything on the surface of the Earth is related, but closer things are more related than things farther apart.

If spatial autocorrelation in the data is significant but not recognized, the results of statistical analysis will be incorrect. For example, the regression parameter estimates in a classical regression model may not be biased, but when the significance of the regression parameter estimates is tested, the probabilities derived from the spatial data will likely be incorrect because of the spatial dependency among the observations. As a result, regression parameter estimates that are not statistically significant may turn out to be significant due to the presence of spatial autocorrelation. In other words, relationships that do not exist may be assessed as statistically significant, leading to inappropriate decisions.

The need to recognize spatial autocorrelation in spatial data is important because when spatial autocorrelation exists, observations in the same geographic area will capture similar information and reduce the so-called effective sample size. The relationships between variables are "repeated" or inflated by the similar or autocorrelated observations. Though the true relationships between variables may not be strong when the duplicated observations are removed, the relationships as reflected by the parameter estimates will tend to be stronger and more significance when "duplicated" observations are included. Therefore, measuring and testing the significance of spatial autocorrelation is essential before any statistical analysis is conducted.

Several statistics can be used to measure levels of spatial autocorrelation in point and polygon data. Since the methods for measuring spatial autocorrelation of point data have been covered in Chapter 6, this chapter will focus on the methods suitable for polygon data only. In terms of attribute values, they can be of nominal or interval-ratio scale. The spatial autocorrelation statistics discussed in this chapter can handle these scales of measurement.

Specifically, **joint count statistics** can be used to measure the magnitude of spatial autocorrelation among polygons with binary nominal data. For interval or ratio data, we may use **Moran's I** Index, **Geary's Ratio *C*,** and the ***G*-statistic.** These measures, which are considered *global measures,* assume that the magnitude of the spatial autocorrelation is reasonably stable across the study region. In addition, the variability of spatial autocorrelation is relatively constant over space. These assumptions imply the existence of a spatially homogeneous process in the data.

But in reality, the variability of spatial autocorrelation may not be stable over the region. Therefore, it is necessary to use *local* measures to depict the spatial variability of spatial autocorrelation. The local spatial autocorrelation measures included in this chapter are the local version of the ***G*-statistic** and the **Local Indicators of Spatial Autocorrelation** (LISA). In addition, a graphical tool, the **Moran scatterplot,** can be used to identify local variability of spatial autocorrelation visually. The concept of spatial autocorrelation can be expanded to describe not just how *values* across space are similar to each other, but also how *observations* are similar. Observations often possess more than one attribute or variable, such as median household incomes and median housing values in the example of the seven Ohio counties. Therefore, at the end, we will discuss a **bivariate spatial autocorrelation** measure that can evaluate the magnitude of spatial autocorrelation of two variables.

8.4 SPATIAL WEIGHTS MATRICES

For measuring spatial autocorrelation in a set of geographic objects, we must discuss methods for capturing spatial relationships among areal units, or polygons. Recall that spatial autocorrelation measures the degree of sameness of attribute values among areal units within their neighborhood. But the concept of a *neighborhood* must be quantified first so that it can be applied in the calculation of spatial autocorrelation statistics. In other words, the neighborhood relationship among areal units must be quantified before we can proceed to calculate the statistics.

Assume that we have n areal units in our study area. Given any predefined method to determine the neighborhood relation of these areal units, we have n^2 ($= n \times n$) pairs of relationship to be captured. Conventionally, we use a matrix to store and organize the spatial relationship among these areal units. Since there are n areal units, this matrix will have n rows by n columns, or $n \times n$ dimension.

Each areal unit is represented by a row and a column. Each value in the matrix indicates the spatial relationship between the geographic features represented by the corresponding row and column. For example, a binary value of 0 or 1 can be assigned to a cell to indicate whether the two corresponding polygons are adjacent (1) or not (0). However, since different criteria are used to define a neighborhood relationship, we may use these criteria to construct different matrices (Griffith, 1996). For example, in addition to the binary matrix mentioned above, we can use the actual distances between centroids of polygons to represent the spatial relationships between polygon pairs in the matrix.

In the following subsections, we discuss in more detail different ways of specifying spatial relationships and the resulting matrices. In general, these matrices are called **spatial weights matrices** because elements within them are often used as weights in the calculation of spatial autocorrelation statistics or in the spatial regression models (Getis and Aldstadt, 2004).

8.4.1 Neighborhood Definitions

There are many ways to define a spatial relationship. Even if we are concerned with only the immediate neighbors of an areal unit, there are at least two common methods for defining a spatial relationship. Figure 8.4 illustrates the rook's case and the queen's case of neighbor relationships in a grid. In a highly simplified polygon structure similar to a set of grid cells, there are nine areal units.

For the purpose of this discussion, let's focus on the center cell, X. Using the rook's case as the criterion to define its neighbors, only B, D, E, and G are regarded as neighbors because each of them shares a boundary with X. If we adopt the queen's case, then all surrounding areal units can be regarded as neighbors of X as long as they touch each other even at a point. In this case, X's neighbors will include A, B, C, D, E, F, G, and H.

The neighborhood definitions described in Figure 8.4 correspond to the adjacency criterion. If any two areal units are next to each other (juxtaposed), then they are neighbors of each other. In the rook's case, the neighboring units must share a boundary with length greater than 0. In the queen's case, the shared boundary can be a section of the boundary or just a point anywhere along the boundary. In other words, when the spatial system involves highly irregularly shaped polygons, the definition of a neighbor is based on whether the rook's case or the queen's case is adopted and the length of the shared boundary.

The neighboring polygons of X identified by these two criteria are its immediate or *first-order neighbors*. The adjacency measure can be extended to identify neighbors of the immediate neighbors or, more technically, *second-order neighbors*. The concept can be further extended to identify *higher-order neighbors*, which will be discussed again briefly in the next section, even though computationally it may sometimes be incomprehensible.

Besides using adjacency as the criterion to define a neighborhood, another common measure is the distance between polygon centroids. Distance is the most general measure of a spatial relationship. Corresponding to the geographic concept of distance decay, the intensity of spatial relationship between

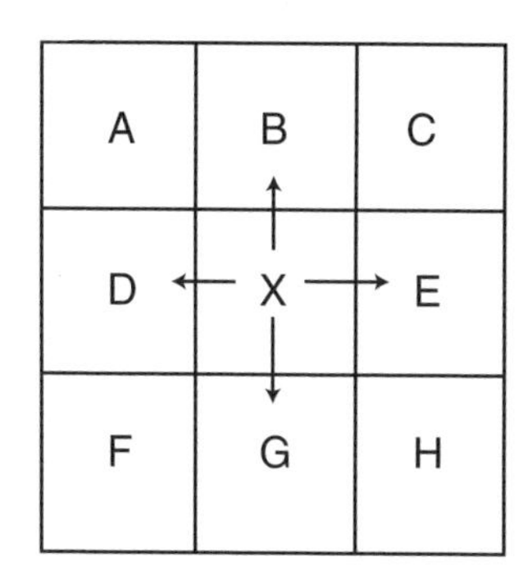

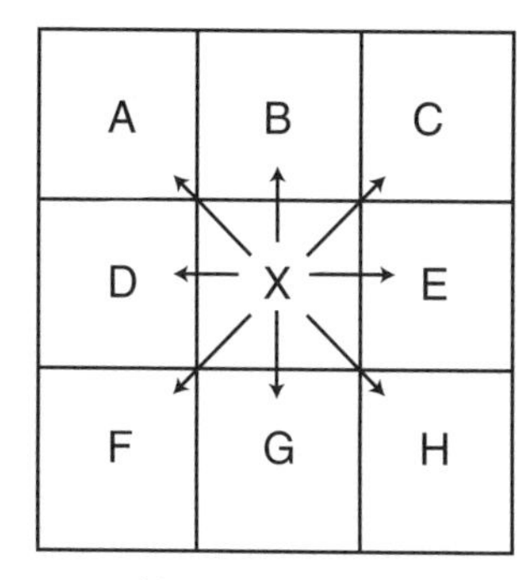

Figure 8.4 Neighbor definitions: rook's and queen's cases.

two distant places is said to be less than the intensity of relationship between places that are closer together, assuming that all places have similar characteristics. Consequently, the distance between any two geographic features can be used as an indicator of their spatial relationship. The distance measure can also be used to define a neighborhood.

If, for a household, neighbors mean those whose houses are less than 1 mile away, then its neighborhood can be quantitatively defined as all houses within a radius of 1 mile from the household. With this definition, we can check all houses to see if they fall within this area. The results can be displayed in a binary form (inside or outside of the defined neighborhood). In other words, using the concepts of the measurement scales discussed in Chapter 1, the distance measure for a neighborhood definition is in ratio scale, while the adjacency measure is just a binary representation of the distance measure, with 1 indicating zero distance between two neighboring units.

8.4.2 Binary Connectivity Matrix

With different ways of defining neighbors, different spatial weights matrices can be constructed to capture the spatial relationships among geographic features. Using the simplest adjacency definition of neighborhood, areal units sharing a common boundary, either in the rook's or the queen's case, are neighbors. In the matrix capturing the spatial relationship, we can put a 1 in the cell corresponding to the two geographic features or polygons if they are next to each other. If any two polygons are not adjacent to each other, the corresponding cell will have a value of 0.

In the entire matrix, the cell will either be 0 or 1. Therefore, this type of matrix is sometime called a **binary matrix.** Because this matrix shows how pairs of polygons are connected, and because a similar concept has been applied to linear features in a network (Chapter 7), this binary matrix is also called a *connectivity matrix*. Formally, let C denote such a matrix, and let c_{ij} denote the value (either 0 or 1) in the matrix in the ith row and the jth column.

$$c_{ij} = 1 \text{ when the } i\text{th polygon is adjacent to the } j\text{th polygon and}$$
$$c_{ij} = 0 \text{ when } i\text{th polygon is not adjacent to the } j\text{th polygon, } i \neq j.$$

A binary matrix has several interesting characteristics. First, all elements along the major diagonal, c_{ii} (the diagonal from upper left to lower right), are generally defined as 0 because it is assumed that an areal unit is not a neighbor of itself. However, in some specific applications, c_{ii} may equal to 1 (Wong, 2002). Second, this matrix is symmetrical, that is, the upper-right triangle of the matrix divided along the major diagonal is a mirror image of the lower-left triangle. Using our notation, $c_{ij} = c_{ji}$. This symmetrical property of the matrix basically reflects the reciprocal nature of the spatial relationship: if areal unit A is a neighbor of B, then B is also a neighbor of A. The symmetrical structure of the matrix requires information on the spatial relationship to be stored twice—a redundancy in this type of matrix. Finally, a row in the matrix represents how one areal unit is related spatially to all other units. Therefore, by summing the cell values for each row across all columns, we can obtain the row sum, indicating the number of neighbors that the areal unit has. Using our notation, the row sum for areal unit i is

$$c_{i.} = \sum_j c_{ij}. \tag{8.1}$$

Similarly, by summing the cell values for each column across all rows, we can obtain the column sums. In our notation, a column sum for areal unit j is

$$c_{.j} = \sum_i c_{ij}. \tag{8.2}$$

If we sum all the cell values, the notation is

$$c_{..} = \sum_i \sum_j c_{ij}. \tag{8.3}$$

Table 8.1 is a binary connectivity matrix of the seven counties in northeast Ohio, using the queen's case. As can be seen in Figure 8.1, Cuyahoga and Portage counties are neighbors. Similarly, Geauga and Summit counties are neighbors in the matrix. The bottom row indicates the column sums, and the rightmost column gives the row sums. According to the row sums, Geauga County has the most neighbors (six) among the seven counties, mainly due to its central location in regard to the other counties in the region. On the other hand, Trumbull, Summit, Ashtabula, and Lake counties all have the fewest neighbors (three). We probably should not be surprised by these re-

TABLE 8.1 Binary Connectivity Matrix (Seven Ohio Counties)

ID	Geauga	Cuyahoga	Trumbull	Summit	Portage	Ashtabula	Lake	Row Sum
Geauga	0	1	1	1	1	1	1	6
Cuyahoga	1	0	0	1	1	0	1	4
Trumbull	1	0	0	0	1	1	0	3
Summit	1	1	0	0	1	0	0	3
Portage	1	1	1	1	0	0	0	4
Ashtabula	1	0	1	0	0	0	1	3
Lake	1	1	0	0	0	1	0	3
Col Sum	6	4	3	3	4	3	3	26

sults, as these counties are all located along the edge of the seven-county region. There is a total of 13 shared boundaries, or joints. Let's use J to represent the number of joints. If we sum all the cell values (26), it is twice the number of shared boundaries or joints ($2J$).

In the previous section, we briefly mentioned that the binary adjacency concept in identifying immediate neighbors can be extended to identify higher-order neighbors. For instance, we know that, based on Table 8.1, Portage County and Lake County are not immediate neighbors. But could they be second-order neighbors? That is, could Portage County be a neighbor of an immediate neighbor of Lake County? If one looks at the map in Figure 8.1, it is obvious that the two counties are second-order neighbors through Geauga County. Can we derive the second-order neighbor information from the connectivity matrix? The answer is yes. The concept is similar to that of the network analysis described in Chapter 7. The major difference is that in Chapter 7, the objects are network links, whereas the objects here are polygon units.

Without going into the details of matrix operations, the connectivity matrix C can multiply by itself to obtain C^2. This second-power matrix is not a binary matrix, however. Using the seven-county example again (Figure 8.1), the C^2 matrix is shown in Table 8.2. In general, a cell value in this matrix indicates the number of ways the two corresponding areal units can be connected to each other through an intermediate neighbor. Therefore, the cell corresponding to Lake County and Portage County is 2, indicating that there are two ways the two counties can be regarded as second-order neighbors. First, Geauga County is an immediate neighbor of both Lake County and Portage County. The second linkage between the two counties is through Cuyahoga County, which is an immediate neighbor of both Lake County and Portage County. Though in this chapter we will not utilize the information on higher-order neighbors, it is important to realize the power of this intuitively simple binary connectivity matrix.

Apparently, the binary matrix is not a very efficient format to store spatial relationships based on adjacency. Not only does the upper triangle of the

TABLE 8.2 Square of the Binary Connectivity Matrix for Seven Ohio Counties

ID	Geauga	Cuyahoga	Trumbull	Summit	Portage	Ashtabula	Lake
Geauga	6	3	2	2	3	2	2
Cuyahoga	3	4	2	2	2	2	1
Trumbull	2	2	3	2	1	1	2
Summit	2	2	2	3	2	1	2
Portage	3	2	1	2	4	2	2
Ashtabula	2	2	1	1	2	3	1
Lake	2	1	2	2	2	1	3

matrix duplicate the lower triangle; in most spatial configurations, most of the cell values in the matrix are 0's, indicating that the two areal units are not neighbors. In the example of the seven Ohio counties, this characteristic of the matrix may not be obvious. But still, among these seven areal units with a matrix of 49 cells, there are 23 0's. Imagine creating a binary matrix for the more than 3000 counties in the entire United States. The matrix would not only be gigantic ($3000 \times 3000 = 9$ million cells), but most of the cell values would be 0's. Most of the time, we are only interested in neighbors. Using a lot of space to store nonneighbor information is inefficient.

Instead of using a square matrix ($n \times n$) to record areal units that are neighbors and nonneighbors, we use a more compact format of the matrix that includes only areal units that are neighbors of a given unit. In general, each areal unit has a unique identification number, such as the Federal Information Processing Standards (FIPS) code of a county (which combines the state and county FIPS codes), or a unique label such as the county name (unique only within the state). Using the unique identifier, we can construct a much more compact (sparse) matrix that lists only the identifiers of neighbors. Table 8.3 captures the same information as Table 8.1, but its size is slightly reduced. Instead of 49 cells (7 by 7) with many 0's, the dimension of the matrix is now only 7 by 6, with 26 cells containing the same information. In a much larger area with more polygons, such as the county-level study of Griffith et al. (2003), the matrix dimension was reduced from approximately 3000 by 3000 to approximately 3000 by 60, a much more dramatic reduction in the matrix size compared to the Ohio seven-county region example. Imagine how much reduction can be achieved in the nationalwide study at the census tract or block group level (Griffith et al., 2003).

8.4.3 Stochastic or Row-Standardized Weights Matrix

The binary matrix basically gives a weight of 1 if the areal unit is a neighbor. Mathematically, this unit weight may not be very effective in modeling spatial relationships. For instance, suppose that we want to analyze how the value of a house is affected by the values of its surrounding houses. Following the general practice of realtors, we can think of the house value as receiving a fractional influence from each of its neighbors. If there are four neighbors for that house, its value may receive 0.25 influence from each neighboring house.

Recall that a binary matrix typically consists of 1's and 0's. A 1 indicates that the corresponding areal units represented by the row and column are neighbors. Therefore, for any given row, the row sum, $c_{i.}$, indicates the total number of neighbors that the areal unit i has (refer to Table 8.1). In general, we can assume that each of its neighbors exerts the same amount of influence on that areal unit. To find out how much each neighbor contributes to the value of the target areal unit, we calculate the ratio of the weight of each neighbor unit with regard to the total influence (or weight). This gives the weight (w_{ij}) of each neighboring unit:

TABLE 8.3 A Sparse Matrix Capturing the Same Information as the Binary Matrix

ID	Neighbor 1	Neighbor 2	Neighbor 3	Neighbor 4	Neighbor 5	Neighbor 6
Geauga	Cuyahoga	Trumbull	Summit	Portage	Ashtabula	Lake
Cuyahoga	Geauga	Summit	Portage	Lake		
Trumbull	Geauga	Portage	Ashtabula			
Summit	Geauga	Cuyahoga	Portage			
Portage	Geauga	Cuyahoga	Trumbull	Summit		
Ashtabula	Geauga	Trumbull	Lake			
Lake	Geauga	Cuyahoga	Ashtabula			

$$w_{ij} = c_{ij}/c_{i.}. \quad (8.4)$$

Figure 8.5 uses the seven Ohio counties to illustrate how a row-standardized matrix or a **stochastic matrix,** which is usually denoted as W, can be derived from the binary matrix C. Note that the stochastic matrix still has a major diagonal of 0's. However, this matrix is no longer symmetrical. Specifically, each cell value in the stochastic matrix is calculated by using Equation 8.4. For example, the row sum of Cuyahoga County is 4. This means that Cuyahoga has four immediate neighbors. For each of the four cell values in the stochastic matrix for its neighbors, each cell in the row of Cuyahoga is assigned a value of 0.25, which is $= 1/4 = c_{ij}/c_{i.} = w_{ij}$. Nonneighbor cells are assigned a value of 0, which is $w_{ij} = 0/4$. Similarly, cells with neighbors in Geauga County's row are assigned a value of 0.17, which is $1/6 = c_{ij}/c_{i.}$.

8.3.4 Centroid Distances

Besides using adjacency as a measure to describe the spatial relationship among a set of geographic features and to define a neighborhood among them, one can use another common measure of spatial relationship, the distances between these geographic features. Using the distance between two geographic features as a weight in describing a spatial relationship is very powerful. Recall the First Law of Geography, the relationship between any two geographic features can be structured as an inverse function of the distance between them. Generally, we expect features close to each other to be more related than distant features. Therefore, using the distances between geographic features as the weights to depict a spatial relationship is theoretically sound.

The distance between any two point features is rather easy to define, putting aside difference ways to calculate distance, such as the *great circle distance,* the *Euclidean distance,* or the *Minkowskian distance* (Batchelor, 1978). However, there are several ways to measure the distance between any two polygons. A very popular practice, especially in transportation studies, is to use the centroid of the polygon to represent that polygon. The centroid is the geometric center of a polygon. There are different ways of determining the centroid of a polygon, and each of them identifies a centroid differently.

In general, the shape of the polygon affects the location of its centroid. Polygons with unusual (geometrically irregular) shapes may generate centroids that are located in undesirable locations. For instance, certain methods for calculating centroids of polygons may generate the centroid of the State of Florida located in the Gulf of Mexico or the centroid of the State of Louisiana located in the neighboring State of Mississippi. Still, in most cases and with the advances in algorithms in computational geometry, it is now common and feasible, using the distance between the centroids of two polygons, to represent the distance between the polygons themselves.

Because distances are used as weights, the spatial weights matrix using distances to represent spatial relationships is sometimes labeled D, and the

Binary Connectivity Matrix

ID	GEAUGA	CUYAHOGA	TRUMBULL	SUMMIT	PORTAGE	ASHTABULA	LAKE	Row Sum
Geauga	0	1	1	1	1	1	1	6
Cuyahoga	1	0	0	1	1	0	1	4
Trumbull	1	0	0	0	1	1	0	3
Summit	1	1	0	0	1	0	0	3
Portage	1	1	1	1	0	0	0	4
Ashtabula	1	0	1	0	0	0	1	3
Lake	1	1	0	0	0	1	0	3

Stochastic or Row-Standardized Matrix

ID	GEAUGA	CUYAHOGA	TRUMBULL	SUMMIT	PORTAGE	ASHTABULA	LAKE
Geauga	0	0.17	0.17	0.17	0.17	0.17	0.17
Cuyahoga	0.25	0	0.00	0.25	0.25	0.00	0.25
Trumbull	0.33	0.00	0	0.00	0.33	0.33	0.00
Summit	0.33	0.33	0.00	0	0.33	0.00	0.00
Portage	0.25	0.25	0.25	0.25	0	0.00	0.00
Ashtabula	0.33	0.00	0.33	0.00	0.00	0	0.33
Lake	0.33	0.33	0.00	0.00	0.00	0.33	0

Figure 8.5 Deriving the stochastic weights matrix from the binary connectivity matrix.

elements of D are labeled d_{ij}, representing the distance between (the centroids of) areal units i and j. Table 8.4 shows the D matrix using the distances between the centroids of the seven Ohio counties. In modeling spatial processes, the distance weight is often used in an inverse manner, as the strengths of most spatial relationships between polygons diminish when the distance between them increases. Therefore, when the distance matrix is used, the weight,

$$w_{ij} = \frac{1}{d_{ij}}, \tag{8.5}$$

is an inverse of the distance between features i and j. In other words, the weight is inversely proportional to the distance between the two features. Based on empirical studies of spatial processes, however, the strength of many spatial relationships has been found to diminish more than just proportionally to the distance separating the features. Therefore, the squared distance is sometimes used in the following format:

$$w_{ij} = \frac{1}{d_{ij}^2}. \tag{8.6}$$

The square power applied to the distance is also known as the **distance-decay** parameter in spatial interaction literature (Fotheringham and O'Kelly, 1989). This is because the square power shows how fast the relationship diminishes when the distance increases. In other words, the relationship between weights and distances is not necessarily linear. To generalize the possibility of nonlinear relationships and to allow for a distance-decay parameter value other than 2, the power for the distance is often denoted β. These weighting schemes of inverse distances will be used later for different measures of spatial autocorrelation.

8.4.5 Nearest Distances

With the advances in GIS software and computation algorithms, it is now fairly easy to determine the distance between any two geographic features besides their centroid distance. One alternative method to determine the distance between any two features is based on the distance of their nearest parts. Selecting the two farthest counties among the seven Ohio counties, the nearest parts between Summit and Ashtabula are the northeastern corner of Summit and the southwestern corner of Ashtabula (Figure 8.1). This conceptualization of distance between geographic features is potentially useful if the investigator is interested in spatial contacts between areal units or the diffusion of geographic phenomena over areal units.

An interesting situation involving the distance of nearest parts occurs when the two features are adjacent to each other. When this is the case, the distance

TABLE 8.4 Spatial Weights Matrix in Miles Using Centroid Distances Among Seven Ohio Counties

ID	Geauga	Cuyahoga	Trumbull	Summit	Portage	Ashtabula	Lake
Geauga	0.0000	25.1970	26.6389	32.7420	25.0076	26.5672	12.6331
Cuyahoga	25.1970	0.0000	47.7889	23.4420	31.5755	50.8241	28.2636
Trumbull	26.6389	47.7889	0.0000	41.8271	24.4707	29.5520	36.7000
Summit	32.7420	23.4420	41.8271	0.0000	17.7783	58.0720	42.7222
Portage	25.0076	31.5755	24.4707	17.7783	0.0000	45.5305	37.4669
Ashtabula	26.5672	50.8241	29.5520	58.0720	45.5305	0.0000	24.7187
Lake	12.6331	28.2636	36.7000	42.7222	37.4669	24.7187	0.0000

between two features is 0. In other words, in this distance measurement scheme, 0 distance means that the corresponding features are immediate neighbors. The 1's in a binary connectivity matrix also capture this situation. By extracting all the nondiagonal 0 cells from the distance matrix and replacing them with 1's, we can also derive a binary matrix. Since this binary connectivity matrix can be seen as a simplified distance weights matrix, it is sometimes said that the binary matrix is a *derivative* of the distance matrix based on nearest parts. Note that the above discussions on inverse distance weighting schemes are also applicable to the current distance measure based on nearest parts.

Table 8.5 is the distance matrix based on nearest parts for the seven Ohio counties. Compare this matrix with the binary matrix in Table 8.1. The nondiagonal cells in the distance matrix with 0 values correspond to the cells with 1's in the binary matrix.

8.4.6 ArcView Example: Spatial Weights Matrices

It is critical to recognize the importance of spatial coordinate systems and projections when distance is involved in the creation of spatial weights matrices. If data are not projected (i.e., in longitude-latitude), the ArcView tool can still compute the distances between observations, but the computed distances will not be precise due to limitations of the distance unit conversion tools embedded in ArcView. The best way to deal with this issue is to convert the data into projections that provide the least distortion of distance before creating the distance-based spatial weights matrices. For contiguity- or adjacency-based matrices, projection is not an issue.

In preparing the calculation of various spatial autocorrelation measures, we have developed a suite of tools for creating spatial weights matrices. These tools are included in Ch8.avx. ArcView Example 8.1 demonstrates how these tools can be used.

ArcView Example 8.1: Spatial Weights Matrices

Step 1 Data and ArcView extension

Using the accompanying ArcView extension, Ch8.avx, several types of spatial weights matrices can be calculated.

As in previous chapters, data should be in the C:\Temp\Data\Ch8_data folder.

Like other extensions, Ch8.avx should have been pasted into the EXT32 folder inside the installation folder of ArcView GIS.

The data theme used in this example is OhCounties.shp, a polygon theme that contains county data and boundaries of the 88 counties in Ohio.

TABLE 8.5 Spatial Weights Matrix in Miles Based upon Nearest Parts (Seven Ohio Counties)

ID	Geauga	Cuyahoga	Trumbull	Summit	Portage	Ashtabula	Lake
Geauga	0.0000	0.0000	0.0000	0.0000	0.0000	0.0000	0.0000
Cuyahoga	0.0000	0.0000	18.4815	0.0000	0.0000	18.9278	0.0000
Trumbull	0.0000	18.4915	0.0000	19.2631	0.0000	0.0000	10.7778
Summit	0.0000	0.0000	19.2631	0.0000	0.0000	21.9251	15.0322
Portage	0.0000	0.0000	0.0000	0.0000	0.0000	10.4689	15.0353
Ashtabula	0.0000	18.9278	0.0000	21.9251	10.4689	0.0000	0.0000
Lake	0.0000	0.0000	10.7778	15.0322	15.0353	0.0000	0.0000

To prepare the data theme for the creation of spatial weights matrices, we must first define the data unit and distance unit.

- Use the **Add Theme Button** to add `OhCounties.shp` to a View document.
- From the **View** menu, choose **Properties.**
- In the **View Properties** dialog box, click the **Projection** button.
- In the **Projection Properties** dialog box, set the projection **Category** to `UTM-1983` and the projection type to `Zone 17`.

Click OK to close the Projection Properties dialog box.
Make sure that the **Map Units** and **Distance Units** in the **View Properties** dialog box have been set to `meters` and `miles`.

- Click **OK** to close the **View Properties** dialog box.

Now the data theme has been projected to UTM 1983, Zone 17. Note that the shape of the state's outline has been changed slightly to reflect the newly defined projection but the original data, OhCounties.shp, have not be altered.

Step 2 Set up an ID field
The creation of a spatial weights matrix has several options in Ch8.avx.

- From the **Ch.8** menu, select **Create Spatial Weights Matrix (Polygon).**

Ch.8
Create Spatial Weights Matrix (Polygon)
Joint Count Statistics
Global Moran _Geary Statistics
General G-Statistic (Global)
Local Indicators of Spatial Association (LISA)
Local Gi(d) Statistic
Moran Scatterplot
Bivariate Sp. Association Statistic

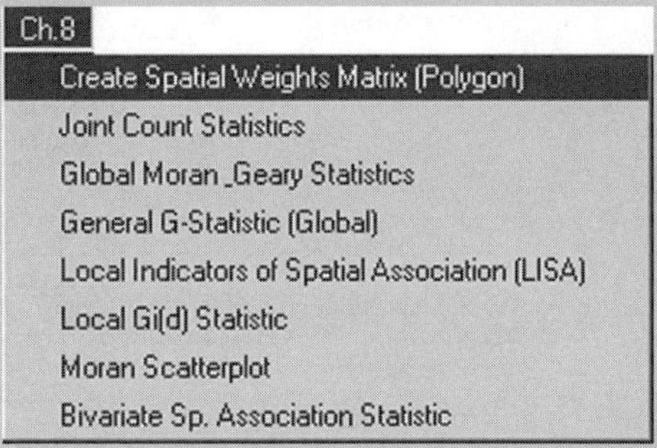

In the Create Spatial Weights Matrix dialog box:

- First, click the `Ohcounties.shp` theme to select it.
- Click the **Choose the Theme** button to list all attribute fields in the data theme.
- Scroll down and choose `Cnty_fips` as the ID field.
- Make sure that the **All** option is checked for all records to be included.
- Click the **Calculate** button to proceed to the next step.

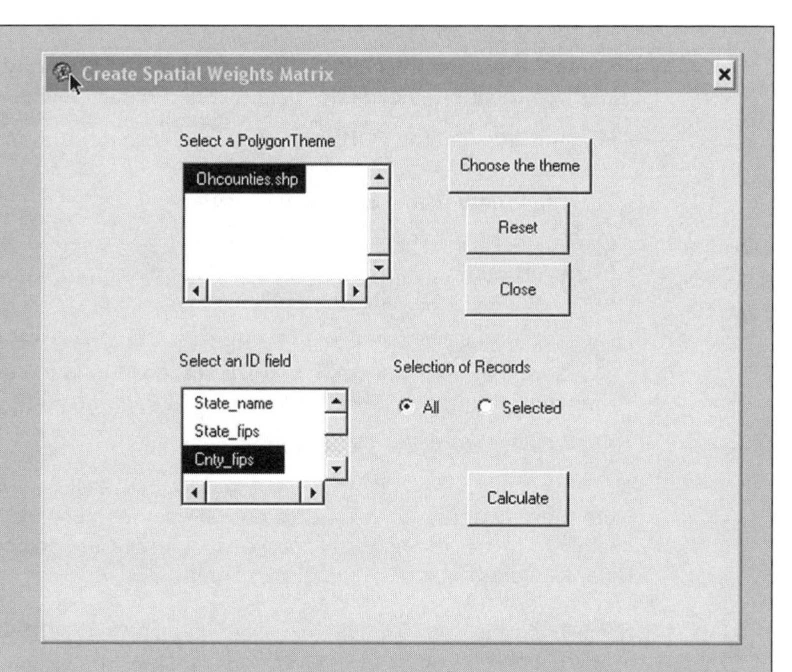

Step 3 Choose the type of spatial weights matrix
In the **Distance Definition** dialog box, a total of four options are available for creating different types of spatial weights matrices. Click the button beside the `Binary Connectivity` option to bring out a drop-down list.

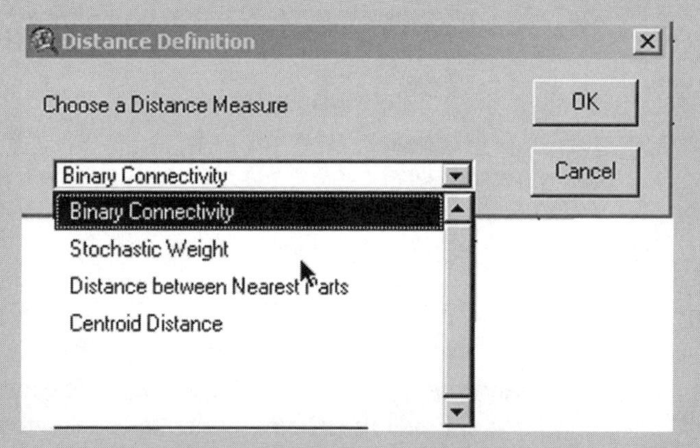

Notice that the other options include `Stochastic Weight`, `Distance between Nearest Parts`, and `Centroid Distance`. Refer to the text discussion on how each option is defined.

- Choose the `Binary Connectivity` option and then click **OK** to proceed.

- In the **Output FileName** dialog box, navigate to a folder and give a name for the matrix to be created. For this matrix, name it `distbin.dbf`.

The default name is `distmatrix.dbf`. Note that the matrix will be created and stored in DBF format, so it is possible to use other spreadsheet or database management programs to open and manipulate the matrix if necessary. Clearly, the ArcView Ch8.avx extension will use the matrix in DBF format for subsequent operations.

- Click OK to close the Info dialog box that reports the completion of this task.

Step 4 Additional spatial weights matrices
To create spatial weights matrices with other definitions, repeat Steps 2 and 3 but choose a different option in the **Distance Definition** dialog box to create and name the following matrices:

Stochastic weight:	diststoch.dbf
Distance between nearest parts:	distnear.dbf
Centroid distance:	distcent.dbf

Step 5 End

8.5 SPATIAL AUTOCORRELATION STATISTICS AND NOTATIONS

This section provides an overview of different types of spatial autocorrelation statistics. Most of these statistics can use one or more of the spatial weights matrices discussed in the previous section. Various statistics have been introduced to deal with data values measured at different scales, such as those measured at nominal or interval/ratio scales. Moreover, different statistics capture different aspects of spatial autocorrelation in the data. In this section, we classify these statistics according to the measurement scales of the data that can be handled by the statistics and according to the geographic scale of the statistics, that is, whether they are applicable to the entire region (global) or to the neighborhood (local).

First, spatial autocorrelation statistics can be categorized according to the measurement scales of the attribute data used in the analysis. If the attribute or variable contains nominal data and is binary (i.e., the attribute or variable has only two possible values or outcomes), then **joint count statistics** are most appropriate. If the variable is in interval or ratio scale, the appropriate spatial autocorrelation statistics would be **Moran's I** and **Geary's Ratio.** Moran's I uses the mean of the attribute's data values as the benchmark for comparison when neighboring values are evaluated; Geary's Ratio is based on a direct comparison of neighboring values. Another possible choice for interval/ratio data is the general *G*-statistic, which is based on the concept of spatial association or cross-product statistics. All these measures can be regarded as **global measures** of spatial autocorrelation or spatial association because one statistic or value is derived for the entire study area, describing the overall spatial relationship of all the areal units. In this case, the spatial variation of spatial autocorrelation within the study region is not a major concern.

However, it is the norm that the magnitude of spatial autocorrelation varies over space. In that case, global measures will be inadequate to capture and describe the spatial variability of spatial autocorrelation. In addition, for many geographic phenomena or events, there is no reason to believe that the spatial variability of spatial autocorrelation is constant over the region, a situation that statisticians refer to as *spatial homogeneity.* Alternatively, if the variance of spatial autocorrelation does change significantly across the study area, it is referred to as *spatial heterogeneity.*

To better describe how geographic phenomena or events distribute and change over space, we need to document the change and the variability of spatial autocorrelation over the study region. To do this, we must rely on **local measures** that can detect spatial autocorrelation at the local scale—that is, provide a spatial autocorrelation value for each areal unit. LISA, which are local versions of Moran's I and the Geary Ratio, can serve the same purposes. In additional, the local version of the *G*-statistic can be used to evaluate spatial association at the local scale.

All the above statistics are univariate in nature. They handle only one variable and evaluate how that variable is correlated over space. But in reality, we are often interested in more than one variable. For instance, besides knowing how much housing values are correlated over space, we may want to find out how the housing value of one house is correlated with the income levels of its neighboring residents. This is a bivariate correlation analysis since two variables are involved. But the correlation crosses over to the neighboring units. Therefore, we will also review a bivariate spatial autocorrelation measure, though it is still limited to a global measure.

In deriving various spatial autocorrelation statistics and related statistics to test for their significance, several notations that are commonly used in association with spatial weights matrices will be used repeatedly. Let's introduce and define them all at once here before we discuss those spatial autocorrelation statistics.

Even though the notation w_{ij} is often used to represent the cell value in a stochastic weights matrix, W for row i and column j, it is also quite common to use w_{ij} to generically represent the cell value of any spatial weights matrix. For any given weights matrix, summing up all the cell values of a given row i across all columns (row sum) is denoted

$$w_{i.} = \sum_j w_{ij}. \tag{8.7}$$

Similarly, a column sum is the sum for a given column j across all rows:

$$w_{.j} = \sum_i w_{ij}. \tag{8.8}$$

Sometimes W or $w_{..}$ also represents the sum of all cell values of the weights matrix:

$$W = \sum_i \sum_j w_{ij}. \tag{8.9}$$

In testing the significance of spatial autocorrelation statistics, the structure of weights has to be summarized by several parameters, including S_1 and S_2. These two terms are defined as

$$S_1 = \frac{1}{2} \sum_i \sum_j (w_{ij} + w_{ji})^2 \tag{8.10}$$

and

$$S_2 = \sum_i \left(\sum_j w_{ij} + \sum_j w_{ji} \right)^2 \tag{8.11}$$

Note that conventionally, S stands for the sample standard deviation. Here S with the subscript 1 or 2 refers to the specific parameter pertaining to the spatial weights matrix. Assume that we can divide the W matrix into upper and lower triangles along the major diagonal. Then each w_{ij} has a mirror cell, w_{ji}, on the other triangle.

In the first term, S_1, the weight, w_{ij}, and its mirror cell value, w_{ji}, are first added together and then squared. This process is performed on all i-j pairs. The total is then halved to obtain S_1. If the weights are binary and the matrix is symmetric (i.e., the C matrix), then $(w_{ij} + w_{ji})^2$ is equal to 4. Thus, S_1 is four times the total number of joints or shared boundaries in the entire study area.

The second term, S_2, is based on the sums of the weights associated with each areal unit first, but in both directions (i.e., for both w_{ij} and w_{ji}). The sums are then added, squared, and summed over all areal units. For a binary matrix,

$$S_2 = 4 \sum_i \left[\left(\sum_j w_{ij} \right)^2 \right]. \tag{8.12}$$

Equation 8.12 can be obtained by first counting the number of joints for each areal unit, i (i.e., $\Sigma_j\, w_{ij}$ across all j's), then squaring this sum for each i, totaling the square of the sums, and multiplying the total by 4.

If there are two groups of areal units, we need two sets of notation. Let's use n to denote the number of areal units in the entire study area. If an attribute carries two values, x and y, and the attribute is used to divide all areal units into two groups, then we will use n_x and n_y to indicate the number of areal units in the two groups.

Similar in notation but very different in meaning, we sometimes use $n^{(x)}$ to represent the product of multiplying n by all subsequent values of n until $(n - (x - 1))$. Specifically,

$$n^{(x)} = n \times (n - 1) \times (n - 2) \times (n - 3) \times \ldots \times (n - x + 1), \tag{8.13}$$

where $n > x$. For example, if $n = 5$,

$$5^{(3)} = 5 \times (5 - 1) \times (5 - (3 - 1)) = 5 \times 4 \times 3 = 60.$$

Finally, $n^{(1)} = n$.

If x_i is the attribute value for areal unit i, a new parameter, m_j, can be derived from x_i:

$$m_j = \sum_{i=1} x_i^j, \tag{8.14}$$

where $j = 1, 2, 3$, or any integer. Therefore, if $j = 1$, m_j is the sum of x_i for all i, or $m_1 = \Sigma_{i=1}\, x_i$. If $j = 2$, m_j is the sum of all the squares of x_i, or $m_2 = \Sigma_i\, x_i^2$.

All these terms will be used later in this chapter in discussing various spatial autocorrelation measures. Readers should refer back to this section later as needed. In the rest of this chapter, we will discuss different spatial autocorrelation measures, starting with joint count statistics for binary attribute data, then global measures (including Moran's I, the Geary Ratio, and the general G statistic), and local measures (including LISA and the local $-G$ statistic) for interval and ratio data. Next, we will discuss the Moran scatterplot, a graphical technique used to visualize the spatial heterogeneity of spatial autocorrelation. Finally, we will discuss the basic formulation of the bivarite spatial autocorrelation measure.

8.6 JOINT COUNT STATISTICS

Joint count statistics provide a simple and quick method of quantifying the degree of clustering or dispersion among a set of contiguous polygons. This method is applicable only when the attribute of the polygons is in *nominal* scale but has binary outcomes. Binary data are those with only two possible values, such as arable/nonarable land parcels, high/low income groups, urban/rural counties, and so on. In the convention of joint count statistics, we use black and white as the binary attribute values. With these two possible attribute values, the polygons can be classified as black or white. In joint count statistics, we treat the shared boundary between any two polygons as a joint. Consequently, all boundaries can be classified as black-black (BB), white-white (WW), and black-white (BW) joints, representing the values of the two neighboring polygons.

If the polygons exhibit a pattern such as that of Case 1 in Figure 8.1, where white polygons cluster together in certain areas and black polygons cluster in other areas, then we would expect to find more WW and BB joints together than WB joints. This situation is also known as a *positive spatial autocorrelation,* a pattern in which similar values are close to each other. On the other hand, a dispersed pattern, or a *negative spatial autocorrelation,* has more BW than BB or WW joints. In that situation, dissimilar values tend to be close together. Case 2 in Figure 8.1 is an example of a dispersed pattern. Of course, for a random pattern, we would expect the actual counts of various types of joints to be fairly close to those of a typical random pattern with the same number of polygons. The basic idea of the joint count statistics is to compare the actual or observed numbers of joints of various types (BB, WW, and BW) with those expected from a random pattern.

While the joint count statistics method is limited to binary nominal data, data in other measurement scales can be easily downgraded or converted to binary form. For a set of ordinal data values, one can set a rank as the cutoff level so that ranks above the cutoff level are counted as one type and those below it are counted as another type. For example, in ranking major U.S. cities by population size, those ranked larger than Cleveland are assigned one value and those ranked smaller than Cleveland are assigned another value. Similarly, downgrading interval/ratio data is only a matter of setting a cutoff point to give the data above and below the cutoff level two binary values. Take average family income as an example. One can consider those cities with an average family income above the national average as one type and those cities with an average family income below the national average as another type. Therefore, the two essential conditions for using the joint count statistics are:

1. Data are related to area (polygons) or raster cells, and
2. Data are measured in binary form (i.e., only two possible values exist).

Note that although it is feasible to convert data from interval/ratio and ordinal scales into binary form, this process also reduces the amount of information captured by the data. We normally try to avoid this process whenever possible. Especially if interval or ratio data are converted into nominal scale, the precision of the original data is lost.

The joint count statistics are essentially the number of BB, WW, or BW joints in the observed region. With small spatial systems, it is probably feasible to manually count the number of the three different types of joints as the statistic. But if the cartographic data representing the study area and the associated attributes are already in digital formats and the spatial systems are too large for manual enumeration, the joint counting processes have to be automated. Below are the general steps used to derive the three joint count statistics in a computational environment.

1. Let $x_i = 1$ if polygon i is black and $x_i = 0$ if it is white.
2. Then, for BB joints,

$$O_{BB} = \frac{1}{2} \sum_i \sum_j (w_{ij} x_i x_j). \tag{8.15}$$

3. For WW joints,

$$O_{ww} = \frac{1}{2} \sum_i \sum_j [w_{ij}(1 - x_i)(1 - x_j)]. \tag{8.16}$$

4. For BW or WB joints,

$$O_{Bw} = \frac{1}{2} \sum_i \sum_j [w_{ij}(x_i - x_j)^2]. \tag{8.17}$$

The weight, w_{ij}, can be either binary or row-standardized. The three statistics above are the observed joint counts describing the actual pattern. If we observe a large number of O_{BB} or O_{ww} joints, or both, we may postulate that the observed pattern may exhibit positive spatial autocorrelation or clustering. However, we cannot conclude that positive spatial autocorrelation exists until we demonstrate that the observed pattern is different from a random pattern and that the difference is probably not due to chance or coincidence. That is the concept of *likelihood.*

Users of joint count statistics must know how to estimate the likelihood that each polygon has a white or a black value (the attribute value). Different ways of estimating attribute values for the polygons will affect the outcomes of testing the significance of the joint count statistics.

If the probability of a polygon's being black or white is based on known theories or a trend derived from a larger region, the method by which the attribute values are estimated is known as *free sampling*. This means that the probability of a polygon's being white or black is not limited or affected by the total number black or white polygons in the group. Consequently, this approach is sometimes referred to as *normality sampling,* implying that the attribute values are selected from a normal distribution. Adopting this sampling assumption implies that we treat the observations as a sample of a larger population, and the purpose of the study is to infer what happens to the larger population. Alternatively, if the probability of a polygon's being black or white is limited by or dependent on the total number of black or white polygons in the study area, the method by which the attribute values are estimated is known as *nonfree sampling* or *randomization sampling.* Adopting this assumption implies that the observed pattern is one of many possible patterns given the set of values. These assumptions are the essence of the debates between Constanzo (1983) and Summerfield (1983) on whether geographic data should be treated as populations or samples in Chapter 2.

In our seven-county Ohio example, the nonfree sampling case can only have three black polygons and four white polygons, no matter how they are rearranged. Since the total number of black and white polygons is fixed, the method is nonfree sampling, or *sampling without replacement.* That is, each observation can be used only once. Compared to nonfree sampling, free sampling does not limit how many polygons can be black or white; the method is also known as *sampling with replacement.*

When using joint count statistics, the choice between normality and randomization sampling is important. As a rule of thumb, randomization sampling should be used whenever references to trends from larger regions or those outside of the study area cannot be used with certainty. This is because randomization sampling requires less rigorous assumptions than free sampling. Normality or free sampling, on the other hand, should be used if the relationship between the study area and the national trend or a trend from a larger region can be established with known theories or by experience.

8.6.1 Free Sampling

In both normality sampling and randomization sampling assumptions, calculating joint count statistics involves the estimation of expected numbers of BB, WW, and BW joints and their standard deviations. The expected numbers of these joints reflect a random pattern, or a pattern with no significant spatial autocorrelation of any type under the chosen sampling assumption. The number of BB and WW joints indicates the magnitude of positive spatial autocorrelation, while the number of BW or WB joints indicates the magnitude of negative spatial autocorrelation.

These observed values are compared with their expected counterparts from a random pattern to derive their differences. These differences are then standardized by their corresponding standard deviations in order to obtain standardized scores. Using the standardized scores, we can decide if there is a significant positive or negative spatial autocorrelation in the pattern. In other words, three pairs of comparisons have to be conducted. For illustrative purpose, we will only show in detail how the negative spatial autocorrelation statistic (BW joints) can be tested for its significance. With this example, the other two situations can be repeated easily or derived using the accompanied ArcView extension.

Using the normality assumption, the equations for the expected number of BB and WW joints are

$$E_{BB} = \frac{1}{2} Wp^2 \tag{8.18}$$

and

$$E_{WW} = \frac{1}{2} Wq^2. \tag{8.19}$$

Then the equation for expected BW joints is

$$E_{BW} = Wpq, \tag{8.20}$$

where

E_{BB}, E_{WW}, E_{BW} are the expected numbers of BB, WW, and BW joints, respectively,

p is the probability that an area will be black, and

q is the probability that an area will be white, or $(1 - p)$.

The two probabilities must sum to 100%, or $(p + q = 1.0)$. If no other information is available, a common method used to estimate p and q is to set, $p = n_B/n$, where n_B is the number of black polygons and $q = 1 - p$. But there are other considerations in determining p, which will be discussed later.

If the spatial weights matrix is a binary matrix, the expected values can be simplified to

$$E_{BB} = Jp^2, \tag{8.21}$$

$$E_{ww} = Jq^2, \tag{8.22}$$

and

$$E_{BW} = 2Jpg, \tag{8.23}$$

where J is the total number of joints in the study area, or $1/2 \Sigma_i \Sigma_j c_{ij}$ of the binary matrix.

To test if the observed pattern is statistically significantly different from a random pattern (as indicated by the expected values), we can apply the z-test discussed in previous chapters. To perform this test, the standard deviations of the expected joints are also needed. When a stochastic weights matrix is used, the three standard deviations are

$$\sigma_{BB} = \sqrt{\frac{1}{4} p^2 q[S_1 q + S_2 p]} \tag{8.24}$$

$$\sigma_{WW} = \sqrt{\frac{1}{4} q^2 p[S_1 p + S_2 q]} \tag{8.25}$$

$$\sigma_{BW} = \sqrt{\frac{1}{4} \{4S_1 pq + S_2 pq \, [1 - 4\, pq]\}}. \tag{8.26}$$

If a binary matrix is used, these formulas are reduced to

$$\sigma_{BB} = \sqrt{p^2 J + p^3 K - p^4(J + K)} \tag{8.27}$$

$$\sigma_{BB} = \sqrt{q^2 J + q^3 K - q^4(J + K)} \tag{8.28}$$

$$\sigma_{BW} = \sqrt{2pqJ + pqK - 4p^2 q^2(J + K)}, \tag{8.29}$$

where

σ are the standard deviations of the corresponding types of joints,
J, p, and q are defined previously,
K is $\Sigma_{i=1}^{n} L_i(L_i - 1)$, and
L_i is the number of joints between polygon i and all of its adjacent polygons. Therefore, it is the same as the row sum, as described in Table 8.1, and n is the total number of polygons.

Let us focus on testing the negative spatial autocorrelation statistic (i.e., BW joints) and using the binary matrix (and its corresponding formulas). When calculating the expected number of BW joints, the probabilities of any polygon's being black or white must be known, namely, p and q. If these probabilities are known, the only step needed to complete the task is to count the total number of joins, J.

Calculating J is quite straightforward since it only involves counting the number of joints between the polygons or summing up all values of the binary connectivity matrix and then dividing this number in half. For the example in Figure 8.6, the number of joints is 11. Consequently, we will have $J = 11$. Note that all joints are to be counted for J, including all BB, BW, and WW joints. Therefore, a good way to check the correctness of the calculation is to verify that

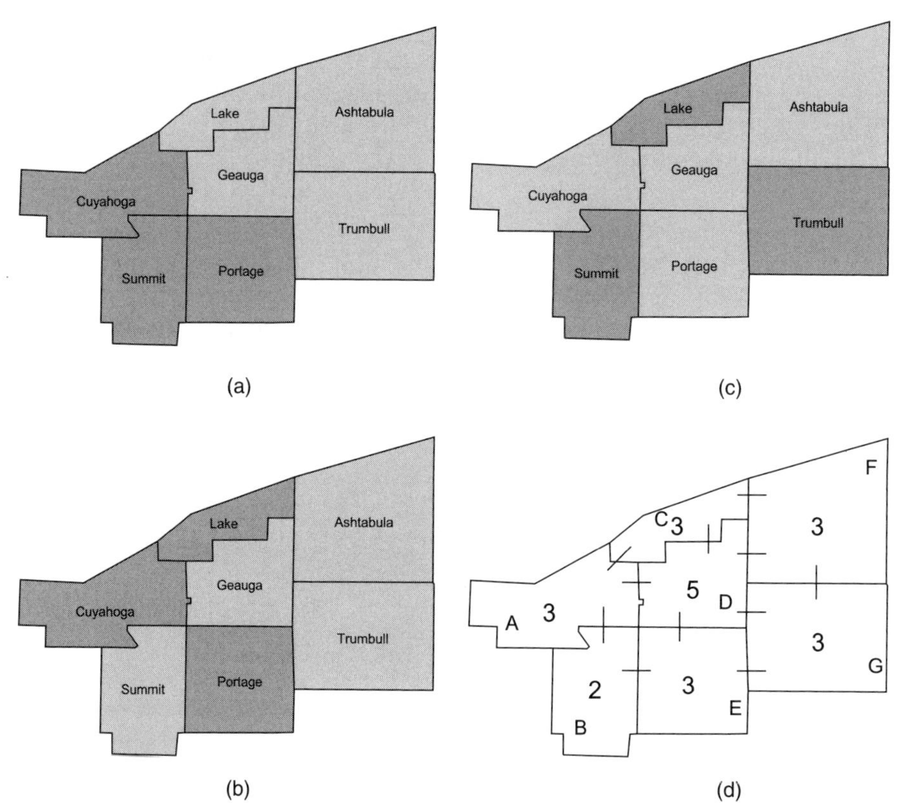

Figure 8.6 Joint structure. (a) Clustered: 4 BW joints. (b) Random: 6 BW joints. (c) Dispersed: 8 BW joints. (d) Number of joints.

$$J = O_{BB} + O_{WW} + O_{BW}. \tag{8.30}$$

Two or more polygons may be contiguous at only one point rather than sharing borders. In this case, it is up to the analyst to decide if this point should be counted as a joint. That is, the analyst must decide if the rook's case or the queen's case should be used in constructing the spatial weights matrix. If one joint of this type is included, it is imperative that all such joints are counted for both the total number of joints and the number of BW joints and other types of joints. In this illustrative example, we adopt the rook's case in defining neighbors. Therefore, Cuyahoga and Portage counties are not treated as neighbors, nor are Geauga and Summit counties. Using different criteria to construct the spatial weights matrix (such as the rook's case vs. the queen's case) will give different results and lead to different conclusions. One may even want to experiment with different neighborhood criteria to evaluate how robust or sensitive the results are to the changing criteria.

With regard to the p and q ratios, the analyst may derive them empirically based on the observed number of black and white polygons, as described above. But it is often desirable to research or look for proper values from theories, literature, historical documents, or based on past experience with the subject. For example, if the subject matter shown by maps in Figure 8.6 is the ratios of the infant mortality rate exceeding a preset threshold at the county level, it would make sense to use the statewide trend to derive proper values for p and q. Similarly, if the maps in Figure 8.6 show the preferences for either the Democratic Party or the Republican Party at the county level, it may be appropriate to find out how the two parties fare across the entire state and use those statewide ratios to determine the values for p and q. To illustrate how to calculate E_{BW} and σ_{BW}, let's assume that $p = 0.3$ and $q = 0.7$. The constraint of $p + q = 1$ should hold for all cases.

The next step is to calculate the value of $\Sigma\, L_i(L_i - 1)$. While there may be several ways in which this number can be calculated, we suggest that a table be set up to calculate the values of L_i, then $L_i - 1$, and then $L_i(L_i - 1)$ for each polygon, i. In this manner, the calculation of $\Sigma\, L_i(L_i - 1)$ will be less confusing and easier to manage. For convenience, we will drop the subscript i in the subsequent discussion of L.

In Figure 8.6d, the polygons representing the seven Ohio counties are labeled A, B, C, D, E, F, and G. The number of joints for each polygon is its L. As shown in Table 8.6, summing the number of all joints yields $\Sigma_{i=A}^{G} L_i = 22$. When each L is reduced by 1, the value of $L - 1$ can be easily calculated. The value of $\Sigma\, L(L - 1)$ is derived by adding up all $L(L - 1)$ values after multiplying each L and $L - 1$ for all rows. With the seven polygons in Figure 8.6, $\Sigma\, L(L - 1) = 52$, as shown in Table 8.6.

All the necessary values have now been calculated and can be substituted into the following equations:

$$E_{BW} = 2 \times 11 \times 0.3 \times 0.7 = 4.62$$

$$\sigma_{BW} = \sqrt{[2 \times 11 + 52] \times 0.3 \times 0.7 - 4 \times [11 + 52] \times 0.3^2 \times 0.7^2}$$

$$= \sqrt{15.54 - 11.11} = \sqrt{4.43} = 2.1.$$

In this example, the expected value of BW joints, E_{BW}, indicates that given 3 black polygons and 4 white polygons, we can expect 4.62 BW joints if the black and white polygons are randomly arranged. This expected value is very important if we want to determine whether the pattern of a number of BW polygons is statistically significantly different from a random pattern. To do that, we would need to take the difference between the observed number and the expected number of BW joints. We have already defined O_{BW} to be the observed number of BW joins in a polygon pattern. This number can be obtained by manually counting the number of BW joints in a simple spatial

TABLE 8.6 Joint Count Statistics: Calculating $\Sigma\ L(L - 1)$

	L	$L - 1$	$L(L - 1)$
A	3	2	6
B	2	1	2
C	3	2	6
D	5	4	20
E	3	2	6
F	3	2	6
G	3	2	6
	$\Sigma\ L - 11$		$\Sigma\ L(L - 1) = 52$

configuration or by using the computational procedure described at the beginning of Section 8.6 in more complicated spatial systems.

In our seven-county example, $O_{BW} = 4$, 6, and 8 for the configurations depicted in Figures 8.6a, 8.6b, and 8.6c, respectively. With these numbers, we know that Figure 8.6a is more clustered than a random pattern because it has a number of joints (4) slightly less than the number of joints for a random pattern (4.62) and that $4 < 4.62$. Figures 8.6b and 8.6c have more BW joints than a random pattern—since $6 > 4.62$ and $8 > 4.62$, respectively. Therefore, they seem to be more dispersed than a random pattern.

The difference between the expected number and the observed number of BW joints can help us assess a polygon pattern in terms of its clustering or dispersion tendency. However, we still do not know how different each pattern is from a random pattern. Measuring this requires that the calculated difference $O_{BW} - E_{BW}$ be standardized.

To standardize the comparison, we will use the z-score in this procedure. Like the z-score described in previous chapters, this z-score (or standardized score) is the difference between the observed and expected statistics divided by the standard deviation, σ_{BW}:

$$z = \frac{O_{BW} - E_{BW}}{\sigma_{BW}}. \tag{8.31}$$

For the examples in Figure 8.6, we have

$$z = \frac{4 - 4.62}{2.1} = -0.29 \quad \text{for Figure 8.6a,}$$

$$z = \frac{6 - 4.62}{2.1} = 0.65 \quad \text{for Figure 8.6b,}$$

and

$$z = \frac{8 - 4.62}{2.1} = 1.61 \quad \text{for Figure 8.6c.}$$

According to the probability distribution of the z-score, any z value that is less than -1.96 or greater than 1.96 is less likely to occur by chance than 5 out of 100 cases ($\alpha = 0.05$). But in this specific significance testing situation, because we test for the significance of negative spatial autocorrelation, and only if O_{BW} is larger than the expected value, E_{BW}, and reflects negative spatial autocorrelation, we should perform a one-tail test. If we still use $\alpha = 0.05$, then the critical value will be 1.645. With this information, we may conclude that none of the patterns in Figure 8.6 has a statistically significant negative spatial autocorrelation or dispersion pattern, even though the pattern in Figure 8.6c has a z-score close to the critical value. If α is changed to 0.1, an increase in the area of the critical region on the right-hand tail, the pattern in Figure 8.6c would be significantly different (statistically) from a random pattern.

The above framework for comparing observed and expected numbers of joints is also applicable to the testing of positive spatial autocorrelation statistics (BB and WW joints). However, in using joint count statistics, there are two statistics depicting positive spatial autocorrelation: O_{BB} and O_{WW}. In order to confirm the presence of positive spatial autocorrelation, we must test both O_{BB} and O_{WW} for significance. That is, O_{BB} has to be compared with its expected value, E_{BB}, and O_{WW} has to be compared with E_{WW} based on their corresponding z-scores. Only if both comparisons turn out to be significant (i.e., both z-scores for the BB and WW comparisons have to be larger than 1.645 if $\alpha = 0.05$ in a one-tail test) can we conclude that there is a significant positive spatial autocorrelation in the pattern.

8.6.2 Randomization Sampling

In randomization sampling, the probability of a polygon's being black or white will depend on the number of black polygons and the number of white polygons in the study area. Using Figure 8.6 as an example again, each of the three possible arrangements has exactly three black polygons and four white polygons. The probability of a polygon's being black is 3/7, and the probability of a polygon's being white is 4/7. In other words, in randomization sampling, the probability of a polygon in a group is constrained by the proportions of the two types of polygons in the study area.

The equations needed to estimate the expected number of joints of all types in nonfree or randomization sampling are

$$E_{BB} = \frac{1}{2} W \frac{n_B(n_B - 1)}{n(n - 1)} \tag{8.32}$$

$$E_{WW} = \frac{1}{2} W \frac{n_W(n_W - 1)}{n(n - 1)} \quad (8.33)$$

$$E_{BW} = W \frac{n_B(n - n_B)}{n(n - 1)} \quad (8.34)$$

with standard deviations equal to

$$\sigma_{BB} = \sqrt{\frac{1}{4}\left(\frac{S_1 n(n_B - 1)}{n(n - 1)} + \frac{(S_2 - 2S_1)n_B^{(3)}}{n^{(3)}} + \frac{[W^2 + S_1 - S_2]n_B^{(4)}}{n^{(4)}}\right] - [E_{BB}]^2} \quad (8.35)$$

and

$$\sigma_{BW} = \sqrt{\frac{1}{4}\left[\frac{2S_1 n_B n_W}{n(n - 1)} + \frac{(S_2 - 2S_1)n_B n_W(n_B + n_W - 2)}{n^{(3)}} + \frac{4[W^2 + S_1 - S_2]n_B^{(2)}n_W^{(2)}}{n^{(4)}}\right] - [E_{BW}]^2} \quad (8.36)$$

if the stochastic row-standardized matrix is used. All the terms were defined in the previous sections, except n_B and n_W. Intuitively, n_B is the number of polygons that are labeled black, and n_W is the number of polygons that are labeled white. We did not present the equation for σ_{WW} because it is similar to σ_{BB} as defined above, except that n_W is used instead of n_B. If the binary connectivity matrix is used instead, the expected value for BW joints and the corresponding standard deviation are

$$E_{BW} = \frac{2Jn_B n_W}{n(n - 1)} \quad (8.37)$$

and

$$\sigma_{BW} = \sqrt{E_{BW} + \frac{\Sigma L(L-1)n_b n_w}{n(n-1)} + \frac{4[J(J-1) - \Sigma L(L-1)]n_b^{(2)}n_w^{(2)}}{n^{(4)}} - E_{BW}^2}. \quad (8.38)$$

The values of J and $\Sigma\, L(L - 1)$ are the same as those for the free sampling or normality case. For the polygon patterns in Figure 8.6, the values of E_{BW} and σ_{BW} are

$$E_{BW} = \frac{2 \times 11 \times 3 \times 4}{7 \times 6} = \frac{264}{42} = 6.286$$

and

$$\sigma_{BW} = \sqrt{6.286 + \frac{52 \times 3 \times 4}{7 \times 6} + \frac{4 \times [(11 \times 10) - 52] \times 3 \times 2 \times 4 \times 3}{7 \times 6 \times 5 \times 4} - 6.286^2}$$
$$= \sqrt{6.286 + 14.857 + 19.886 - 39.514} = \sqrt{1.515} = 1.23.$$

Similarly, we can use the z-score, $z = (O_{BW} - E_{BW})/\sigma_{BW}$, to test how each of our sample polygon patterns compares to a random pattern under the condition of three black polygons and four white polygons:

$$z = \frac{4 - 6.286}{1.23} = -1.85 \quad \text{for Figure 8.6a,}$$

$$z = \frac{6 - 6.286}{1.23} = -0.23 \quad \text{for Figure 8.6b,}$$

and

$$z = \frac{8 - 6.286}{1.23} = 1.39 \quad \text{for Figure 8.6c.}$$

Given these z-scores and the critical value of 1.645 with $\alpha = 0.05$ in a one-tail situation, we can conclude, as in the normality sampling framework, that none of the patterns in Figure 8.6 has a significant negative spatial autocorrelation. This is because none of the z-scores of the negative spatial autocorrelation joint count statistics (O_{BW}) is larger than the critical value. Though the conclusion from the randomization sampling is the same as the conclusion from the normality assumption, note that the corresponding z-scores based on the two sampling assumptions are different. The differences in the z-scores imply that one may not have the same conclusions when different sampling assumptions are adopted.

ArcView Example 8.2: Joint Count Statistics

Joint count statistics are the simplest form of spatial autocorrelation measures. They are easy to understand and calculate. However, these statistics are only for binary nominal data, and the calculated statistics with their z-scores are the bases used to draw conclusions.

Step 1 Explore data theme

As a demonstration, we will use the Ohcounties.shp shapefile from ArcView Example 8.1.
Let's explore the data theme further.

- Use the **Open Theme Table** button to open the attribute table of `Ohcounties.shp`.
- Scroll to the right until the `SeniorPct` and the `Senior` fields are shown.

The `SeniorPct` field holds the percentages of senior (>=65 years old) population for all counties, while the `Senior` field holds the counties' assignments to 1 if their SeniorPct values are higher than the average of all 88 counties and 2 otherwise.

- In the View document, use the **Legend Editor** to set the display symbology so that counties with Senior = 1 are displayed in one color and those with Senior = 2 in another.

Note that the distribution of senior population shows that the rural counties have higher percentages of senior population.

Step 2 Calculate joint count statistics
To calculate joint count statistics:

- From the **Ch.8** menu, choose **Joint Count Statistics.**

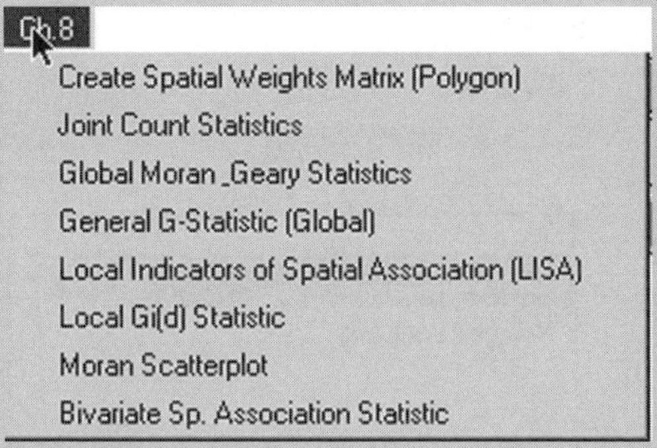

Ch.8
Create Spatial Weights Matrix (Polygon)
Joint Count Statistics
Global Moran _Geary Statistics
General G-Statistic (Global)
Local Indicators of Spatial Association (LISA)
Local Gi(d) Statistic
Moran Scatterplot
Bivariate Sp. Association Statistic

Since no records have been selected prior to this step, we will use all records for the analysis.
In the Joint Count Statistics dialog box:

- Click to highlight the `Ohcounties.shp` shapefile.
- Click the **Choose the Theme** button to list all attribute fields in the two list boxes.
- Check the next box to indicate that the Binary matrix has been created (from ArcView Example 8.1, distbin.dbf).
- Check the box to indicate that all records are to be in the selection.
- For the **ID field,** scroll down the list and click to highlight the `Cnty_fips` field.
- For the binary variable, scroll down the list and click to highlight the `Senior` field.
- Click the **Calculate** button to proceed.
- In the **Choose the Binary or Stochastic Matrix File** dialog box, navigate to the folder that stores distbin.dbf and click to choose it.
- Click **OK** to proceed.

A few intermediate Info dialog boxes will report statistics as the calculation progresses.
First, S_1 is reported as 924. Next, S_2 is reported as 10304.
The calculation will take a few minutes, as it involves a number of iterations in counting various types of polygon joints. When the first part of the calculation is completed, your will be prompted for the assumptions under which z-scores are to be calculated. Since we cannot change the fact that there are 88 counties in Ohio, let's use the nonfree sampling assumption.
In the Sampling dialog box:

- Choose `Nonfree sampling` and then click **OK.**

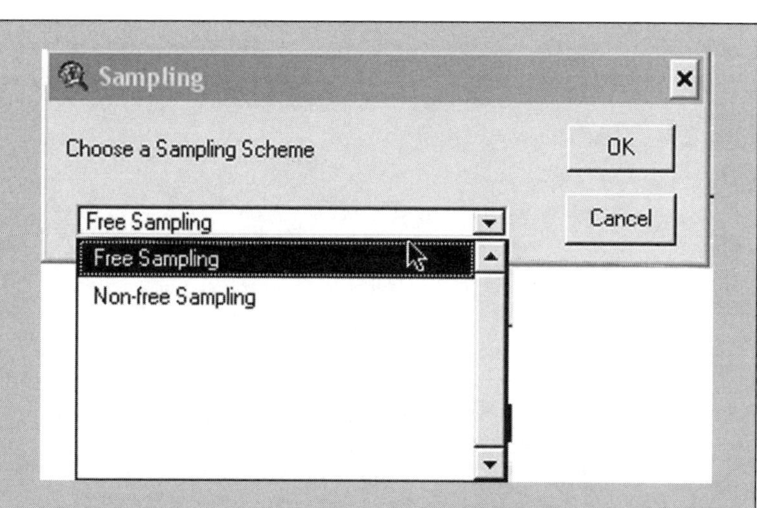

Step 3 Analysis
Base on the calculated statistics and the *z*-scores, we can reach the following conclusions:
For the nonfree sampling assumption, we have

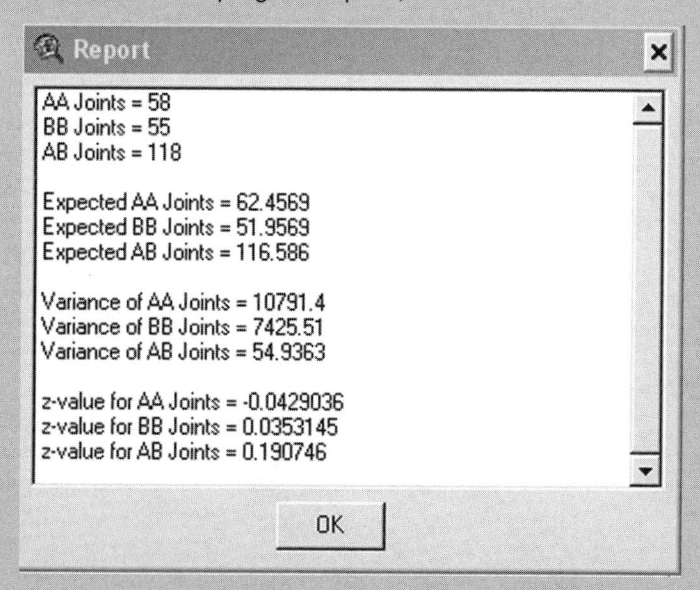

Judging by the values of the *z*-scores, we have to conclude that we cannot reject the null hypothesis that stipulates no significant difference between the observed pattern and a random pattern of the 46/42 division among the 88 counties.
Note that the results can be written to and saved in a text file for future use.

Step 4 End

8.7 GLOBAL SPATIAL AUTOCORRELATION STATISTICS

Joint count statistics are useful global measures of spatial autocorrelation for variables with only two outcomes. In real life, we do have situations resembling the binary variable framework, such as the election of mayors or presidents, if there are only two candidates. Still, the binary outcome situation is relatively uncommon and rather restrictive, as many real-world cases deal with variables at interval or ratio scales. In these cases, Moran's I and Geary's Ratio, *C*, can be used.

Moran's I and Geary's Ratio have some common characteristics, but their statistical properties are different. Most analysts favor Moran's I mainly because the characteristics of its numeric distribution are more desirable than those of Geary's *C* (Cliff and Ord, 1973, 1981). Nevertheless, both statistics are based on comparing attribute values of neighboring areal units. If neighboring units over the entire study area have similar values, then the statistics should indicate a strong positive spatial autocorrelation. If neighboring areal units have dissimilar values, then the statistics should show a strong negative spatial autocorrelation. The two statistics, however, adopt different approaches to compare neighboring values.

8.7.1 Moran's I

Moran's I can be defined simply as

$$\mathrm{I} = \frac{n \Sigma \Sigma w_{ij} \, (x_i - \bar{x})(x_j - \bar{x})}{W \, \Sigma \, (x_i - \bar{x})^2}, \tag{8.39}$$

where

x_i is the value of the interval or ratio variable in areal unit i and
W is the sum of all elements of the spatial weights matrix, which can be of any form, including the binary, stochastic, or distance-based matrix.

Other terms have been defined previously.

The value of Moran's I ranges from -1 for an extremely negative spatial autocorrelation to 1 for an extremely positive spatial autocorrelation. Even though the theoretical range of Moran's I is $(-1, 1)$, empirical studies may occasionally obtain a value slightly beyond this range. Nevertheless, this issue does not pose a significant problem with the use of Moran's I.

If no spatial autocorrelation exists, the expected value for Moran's I is

$$E_I = \frac{-1}{n - 1}. \tag{8.40}$$

Note that E_I is always negative. Because E_I is inversely related to n, the number of areal units, the larger the size of the spatial system, the smaller will be the absolute value of E_I, but E_I will approach 0 only from the negative side. This property of E_I has two implications. First, when the spatial system (n) is small, E_I can be very negative. Therefore, when we observe a negative Moran's I in a small system, we should not assume or develop an impression that there is a strong negative spatial autocorrelation. A relatively negative Moran's I value for a small system may still represent no spatial autocorrelation or even positive spatial autocorrelation, as long as the observed negative Moran's I value is larger than the (negative) expected value. Second, we often use zero as the value to indicate a neutral situation, such as no correlation in the case of Pearson's correlation coefficient. But this is not the case with Moran's I. The value zero cannot be used as a reference point to distinguish positive from negative spatial autocorrelation.

When calculating Moran's I, the most commonly used spatial weights matrices are the binary and stochastic matrices. If a binary matrix is used, W in the denominator is basically twice the number of shared boundaries in the entire study region, or $2J$. But note that it is possible to use other types of weights matrices, even the distance-based weights matrix. For our purpose, let us assume that a binary matrix is used.

In the numerator of Equation 8.39, the Moran's I calculation, if i and j are neighbors, then w_{ij} will be 1. Therefore, if areal unit i and areal unit j are not neighbors, w_{ij} will be 0 for that pair of units i and j. If $w_{ij} = 0$, that pair of values will not be compared because units i and j are not neighbors. If areal units i and j are neighbors ($w_{ij} = 1$), their values are first compared with the mean of the variable separately. The deviations from the mean of the two neighboring units are then multiplied together. The products of the deviations from the mean are then summed for all pairs of areal units as long as they are neighbors. If both neighboring values are above the mean, then the product is a large positive number. So is the product if both neighboring values are below the mean (the product of two negative numbers).

These situations reflect the presence of positive spatial autocorrelation where similar situations—that is, both values are larger than the mean or both are smaller than the mean—are next to each other. But if the value of one areal unit is above the mean and the value of its neighboring unit is below the mean, the product of the two mean deviations will be negative, indicating the presence of negative spatial autocorrelation. Therefore, over the entire study region, if similar situations (can be high-high or low-low) are more likely than dissimilar situations between neighbors, Moran's I tends to be positive. Alternatively, if situations are mostly high-low and low-high, we would have negative values for Moran's I.

The numerator of Moran's I is based on the covariance, $(x_i - \bar{x})(x_j - \bar{x})$, which is a cross-product of the differences between neighboring values and the overall mean. This covariance structure is also the basis of Pearson's product-moment correlation coefficient (Chapter 3), which is defined as

$$r = \frac{(x_i - \bar{x})(y_i - \bar{y})}{n\delta_x \delta_y}. \tag{8.41}$$

Equation 8.41 measures how closely the distributions of the two variables, x and y, resemble each other. For a given observation, i, if both x_i and y_i are above their means, the product will be large and positive. A similar result will occur when both x_i and y_i are below their means. Only if one of the two variables has a value above the mean and the other has a value below the mean will the correlation be negative.

In contrast to Pearson's correlation coefficient, the covariance in Moran's I described in Equation 8.39 is the covariance over space for neighboring units and the covariance will not be counted unless areal units i and j are neighbors. Also, only one variable is considered (x) in Moran's I instead of two (x and y) in the Pearson correlation coefficient. The denominator of Moran's I is essentially the sum of the squared deviations scaled by the total weight of the matrix.

Using the seven Ohio counties to illustrate the concepts related to Moran's I, Figure 8.7 shows the variable of the median household income in the Summary File 3 of the 2000 Census. Using the binary connectivity matrix and the queen's case to define neighbors, Moran's I is calculated. Tables 8.7a and 8.7b show values derived from the intermediate steps. In Table 8.7a, the median family income (Med Income), deviations from the mean $(x - \bar{x})$ and the square of the mean deviations $(x - \bar{x})^2$ are reported. The mean of the median

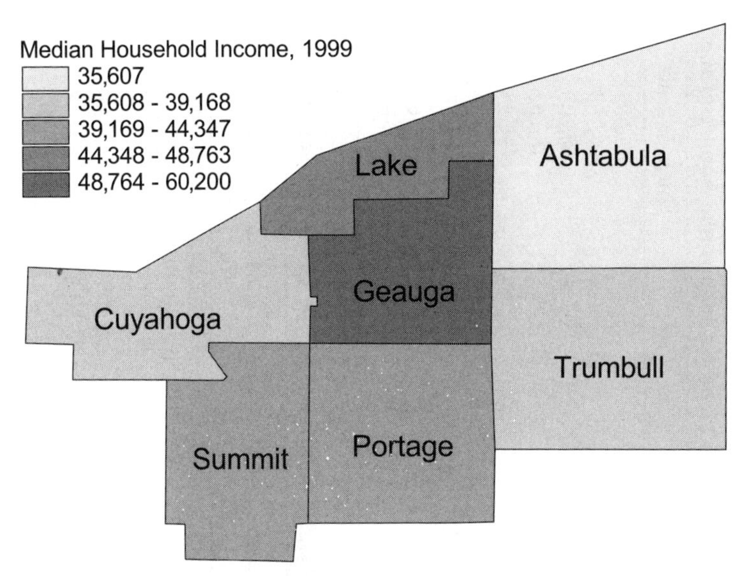

Figure 8.7 Median household income of seven Ohio counties, 2000 Census, Summary File 3.

TABLE 8.7 (a) Mean Deviations and Square of the Mean Deviations

Counties	Med. Income (x)	$x\text{-}\bar{x}$	$(x\text{-}\bar{x})^2$
Ashtabula	35,607	−8,491.14	72,099,507.02
Cuyahoga	39,168	−4,930.14	24,306,308.59
Geauga	60,200	16,101.86	259,269,803.45
Lake	48,763	4,664.86	21,760,892.16
Portage	44,347	248.86	61,929.88
Summit	42,304	−1,794.14	3,218,948.59
Trumbull	38,298	−5,800.14	33,641,657.16
Total	308,687		414,359,046.86
Mean	44,098.14		

(b) Weights Multiplied by the Cross-Product of Mean Deviations

Counties		Geauga	Cuyahoga	Trumbull	Summit	Portage	Ashtabula	Lake	
	Deviations	16.102	−4.930	−5,800	−1,794	249	−8,491	4,665	Total
Geauga	16,102	0	16,102*(−4930)	16,102*(−5800)	16,102*(−1794)	16,102*(249)	16,102*(−8491)	16,102*(4665)	−259,258,302
Cuyahoga	−4,930	−4,930*(16102)	0	0	−4,930*(−1794)	−4,930*(249)	0	−4,930*(4665)	−94,764,460
Trumbull	−5,800	−5,800*(16102)	0	0	0	−5,800*(249)	−5,800*(−8491)	0	−45,588,000
Summit	−1,794	−1,794*(16102)	−1,794*(−4930)	0	0	−1,794*(249)	0	0	−20,489,274
Portage	249	249*(16102)	249*(−4930)	249*(−5800)	249*(−1794)	0	0	0	890,922
Ashtabula	−8,491	−8,491*(16101)	0	−8,491*(−5800)	0	0	0	−8,491*(4665)	−127,084,797
Lake	4,665	4,665*(16102)	4,665*(−4930)	0	0	0	4,665*(−8491)	0	12,506,865
								Grand Total =	−533,787,046

household income is \$44,098. Using the mean, the mean deviations were calculated in the third column. The sum of the squared mean deviation, which is $\Sigma\,(x_i - \bar{x})^2 = 414{,}359{,}046$, is part of the denominator of Moran's I. Another part of the denominator is the sum of the spatial weights, W. Because we use a binary matrix, the sum of all weights is basically $2J$, which is $2 \times 13 = 26$. For the numerator, we have to compute $w_{ij}(x_i - \bar{x})(x_j - \bar{x})$ for each pair of i and j. The binary weight matrix is reported in Table 8.1.

In Table 8.7b, the mean deviations are also listed together with the corresponding counties. As long as the pair of counties has a 1 in the binary matrix, their corresponding mean deviations will be used to form a product reported in Table 8.7b. These cross-products are then summed to give a grand total of −533,787,046.

Therefore, we have Moran's I:

$$\mathrm{I} = \frac{7(-533{,}787{,}046)}{26 \times 414{,}359{,}047} = -0.3468.$$

The calculated value for Moran's I seems to indicate a negative spatial autocorrelation. But as warned before, the expected value for Moran's I, that is, the value of Moran's I for a random pattern, is $-1/(n - 1)$, which is always a negative value, and will be quite negative when n is small (and in this case, we have an $n = 7$ only). Therefore, we have to compute the expected value for Moran's I first and then compare it with the calculated value for Moran's I. In this example, we have the expected value for Moran's I,

$$E(\mathrm{I}) = -1/(7 - 1) = -0.16667.$$

Mathematically we can see that our calculated (observed) value for Moran's I is more negative than the expected value of Moran's I when there is no spatial autocorrelation. Though this simple comparison indicates that the observed pattern is more negatively spatially autocorrelated than a random pattern, we still have to test if this difference between the calculated and expected values for Moran's I occurred by chance or is very unlikely to happen, implying a systematic pattern. To test for the significance of Moran's I, we adopt the same method we used to test the significance of joint count statistics.

The difference between the calculated (observed) and expected values for Moran's I is scaled by the standard error of Moran's I in order to derive the z-score. As mentioned before, the expected value for Moran's I is $E_I = -1/(n - 1)$, that is, the value for Moran's I if there is no spatial autocorrelation. This definition of the expected value for Moran's I applies to both the normality and randomization sampling assumptions, but the estimations of the variance and standard error vary according to the sampling assumptions. Similar to the situation in joint count statistics, the two sampling assumptions of normality and randomization are applicable to the significance testing of Moran's I.

As discussed before, normality sampling assumes that the attribute values of x_i are independently drawn from a normal distribution and therefore are not limited by the existing values of the observed spatial pattern. Thus, the distribution properties of the values for Moran's I are derived from repeated sampling of a set of values drawn from the normal distribution. Different sets of values and their corresponding means are therefore different. The variance of Moran's I under the normality assumption is

$$\sigma^2(\mathrm{I}) = \frac{n^2 S_1 - nS_2 + 3(W)^2}{(W)^2(n^2 - 1)}, \tag{8.42}$$

where all the terms were defined in previous sections.

Under the randomization assumption, however, the set of attribute values (of the areal units) is fixed. What is not fixed is the location associated with each value or the association between attribute values and areal units. In other words, there are many ways to distribute a set of attribute values to a set of areal units in the spatial system. The one we observe is one of many possible spatial patterns given the set of attribute values and the set of areal units. The configurations that yield no significant spatial autocorrelation are those generated by distributing the set of values independently and randomly to the areal units. Therefore, the variance is based on the number of possible permutations of the n data values over the n locations (areal units) and is defined as

$$\sigma^2(\mathrm{I}) = \frac{n[(n^2 - 3n + 3)S_1 - nS_2 + 3W^2] - k[(n^2 - n)S_1 - 2nS_1 + 6W^2]}{(n - 1)(n - 2)(n - 3)(W^2)}, \tag{8.43}$$

where all these terms were defined as before, and k is the kurtosis measure without the "-3" adjustment discussed in Chapters 2 and 6.

Using the seven Ohio counties data and the binary connectivity matrix,

$n = 7$,
$S_1 = 52$,
$S_2 = 416$, and
$W = 26$.

Therefore, under the normality assumption, we have

$$\sigma^2(\mathrm{I}) = \frac{7^2 \times 52 - (7)(416) + 3(26)^2}{(26)^2(7^2 - 1)} = 0.0513.$$

If the randomization assumption is used, $\sigma^2(\mathrm{I})$ is 0.04945. The corresponding z-scores are then

$$z_N(\mathrm{I}) = \frac{-0.3468 - (-0.1667)}{\sqrt{0.0513}} = -0.7952 \quad \text{for the normality assumption}$$

and

$$z_R(\mathrm{I}) = \frac{-0.3468 - (-0.1667)}{\sqrt{0.0471}}$$
$$= -0.8299 \quad \text{for the randomization assumption.}$$

Putting the sampling assumption aside, Moran's I indicates a somewhat negative spatial autocorrelation, even after comparison with the expected values of Moran's I from a random pattern. But regardless of which sampling assumption we adopt, because both z-scores are negative but larger than -1.96, using the traditional criterion ($\alpha = 0.05$ level of significance), we cannot conclude that the observed pattern is significantly different from a random pattern. Though the observed value for Moran's I and the corresponding z-scores based on both sampling assumptions indicate a negative spatial autocorrelation, the magnitudes are not significant enough to conclude that there is a significant negative spatial autocorrelation in the variable of median household income among the seven Ohio counties. The negative Moran's I may be generated by chance; it is not due to a systematic process.

8.7.2 Geary's Ratio

Similar to Moran's I method in measuring spatial autocorrelation, Geary's Ratio, C, also adopts a cross-product term (Getis, 1991). Geary's C is formally defined as

$$C = \frac{(n - 1) \Sigma \Sigma w_{ij}(x_i - x_j)^2}{2W \Sigma (x_i - \bar{x})^2}. \tag{8.44}$$

Like Moran's I, Geary's C can accommodate any type of spatial weights matrices, although the most popular types are the binary and stochastic matrices. In comparing Equation 8.44 with Equation 8.39 for Moran's I, it is apparent that the most significant difference between them is the term in the numerator. In Moran's I, the numerator is based on the deviations from the mean of the two neighboring values. In Geary's C, instead of comparing the neighboring values with the mean, it compares the two neighboring values with each other directly ($x_i - y_i$). To a large degree, the index is not concerned with whether x_i is larger than x_j or vice versa, but rather with how dissimilar the two neighboring values are. Therefore, the differences between neighboring values are squared to remove the directional aspect of the differences.

Geary's C ranges from 0 to 2. Somewhat counterintuitively, a value of 0 in Geary's C indicates a perfect positive spatial autocorrelation when all neighboring values are the same and, therefore, the numerator becomes 0. When Geary's C is 2, it indicates an extremely negative spatial autocorrelation. In contrast to Moran's I, the expected value for Geary's C is not affected by the sample size, n, but is always 1.

By comparing the equations for Moran's I and Geary's C, and the interpretations of the values, one may develop an impression that the two indices are very different. It is true that they have different distributional properties such that Moran's I is more desirable in most situations. However, the two indices are related algebraically in the following manner (Griffith, 1987):

$$C = \left\{ \frac{(n-1) \Sigma \Sigma w_{ij}(x_i - x_j)^2}{2n \Sigma \Sigma w_{ij}(x_i - \bar{x})(x_j - \bar{x})} \right\} \times I. \tag{8.45}$$

In other words, one may compute Geary's C based on Moran's I and vice versa.

Using the Ohio counties example again, we have all the parameters to calculate Geary's C from Moran's I calculation except the sum of the weights times the square of the difference ($\Sigma\Sigma w_{ij}(x_i - x_j)^2$) in the numerator. Table 8.8 shows the derivation of this term. Using thc binary connectivity matrix in Table 8.1 again, only if the corresponding cell in Table 8.1 is 1 (i.e., the weight), the corresponding values of the pair of neighbors are compared in Table 8.8. The difference of the value has to be squared too. The sum of the weights times the square of the difference is 5,158,108,600. Then Geary's C is

$$C = \frac{6 \times 5{,}158{,}108{,}600}{2 \times 26 \times 414{,}359{,}047} = 1.4364,$$

indicating a slightly negative situation, which is consistent with the result from Moran's I.

As in the discussion on Moran's I, we also have to test if the observed value for Geary's C is statistically different from the expected value of C for a random pattern. To derive the z-score, we need to know the expected C and its variance. We know that the expected value of C is 1 regardless of the size of n. For an estimate of the variance under the normality assumption,

$$\sigma^2(C) = \frac{(2S_1 + S_2)(n-1) - 4W^2}{2(n+1)W^2}, \tag{8.46}$$

while under the randomization assumption,

TABLE 8.8 Deriving the Numerator of Geary's C (Weights Multiplied by the Square of the Cross-Product Differences)

Counties		Geauga	Cuyahoga	Trumbull	Summit	Portage	Ashtabula	Lake	
	Med Inc	60,200	39,168	38,298	42,304	44,347	35,607	48,763	Total
Geauga	60,200	0	$(60{,}200-39{,}168)^2$	$(60{,}200-38{,}298)^2$	$(60{,}200-42{,}304)^2$	$(60{,}200-44{,}347)^2$	$(60{,}200-35{,}607)^2$	$(60{,}200-48{,}763)^2$	2,229,247,671
Cuyahoga	39,168	$(39{,}168-60{,}200)^2$	0	0	$(39{,}168-42{,}304)^2$	$(39{,}168-44{,}347)^2$	0	$(39{,}168-48{,}763)^2$	571,065,586
Trumbull	38,298	$(38{,}298-60{,}200)^2$	0	0	0	$(38{,}298-44{,}347)^2$	$(38{,}298-35{,}607)^2$	0	523,529,486
Summit	42,304	$(42{,}304-60{,}200)^2$	$(42{,}304-39{,}168)^2$	0	0	$(42{,}304-44{,}347)^2$	0	0	334,275,161
Portage	44,347	$(44{,}347-60{,}200)^2$	$(44{,}347-39{,}168)^2$	$(44{,}347-38{,}298)^2$	$(44{,}347-42{,}304)^2$	0	0	0	318,903,900
Ashtabula	35,607	$(35{,}607-60{,}200)^2$	0	$(35{,}607-38{,}298)^2$	0	0	0	$(35{,}607-48{,}763)^2$	785,137,466
Lake	48,763	$(48{,}763-60{,}200)^2$	$(48{,}763-39{,}168)^2$	0	0	0	$(48{,}763-35{,}607)^2$	0	395,949,330
									Grand Total = 5,158,108,600

$$\sigma^2(C) = \frac{\begin{array}{l}(n-1)S_1[n^2-3n+3-(n-1)k] \\ -\frac{1}{4}(n-1)S_2[n^2+3n-6-(n^2-n+2)k]+W^2[n^2-3-(n-1)^2k]\end{array}}{n(n-2)(n-1)W^2}. \quad (8.47)$$

In our Ohio example, the variances are 0.0385 and 0.03261 for the normality and randomization assumptions, respectively. Therefore, the respective z-scores, which are based on the observed value minus the expected value (1.4364 − 1) and then divided by the standard deviations ($\sqrt{0.0385}$ and $\sqrt{0.03261}$), are 2.2252 for normality and 2.4166 for randomization.

Remember that large values of Geary's C indicate a negative spatial autocorrelation, 0 indicates a perfectly positive spatial autocorrelation, 2 indicates a perfectly negative spatial autocorrelation, and 1 (expected value) indicates no spatial autocorrelation. Therefore, positive z-scores like the two above from the two sampling assumptions suggest that the observed pattern exhibits negative spatial autocorrelation.

Because the z-scores under both sampling assumptions are larger than 1.96, we can concluded that there is a significant negative spatial autocorrelation in median household income among the seven Ohio counties. Note that in this specific example, the two spatial autocorrelation measures (Moran's I and Geary's C) do not yield consistent conclusions, regardless of which sampling assumption is adopted. Both statistics indicate moderate levels of negative spatial autocorrelation. But values from Moran's I statistics are not significant under the two sampling assumptions. On the other hand, the values from Geary's C statistics are significant. In general, Moran's I and Geary's C should offer similar conclusions, provided that the data are not too skewed in their distribution. In the Ohio case, the data are quite skewed, with Geauga County's income being far higher than the rest. Therefore, the results of significance testing of these statistics are not consistent.

For each of the spatial autocorrelation statistics, the two sampling assumptions yield the same conclusion. In general, we do not expect the two sampling assumptions to offer dramatically different results. However, minor differences in their z-scores and corresponding probability values should be the norm rather than the exception. One should not assume that results based on one sampling assumption will be identical to those based on other assumptions in all cases.

8.7.3 General *G* Statistic

Moran's I and Geary's C both have well-established statistical properties. Each of them offers a summary measure for describing the level of spatial autocorrelation for the entire study region. However, neither was designed to assess localized spatial autocorrelation surrounding individual areal units. Therefore, Moran's I and Geary's C are known as global measures. They are not effective in identifying different types of local clustering patterns. These patterns are sometimes described as *hot spots* and *cold spots.* For instance, if high values are close to each other, Moran's I and Geary's C will indicate relatively high positive spatial autocorrelation—but each with an index value for the entire area. A local cluster of high values may be labeled a hot spot in spatial clustering analysis. When low values are close to each other to form a local cluster, the cluster may be labeled a cold spot. But when two types of clusters are evaluated by Moran's I and Geary's C, both indices will return a high positive spatial autocorrelation. This is because both indices are concerned with only whether neighboring values are similar or not; they cannot determine if the similarity is due to high values or low values. In other words, Moran's I and Geary's C cannot distinguish the two types of spatial autocorrelation (H-H vs. L-L).

The **general *G*-statistic** (Getis and Ord, 1992), which is also a global spatial autocorrelation statistic, has an advantage over Moran's I and Geary's C in that it is capable of detecting the presence of hot spots or cold spots over the entire study area. These hot spots or cold spots can be thought of as spatial concentrations of particular phenomena, such as air pollutants (ozone), chemicals (dioxin), or income levels.

Similar to Moran's I and Geary's C, the general G-statistic is also based on cross-product statistics. The cross-product is often labeled as a measure of *spatial association.* Formally, the general G-statistic is defined as

$$G(d) = \frac{\Sigma\Sigma\, w_{ij}(d) x_i x_j}{\Sigma\,\Sigma\, x_i x_j} \quad \text{for } i \neq j. \tag{8.48}$$

In the numerator of the G-statistic, the weights are defined by a distance function or neighborhood function with distance d. The weight, $w_{ij}(d)$, is 1 if areal unit j is within d from area unit i and 0 otherwise. That is, the two areal units are regarded as neighhors if they are less than d distance from each other. Therefore, the weights matrix is essentially a binary symmetrical matrix that has 1's for neighboring areal units and 0's for areal units that are not neighbors using the neighborhood function of d. As a result, the product of x_i and x_j contributes to the G-statistic only if they are within d distance of each other.

If i and j are point objects, the inclusion of j should be very simple. But when i and j are polygon features, neighborhood definition based on distance becomes tricky. As discussed in Section 8.4, when distance is involved in defining a neighborhood of polygon features, we can use the centroid distance or the nearest distance. Of course, how neighboring polygons should be defined is dependent upon the actual conditions of the areal units and the phenomena being studied.

The sum of the weights matrix is $W = \Sigma_i \Sigma_j w_{ij}(d)$, where $j \neq i$. Because of this nature of the weight, some of the (x_i, x_j) pairs will not be included in the numerator. For example, if i and j are more than d away from each other, the product of $x_i \times x_j$ will not be included in the numerator. On the other hand, the denominator includes all (x_i, x_j) pairs, regardless of how far i and j are, with the exception of (x_i, x_i). In this manner, the denominator is always larger than or equal to the numerator.

Basically, the numerator, which dictates the magnitude of the $G(d)$ statistic, is relatively large if neighboring values are large and small if neighboring values are small. This is a distinctive property of the general G-statistic. A moderate level of $G(d)$ reflects a spatial association of high and moderate values. A relatively low level of $G(d)$ indicates a spatial association of low and below-average values. Therefore, the $G(d)$ statistic has the ability to detect clusters of high values (hot spots) and low values (cold spots).

Before calculating the general G-statistic, one must define a distance, d, within which areal units will be regarded as neighbors. In the seven Ohio counties, we choose 30 miles for demonstration purposes. Referring to Table 8.9, which shows the centroid distances among these seven counties, the 30-mile distance is chosen somewhat arbitrarily, but is large enough for each county to include a least one other county as its neighbor. This distance is also not too large to include all counties as the neighbors for any given county. Based upon the 30-mile criterion to define neighbors, a binary matrix is derived in Table 8.9. Using median household income as the variable again, we can compute the general G-statistic.

The denominator of the G-statistic is essentially the sum of the cross-products of the values in the variable (median household income), except that the square of the variable is not included (when $i = j$). In Table 8.10, the second top row and the second leftmost column are the median household income. Each cell is derived from pulling the median income values from the corresponding row and column to form a product, except for cells along the major diagonal cells (where $i = j$). The denominator is the sum of all of these cell values, which is equal to 81,260,781,498. For the numerator, it is based on the values in Table 8.10, except that the cells with 0 in the binary matrix in Table 8.9 are not counted, as those cells refer to county pairs that are more than 30 miles away from each other. They are not considered neighbors. Therefore, Table 8.11 shows the results.

Based on the results from these two tables, the general G-statistic for the median household income in the seven-county region in Ohio is

$$G(d) = \frac{45{,}307{,}133{,}554}{81{,}260{,}781{,}498} = 0.5576,$$

where $d = 30$. We may consider $G(d)$ as the proportion of the spatial association accounted for within d distance units of each observation. When this

TABLE 8.9 Converting a Distance Matrix into a Binary Matrix Using 30 Miles as the Threshold

Distance Matrix Based upon Centroid Distance

ID	Geauga	Cuyahoga	Trumbull	Summit	Portage	Ashtabula	Lake
Geauga	0	25.1970	26.6389	32.7420	25.0076	26.5672	12.6331
Cuyahoga	25.1970	0	47.7889	23.4420	31.5755	50.8241	28.2636
Trumbull	26.6389	47.7889	0	41.8271	24.4707	29.5520	36.7000
Summit	32.7420	23.4420	41.8271	0	17.7783	58.0720	42.7222
Portage	25.0076	31.5755	24.4707	17.7783	0	45.5305	37.4669
Ashtabula	26.5672	50.8241	29.5520	58.0720	45.5305	0	24.7187
Lake	12.6331	28.2636	36.7000	42.7222	37.4669	24.7187	0

Based upon 30 Miles as Threshold, Distance Matrix Is Converted into a Binary Matrix

ID	Geauga	Cuyahoga	Trumbull	Summit	Portage	Ashtabula	Lake
Geauga	0	1	1	0	1	1	1
Cuyahoga	1	0	0	1	0	0	1
Trumbull	1	0	0	0	1	1	0
Summit	0	1	0	0	1	0	0
Portage	1	0	1	1	0	0	0
Ashtabula	1	0	1	0	0	0	1
Lake	1	1	0	0	0	1	0

TABLE 8.10 Denominator Calculation for the General-G Statistic

Counties		Geauga	Cuyahoga	Trumbull	Summit	Portage	Ashtabula	Lake	
	Med Inc	60,200	39,168	38,298	42,304	44,347	35,607	48,763	Total
Geauga	60200	0	60200*39168	60200*38298	60200*42304	60200*44347	60200*35607	60200*48763	14,958,917,400
Cuyahoga	39168	39168*60200	0	39168*38298	39168*42304	39168*44347	39168*35607	39168*48763	10,556,520,192
Trumbull	38298	38298*60200	38298*39168	0	38298*42304	38298*44347	38298*35607	38298*48763	10,355,357,922
Summit	42304	42304*60200	42304*39168	42304*38298	0	42304*44347	42304*35607	42304*48763	11,269,066,432
Portage	44347	44347*60200	44347*39168	44347*38298	44347*42304	0	44347*35607	44347*48763	11,722,685,980
Ashtabula	35607	35607*60200	35607*39168	35607*38298	35607*42304	35607*44347	0	35607*48763	9,723,559,560
Lake	48763	48763*60200	48763*39168	48763*38298	48763*42304	48763*44347	48763*35607	0	12,674,674,012
								Grand Total =	81,260,781,498

TABLE 8.11 Numerator Calculation for the General-G Statistic

Counties		Geauga	Cuyahoga	Trumbull	Summit	Portage	Ashtabula	Lake	
	Med Inc	60,200	39,168	38,298	42,304	44,347	35,607	48,763	Total
Geauga	60200	0	2,357,913,600	2,305,539,600	0	2,669,689,400	2,143,541,400	2,935,532,600	12,412,216,600
Cuyahoga	39168	2,357,913,600	0	0	1,656,963,072	0	0	1,909,949,184	5,924,825,856
Trumbull	38298	2,305,539,600	0	0	0	1,698,401,406	1,363,676,886	0	5,367,617,892
Summit	42304	0	1,656,963,072	0	0	1,876,055,488	0	0	3,533,018,560
Portage	44347	2,669,689,400	0	1,698,401,406	1,876,055,488	0	0	0	6,244,146,294
Ashtabula	35607	2,143,541,400	0	1,363,676,886	0	0	0	1,736,304,141	5,243,522,427
Lake	48763	2,935,532,600	1,909,949,184	0	0	0	1,736,304,141	0	6,581,785,925
								Grand Total =	45,307,133,554

statistic has a relatively large value, it implies that some types of clustering exist within the neighborhood. It is also expected that as d increases, $G(d)$ will also increase.

But the more detailed interpretation of the general G-statistic has to rely on its expected value and the standardized score (z-score), as in the case of other spatial autocorrelation measures. To derive the z-score and to test for the significance of the general G-statistic, we must know the expected value of $G(d)$ and its variance. The expected value of $G(d)$ is

$$E(G) = \frac{W}{n(n-1)}. \tag{8.49}$$

Therefore, the expected $G(d)$ is a function of the size of the spatial system, n, with the neighborhood configuration defined by d and captured by W. The expected value of $G(d)$ indicates the value of $G(d)$ if there is no significant spatial association among the areal units. In the Ohio seven-county case,

$$E(G) = 22/[7*6] = 0.5238.$$

Intuitively, because the observed $G(d)$ is slightly higher than the expected $G(d)$, or $0.5576 > 0.5238$, we may say that the observed pattern exhibits some positive spatial association. However, we cannot be sure that this level is significant until we formally test it. To test it, we must calculate the z-score of the observed statistic based on the variance of the expected statistic.

According to Getis and Ord (1992), the variance of $G(d)$ is

$$Var(G) = E(G^2) - [E(G)]^2, \tag{8.50}$$

where

$$E(G^2) = \frac{1}{(m_1^2 - m_2)^2 \, n^{(4)}} [B_0 m_2^2 + B_1 m_4 + B_2 m_1^2 m_2 + B_3 m_1 m_3 + B_4 m_1^4], \tag{8.51}$$

where m_j and $n^{(x)}$ are defined the same way as in Section 8.5. The other coefficients are as follows:

$$B_0 = (n^2 - 3n + 3)S_1 - nS_2 + 3W^2, \tag{8.52}$$

$$B_1 = -[(n^2 - n)S_1 - 2nS_2 + 3W^2], \tag{8.53}$$

$$B_2 = -[2nS_1 - (n + 3)S_2 + 6W^2], \tag{8.54}$$

$$B_3 = 4(n - 1)S_1 - 2(n + 1)S_2 + 8W^2, \tag{8.55}$$

and

$$B_4 = S_1 - S_2 + W^2, \tag{8.56}$$

where S_1 and S_2 are defined in the same way as in Section 8.5.

The median household income data of the seven Ohio counties give

$$E(G^2) = 0.2829.$$

Therefore, the variance of $G(d)$ is

$$Var(G) = 0.2831 - (0.5238)^2 = 0.0087$$

and the standardized score is

$$Z(G) = \frac{0.5576 - 0.5238}{\sqrt{0.0087}} = 0.3609,$$

which is smaller than 1.96, our standard marker indicating the $\alpha = 0.05$ level of significance. In other words, the calculated $G(d)$ has a mildly positive spatial association, and the z-score indicates that the counties with high median household income are close to (within 30 miles of) counties with moderate income. However, this relationship is not statistically significant. That is, the pattern is probably created by chance rather than by some systematic process. One should recognize that high values in G-statistic indicate H-H clusters in the study area, and low values for L-L clusters. Thus, both the positive and negative z-scores indicate positive spatial autocorrelation. The negative z-scores do not represent negative spatial autocorrelation.

8.7.4 ArcView Example: Global Statistics for Spatial Autocorrelation

So far in this chapter, we have discussed four statistics for measuring the spatial autocorrelation of an attribute distributed over a set of polygons. With the exception of joint count statistics, which are used with binary nominal data, Moran's I, Geary's C, and the general G-statistic can measure global spatial autocorrelation in interval-ratio scales.

Also using the data for the 88 Ohio counties, as in ArcView Examples 8.1 and 8.2, the following example aims at demonstrating the procedures for using the Ch8.avx extension to calculate the three global statistics of spatial autocorrelation. In the following example, we will use the stochastic weights matrix to calculate these statistics. Even though the three global statistics can work with binary connectivity matrices, the stochastic distance matrices are used here as part of the demonstration.

ArcView Example 8.3: Global Statistics for Spatial Autocorrelation

Step 1 Data and extension
Prepare the Ohcounties.shp data theme and the Ch8.avx extension as described in ArcView Example 8.1.

Step 2 Stochastic weights matrix
To calculate the three global statistics of spatial autocorrelation in this example, we will use the stochastic weights matrix:

- From the **Ch.8** menu, choose **Create Spatial Weights Matrix (Polygon).**
- Click and select `Ohcounties.shp` as the data theme.
- Choose to use all records.
- Click the **Show Variables** button to list all attribute fields.
- Click to select `Cnty_fips` as the ID field.
- Click the **Calculate** button to proceed.

In the **Distance Definition** dialog box:

- From the drop-down menu, select `Stochastic Weight` as the distance measure.
- Click **OK** to proceed.
- When prompted, navigate to a folder and give a name to store the stochastic weights matrix to be created.
- Click **OK** when informed of the completion of the step.

Step 3 Moran's I and Geary's *C*
The first set of global statistics to be computed are Moran's I and Geary's *C*.
From the **Ch.8** menu, choose **Global Moran_Geary Statistics.**
In the Moran's I & Geary's Ratio dialog box:

- Click to highlight the `Ohcounties.shp` data theme.
- Click the **Choose the Theme** button to list the attribute fields in the two list boxes.
- Check the box to indicate that spatial weights matrix has been created (in Step 2).
- Check the option to indicate that all records are to be included in the calculation.
- For the ID field, scroll down and click `Cnty_fips`.
- For the variable, scroll down and click `SeniorPCT`.
- Either enter `1` as the distance decay parameter or leave it blank, as we are not using the distance-based weights matrix.
- Finally, click the **Calculate** button to proceed.

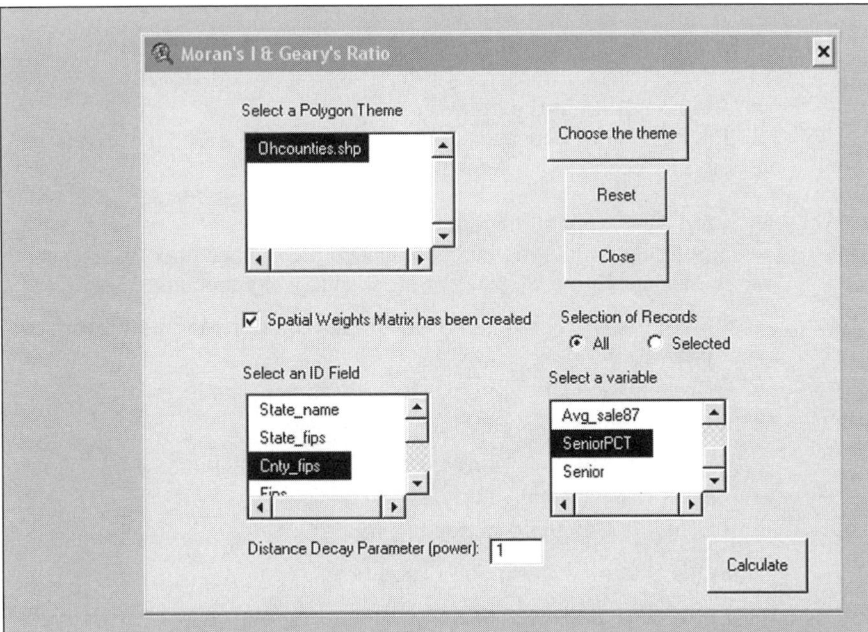

When prompted, navigate to the folder where the spatial weights matrix was created and stored. Select it for the calculation. When prompted, select **Stochastic Weight** as the type of spatial weights matrix.

Step 4 Analysis of Moran's I and Geary's *C*
In the Report information box, the results of the two statistics are reported as

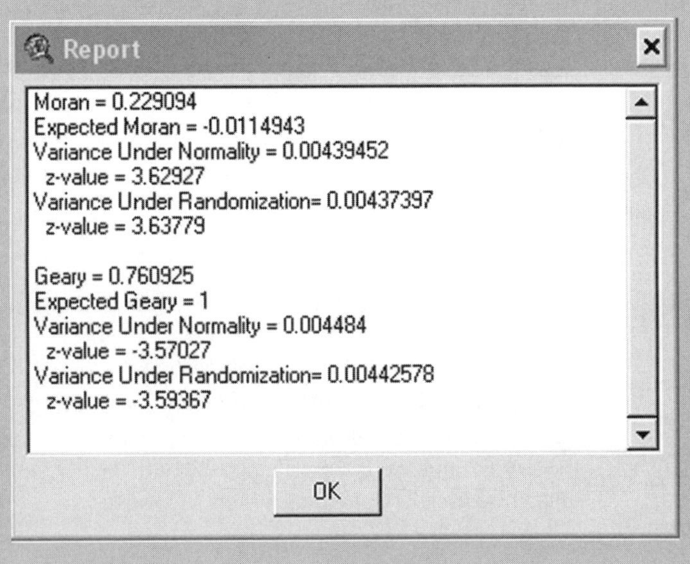

Step 5 General *G*-statistic

To calculate the last of the three global statistics, we must generate a distance matrix first.

- From the **Ch.8** menu, choose **Create Spatial Weights Matrix (Polygon).**
- Click and select `Ohcounties.shp` as the data theme.
- Choose to use all records.
- Click the **Show Variables** button to list all attribute fields.
- Click to select `Cnty_fips` as the ID field.
- Click the **Calculate** button to proceed.

In the Distance Definition dialog box:

- From the drop-down menu, select `Centroid Distance` as the distance measure.
- Click **OK** to proceed.
- When prompted, navigate to a folder and give a name to store the distance weights matrix to be created.
- Click **OK** when informed of the completion of the step.

To compute the general *G*-statistic:

- From the **Ch.8** menu, click the **General *G*(*d*) Statistic** item.
- In the General *G*(*d*) Statistic dialog box:
  - Click `Ohcounties.shp` to select it.
  - Click the **Choose the Theme** button to list the attribute fields in the two list boxes.
  - Check the box to indicate that the distance matrix has been created.
  - Check the **All** option to indicate the inclusion of all records for calculation.
  - For the ID field, scoll down the left list box and click `Cnty_fips`.
- For the variable, scroll down the right list box and click `SeniorPCT`.
- Enter `60` (miles) as the distance threshold defining the neighborhoods.
- When prompted, navigate to the folder where the distance weights matrix was created and added.

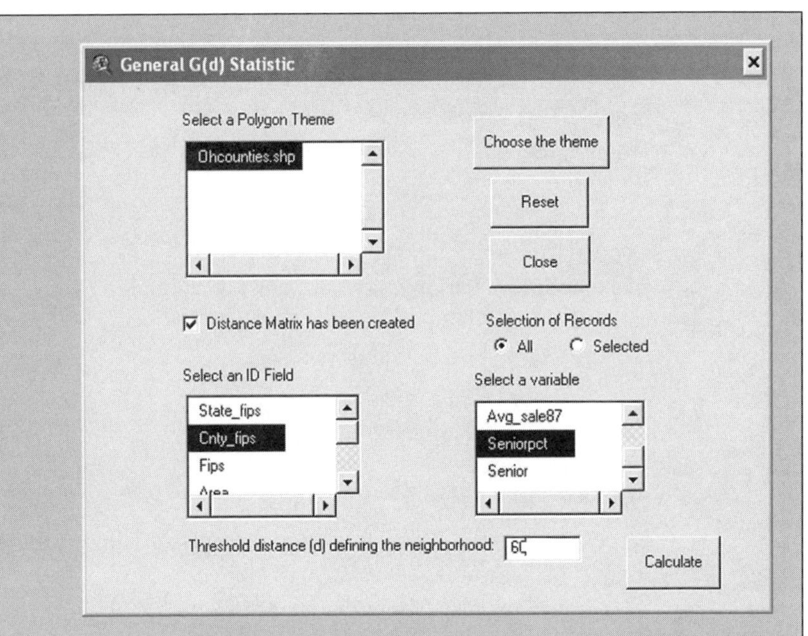

When the calculation is completed, the results are reported in an Info dialog box:

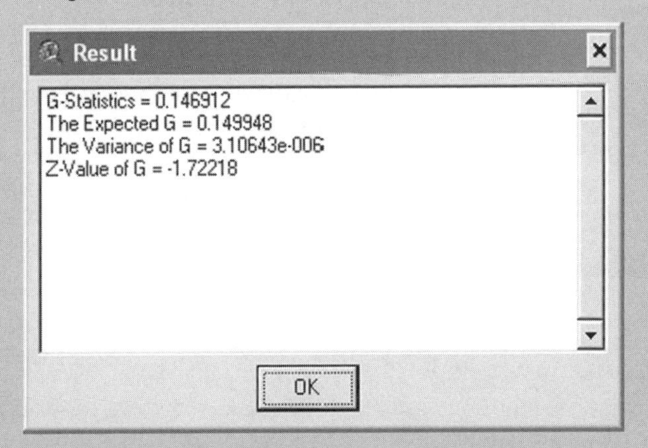

Step 6 Analysis of the global statistics
Given that we cannot change the configuration of the 88 Ohio counties, the distribution of the attribute values, SeniorPCT, over the 88 polygons is appropriate to be tested under the nonfree assumption. Therefore, using the randomization *z*-scores for Moran's I and Geary's *C*, we have the following:

	Statistic	Expected Value	z-Score
Moran's I	0.2291	−0.0115	3.6378
Geary's C	0.7609	1	−3.5937
G-statistic	0.1469	0.1499	−1.7222

Given these results, we reject the null hypothesis for the two spatial autocorrelation statistics, Moran's I and Geary's C, and conclude that the distribution of values of SeniorPCT over the 88 county polygons is not a random pattern. There is a certain degree of clustering, as indicated by the values of these two statistics. The G-statistic indicates a mildly spatial association, of low values as the observed statistic is slightly smaller than the expected value. However, the z-score was not significant. Note that this conclusion is inconsistent with the conclusion we reached in ArcView Example 8.2 for joint count statistics. This is because this example uses original data (at the interval/ratio scale), whereas in ArcView Example 8.2 we used binary nominal data of Senior (1's or 2's) downgraded from SeniorPCT.
This result further demonstrates that unless it is absolutely necessary, downgrading data from interval/ratio scale to either ordinal or nominal scale tends to reduce the information in the data and may lead to wrong conclusions.

Step 7 End

8.8 LOCAL SPATIAL AUTOCORRELATION STATISTICS

All of the spatial autocorrelation statistics discussed so far have a common characteristic: they are global statistics because they are summary measures of the entire study region. It is reasonable to suspect that the magnitude of spatial autocorrelation is not necessarily uniform over the region, but rather varies from one part of it to another. In other words, it is likely that the magnitude of spatial autocorrelation is high in some subregions but low in others. In fact, it is even possible to find positive spatial autocorrelation in one part of the region and negative autocorrelation in another.

In order to capture the spatial variability of spatial autocorrelation, we must rely on another set of measures that can indicate spatial autocorrelation at the local scale. These types of measures are referred to as *local measures*. All local measures covered in this chapter are based on their global counterparts discussed above (Moran's I, Geary's C, and the general G-statistic) but are modified to detect and evaluate spatial autocorrelation at a local scale.

8.8.1 Local Indicators of Spatial Association

Local Indicators of Spatial Association (LISA) refer to the local version of Moran's I and Geary's C (Anselin, 1995). In order to indicate the level of spatial autocorrelation at the local scale, a spatial autocorrelation value must be derived for each areal unit. The local Moran statistic for areal unit i is defined as

$$I_i = z_i \sum_j w_{ij} z_j, \tag{8.57}$$

where z_i and z_j are deviations from the mean for the corresponding x values, or

$$z_i = \frac{x_i - \bar{x}}{\delta}, \tag{8.58}$$

where δ is the standard deviation of the variable x. Therefore, z_i is basically the z-score of x_i. Similar to the interpretation of Moran's I, a high value of the local Moran statistic means a clustering of similar values (which can be all high or all low). A low value of the local Moran statistic indicates a clustering of dissimilar values.

In general, w_{ij} is the row-standardized matrix, but other specifications of spatial weights matrices are also possible. If the weights are in the row-standardized form, the operation $\Sigma w_{ij} z_j$ essentially pulls a fraction of the value from each of the standardized scores in the units surrounding areal unit i. It can also be regarded as the linear combination of the neighborhood z-scores. Result from this operation, an estimate pulled from the neighboring values, is then multiplied by the z-score from i to derive the value of the local Moran statistic for unit i.

Table 8.12a is similar to the row-standardized matrix in Figure 8.5. From Figures 8.1 and 8.5, we can tell that Geauga County has six neighboring counties. If the row-standardized stochastic weights are adopted, the weight for each of Geauga County's neighbors is 1/6, or 0.17. Similarly, the weight for each of Cuyahoga County's neighbors is 0.25, as this county has four neighbors. The number of deviations from the mean of median household income for all seven counties is reported in Table 8.12b. These numbers are different from the mean deviations used in Moran's I calculation in Table 8.7a because the deviations from the mean we use here are the mean deviation $(x - \bar{x})$ divided by the standard deviation.

Among the seven Ohio counties, let us focus on Geauga County. Because Geauga County has six neighbors, the weight for each neighbor is 1/6. Applying this weight to the mean deviations for all of the j neighbors of Geauga County, $w_{ij} \times z_j$ produces the results reported in Table 8.12b. The same step is applied to other counties. In the second to last column of Table 8.12b, the

TABLE 8.12 Computing Local Indicators of Spatial Association Using the Seven Ohio Counties

(a) Derivation of the Row-Standardized Stochastic Matrix

Counties	Geauga	Cuyahoga	Trumbull	Summit	Portage	Ashtabula	Lake
Geauga	0	1/6	1/6	1/6	1/6	1/6	1/6
Cuyahoga	1/4	0	0	1/4	1/4	0	1/4
Trumbull	1/3	0	0	0	1/3	1/3	0
Summit	1/3	1/3	0	0	1/3	0	0
Portage	1/4	1/4	1/4	1/4	0	0	0
Ashtabula	1/3	0	1/3	0	0	0	1/3
Lake	1/3	1/3	0	0	0	1/3	0

(b) Stochastic Weights Matrix Multiplied by Deviations from the Mean and I_j Derivation

Counties		Geauga	Cuyahoga	Trumbull	Summit	Portage	Ashtabula	Lake		
	(*X*-Mean)/std	2.0928	−0.6408	−0.7539	−0.2332	0.0323	−1.1036	0.6063	Sum(WijZj)	Ii
Geauga	2.0928	0	(1/6)*−0.6408	(1/6)*−0.7539	(1/6)*−0.2332	(1/6)*0.0323	(1/6)*−1.1036	(1/6)*0.6063	−0.3488	−0.7300
Cuyahoga	−0.6408	(1/4)*2.0928	0	0	(1/4)*−0.2332	(1/4)*0.0323	0	(1/4)*0.6063	0.6246	−0.4002
Trumbull	−0.7539	(1/3)*2.0928	0	0	0	(1/3)*0.0323	(1/3)*−1.1036	0	0.3405	−0.2567
Summit	−0.2332	(1/3)*2.0928	(1/3)*−0.6408	0	0	(1/3)*0.0323	0	0	0.4948	−0.1154
Portage	0.0323	(1/4)*2.0928	(1/4)*−0.6408	(1/4)*−0.7539	(1/4)*−0.2332	0	0	0	0.1162	0.0038
Ashtabula	−1.1036	(1/3)*2.0928	0	(1/3)*−0.7539	0	0	0	(1/3)*0.6063	0.6484	−0.7156
Lake	0.6063	(1/3)*2.0928	(1/3)*−0.6408	0	0	0	(1/3)*−1.1036	0	0.1161	0.0704

products of the weights and mean deviations for all neighboring counties are summed for each county, i. Then, in the last column, the sum of each county is multiplied by the mean deviation of the concerned county, i. That gives the local Moran statistic for each of the seven counties. The major difference in the calculation among the counties is their weights, which are dependent on the number of neighbors each unit has.

As with other statistics, just deriving the values of local Moran statistic for each county is not very meaningful. High or low local Moran statistic values may occur just by chance. These values must be compared with their expected values and interpreted with their standardized scores. According to Anselin (1995), the expected value under the randomization hypothesis is

$$E[I_i] = \frac{-w_i}{n-1} \tag{8.59}$$

and

$$Var[I_i] = w_{i.}^{(2)} \frac{\left(n - \frac{m_4}{m_2^2}\right)}{(n-1)} + 2w_{i(kh)} \frac{\left(2\frac{m_4}{m_2^2} - n\right)}{(n-1)(n-2)} - \frac{w_{i.}^2}{(n-1)^2}, \tag{8.60}$$

where

$$w_{i.}^2 = \left(\sum_j w_{ij}\right)^2, \tag{8.61}$$

$$w_{i.}^{(2)} = \sum_j w_{ij}^2, \quad i \neq j, \tag{8.62}$$

$$m_2 = \frac{\sum_i z_i^2}{n}, \tag{8.63}$$

and

$$m_4 = \frac{\sum_i z_i^4}{n}. \tag{8.64}$$

The term

$$2w_{i(kh)} = \sum_{k\neq i} \sum_{h\neq i} w_{ik} w_{ih}, \quad k \neq h, \quad j \neq i. \tag{8.65}$$

Table 8.13 reports the local Moran values together with their expected values and variances, as well as the z-scores for all seven counties. The z-score for each unit is obtained by

$$z[I_i] = \frac{I_i - E[I_i]}{\sqrt{Var[I_i]}}. \tag{8.66}$$

Note that m_2 is the second moment statistic and m_4 is the fourth moment statistic (refer to Chapter 2). Also note that because each county has its own local Moran value, each local Moran value has its associated expected value and variance. This is an advantage of the local Moran statistic. A value is derived for each areal unit, and therefore the results can be mapped. Figure 8.8 consists of two maps showing the local Moran statistics and the z-scores for all seven counties.

The local Moran statistic indicates how neighboring values are associated with each other. In Figure 8.7, we can see that the median household income of Geauga County is highest ($60,200) and one of its neighboring counties, Ashtabula, has the lowest value ($35,607). Therefore, the local Moran statistic for Geauga County, as shown in Figure 8.8, is negative and the lowest, indicating a negative spatial autocorrelation relationship of this county with its neighbors. This should be obvious because the median income of Geauga County is much higher than the incomes of the six other counties. The next highest income county is Lake County, at $48,763, still much lower than Geauga County. Because Geauga County has such a high income compared to the rest and is a neighbor of all the remaining counties, most of the other counties have negative local Moran statistics. One of Geauga County's neighbors, Ashtabula County, has the second lowest local Moran statistic because it is very different from its surrounding counties, including Geauga County.

Only two counties, Lake and Portage, have positive local Moran statistics. Though they are positive, they are still on the low side. Lake County has a local value of 0.0704, slightly higher than that of Portage County. These slightly positive values for the local Moran statistic are mainly due to the fact

TABLE 8.13 LISA and Related Statistics

Counties	I_i	$E[I_i]$	$\text{Var}[I_i]$	$z[I_i]$	c_i
Geauga	−0.7300	−0.17	0.0567	−2.3650	6.2767
Cuyahoga	−0.4002	−0.17	0.1142	−0.6910	2.4118
Trumbull	−0.2567	−0.17	0.1717	−0.2173	2.9481
Summit	−0.1154	−0.17	0.1717	0.1237	1.8824
Portage	0.0038	−0.17	0.1142	0.5042	1.3469
Ashtabula	−0.7156	−0.17	0.1717	−1.3246	4.4213
Lake	0.0704	−0.17	0.1717	0.5721	2.2297

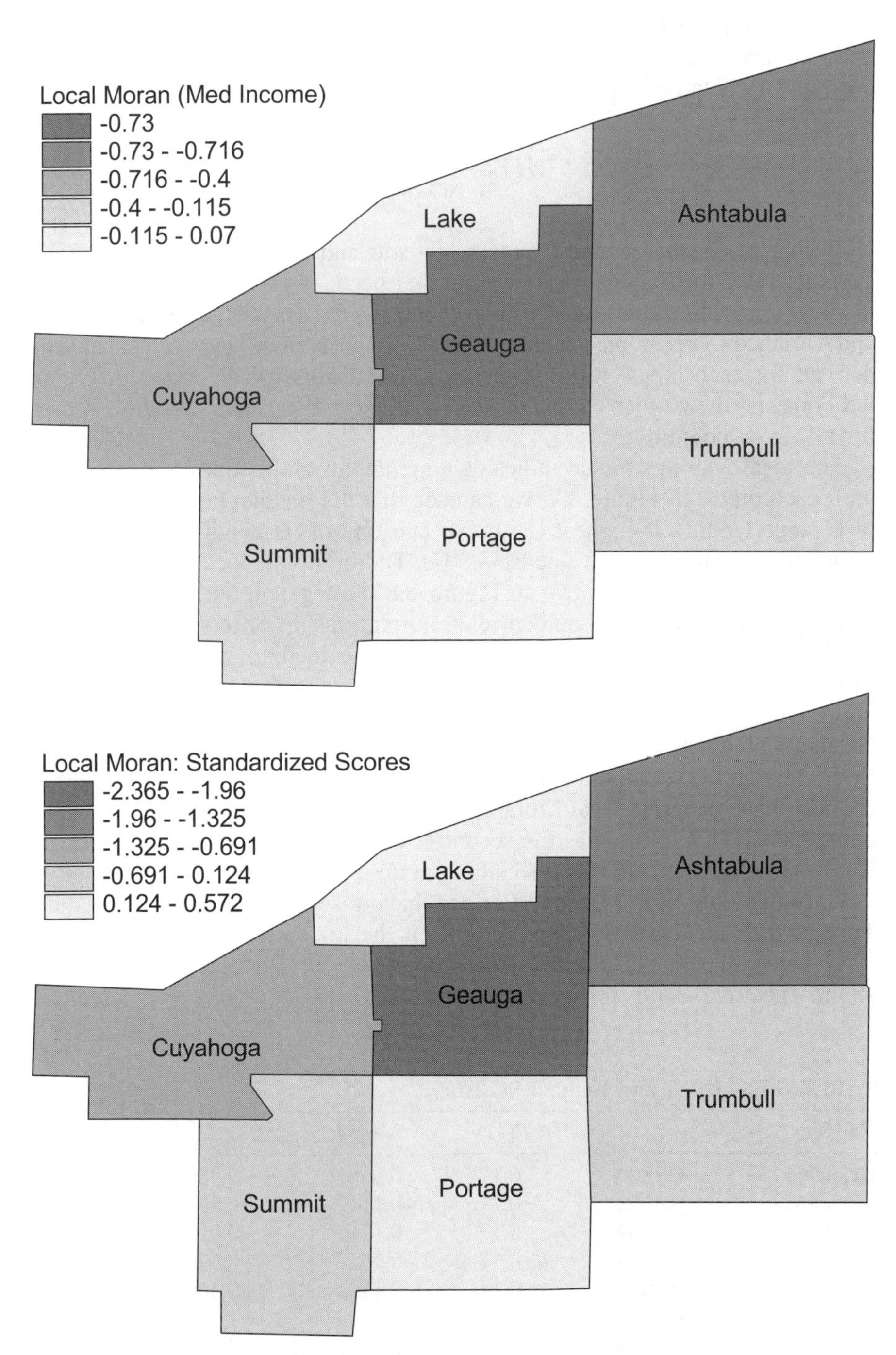

Figure 8.8 Local Moran values and their corresponding standardized scores.

that these two counties have the second and third highest income levels in the region and they are the neighbors of the highest income county—Geauga County. But because their neighboring counties have relatively lower income levels, their values for the local Moran statistic cannot be too high.

One may claim that the two maps in Figure 8.8 show an identical pattern. However, the second map with the z-scores can show which local Moran value is significant. Using the standard marker of -1.96, we see that only the negative value for the local Moran statistic of Geauga County is significant. Its z-score is smaller than -1.96. Thus, the negative spatial association pattern we observed for Geauga County and its surrounding neighbors is not due to a random process.

A local version of Geary's C is also available. Formally, it is defined as

$$c_i = \sum_j w_{ij}(z_i - z_j)^2. \tag{8.67}$$

Very similar to the global version, the z-scores of neighboring values are compared first. As the question is whether they are similar or not, but not which one is larger, the difference in the neighboring z-scores is squared to remove the directional information. Therefore, the larger the c_i value, the more dissimilar it is between the attribute value in areal unit i and the attribute values of its neighboring units.

Unfortunately, the distribution properties of the local Geary statistic are not as desirable as those of the local Moran statistic. Still, mathematically, we can interpret the local Geary statistic in the same way as the global Geary Ratio. Clustering of similar values will create a relatively low value for the local Geary statistic, while clustering of dissimilar values will yield a relatively high value. Similar to the local Moran statistic, a value for the local Geary statistic is computed for each areal unit; therefore, the results can be mapped.

The local Geary statistic values for the seven Ohio counties are reported in the last column of Table 8.13. A causal inspection of the local Geary statistic value and the local Moran statistic reveals that they are inversely related, as expected. The results from the local Geary statistic are consistent with the previous results from the local Moran statistic in terms of local spatial association among the seven counties. Therefore, a map based on the local Geary statistic was not needed.

8.8.2 Local *G*-Statistic

Another local measure of spatial autocorrelation is the local version of the general G-statistic (Getis and Ord, 1992). The **local *G*-statistic** is derived for each areal unit to indicate how the value of the areal unit of concern is associated with the values of the surrounding areal units as defined by a distance threshold, d. Formally, the local G-statistic is defined as

$$G_i(d) = \frac{\sum_j w_{ij}(d)x_j}{\sum_j x_j} \text{ and } j \neq i. \tag{8.68}$$

All other terms are as defined previously in the discussion of the general G-statistic. As with other statistics, it is best to interpret the local G-statistic in the context of the standardized score. To obtain the standardized score, we must know the expected value and the variance of the statistic. The expected value is defined as

$$E(G_i) = W_i/(n - 1), \tag{8.69}$$

where

$$W_i = \sum_j w_{ij}(d). \tag{8.70}$$

The definition of the variance is similar to the definition of the general G-statistic. It is defined as

$$Var(G_i) = E(G_i^2) - [E(G_i)]^2 \tag{8.71}$$

and

$$E(G_i^2) = \frac{1}{\left(\sum_j x_j\right)^2}\left[\frac{W_i(n - 1 - W_i)\sum_j x_j^2}{(n - 1)(n - 2)}\right] + \frac{W_i(W_i - 1)}{(n - 1)(n - 2)}, \tag{8.72}$$

where $j \neq i$. That is, in calculating $E(G_i^2)$ and $E(G_i)$, x_i is excluded when deriving $(\Sigma_j x_j)^2$ and $\Sigma_j x_j^2$ in Equation 8.72.

Given the standardized score of $G_i(d)$ using the above expected value and variance, a high score appears when the spatial clustering is formed by similar but high values. If the spatial clustering is formed by low values, the z-score tends to be highly negative. A z-score around 0 indicates no apparent spatial association pattern. Table 8.14 is extracted from Getis and Ord (1992) to summarize the interpretation of the standardized scores of the $G_i(d)$ statistic. A related statistic is labeled $G_i^*(d)$. This statistic is almost identical to $G_i(d)$, except that it includes cases where $j = i$. Because these two statistics are very similar, we will focus on $G_i(d)$. Readers who are interested in the other statistic can refer to Getis and Ord (1992).

Given the interpretation of the $G_i(d)$ statistic as summarized in Table 8.14, we should expect that, using median household income as the variable, Geauga County should have a somewhat negative standardized score for the

TABLE 8.14 Interpretation of Standard Scores for the $G_i(d)$ Statistic

Situation	$Z(G_i)$
High next to High	Strongly positive
High next to Moderate	Moderately positive
Moderate next to Moderate	0
Random	0
High next to Low	Negative
Moderate next to Low	Moderately negative
Low next to Low	Strongly negative

local $G_i(d)$ statistic because its income is very different from that of the rest of its neighbors. As shown in the map in Figure 8.9 and Table 8.15, the standardized score of the $G_i(d)$ statistic for Geauga County is slightly negative, confirming a high value next to low value situation. Summit County has the lowest standardized $G_i(d)$ statistic, probably because it has a moderate income level, and its neighbors have either lower (Cuyahoga) or higher (Geauga) income levels.

8.9 MORAN SCATTERPLOT

The development of local spatial autocorrelation statistics acknowledges the fact that spatial processes, including spatial autocorrelation, can vary over a geographic region. Apparently, we can adopt one or more of the previously described local measures of spatial autocorrelation to evaluate the entire study area to see if there is any local instability in spatial autocorrelation.

From the statistical visualization and spatial exploratory analysis perspectives, it will be informative and useful if the analyst can identify areas with unusual levels of spatial autocorrelation. Those areas can be regarded as outliers. A very effective visual diagnostic tool is the **Moran scatterplot** based upon a regression framework and Moran's I statistic (Anselin, 1995). Assuming that z is a vector of z_i where z_i is the deviation from the mean or $z_i = (x_i - x_j)/\delta$ and W is the row-standardized spatial weights matrix, we may form a regression of W_Z on z, while the slope of this regression indicates how the neighboring values are related to each other. In other words, the regression is

$$z = a + IWz, \tag{8.73}$$

where a is a vector of the constant intercept term and I is the regression coefficient representing the slope. The slope parameter is not mistakenly la-

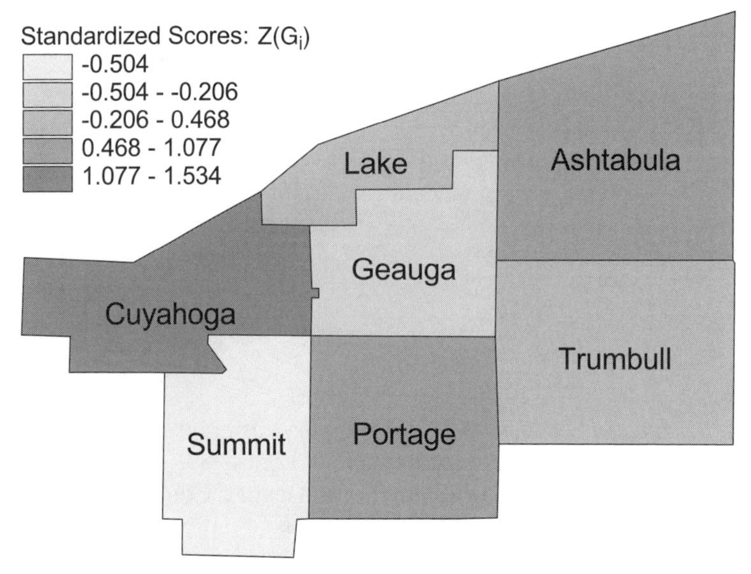

Figure 8.9 Standardized scores for the $G_i(d)$ statistic.

beled *I*. Theoretically, the slope parameter in this spatial regression model is also Moran's I global statistic if the stochastic weights are used. Sometimes, this model is also known as a **spatial average model** (Bailey and Gatrell, 1995).

Moran's I reflects the level of spatial autocorrelation, and the statistic is a global summary statistic. Different observations within the study region, however, may show different levels of spatial autocorrelation with their neighbors. By plotting Wz on z and superimposing the regression line, the scatterplot can potentially indicate outliers in terms of the magnitude of spatial autocorrelation.

If all observations have a similar level of spatial autocorrelation as reflected by the global Moran's I, the scatterplot will show observations distributed rather closely along the regression line. If certain observations show unusually

TABLE 8.15 $G_i(d)$ **and Related Statistics**

County	*Gi*(*d*)	*E*(*Gi*)	Var(*Gi*)	*Z*(*Gi*)
Geauga	0.8298	0.8333	0.0003	−0.2060
Cuyahoga	0.5612	0.5000	0.0016	1.5340
Trumbull	0.5183	0.5000	0.0015	0.4675
Summit	0.3135	0.3333	0.0015	−0.5045
Portage	0.5327	0.5000	0.0018	0.7743
Ashtabula	0.5393	0.5000	0.0013	1.0771
Lake	0.5193	0.5000	0.0017	0.4641

high or low levels of spatial autocorrelation locally in reference to their neighbors, those observations will be plotted far above or below the regression line. This regression line reflects the general trend of spatial autocorrelation in the entire region, and again, the slope parameter is equivalent to Moran's I. In other words, those observations deviating from the general trend of spatial autocorrelation have spatial autocorrelations that are very different from the overall level. Thus, the Moran scatterplot is useful in identifying unusual observations in regard to relationships with neighbors.

In the seven Ohio counties, overall median household income values are moderately negatively spatially autocorrelated, with a Moran's I of −0.3063 when a stochastic weights matrix is used. High-value counties are somewhat close to low-value counties. This is mainly the result of using a few counties for demonstration. As shown in the scatterplot in Figure 8.10, income (z) is inversely related to the product of the spatial weights and income (Wz). The slope is −0.3063, moderately negative. Note that the value of the slope parameter is identical to the Moran's I of the median household income when a row-standardized weights matrix is used.

The R-square (0.87) turns out to be quite high, mainly because only a few observations are involved. The extremes are Geauga County, as shown by the point in the lower right corner, and Ashtabula County, as indicated by the point in the upper left corner. The scatterplot basically shows that the high income of Geauga County (z) is negatively associated with the low incomes of its surrounding counties. By contrast, the low income of Ashtabula County is negatively associated with the high incomes of its surrounding counties. Because only seven observations are involved, the scatterplot does not reveal

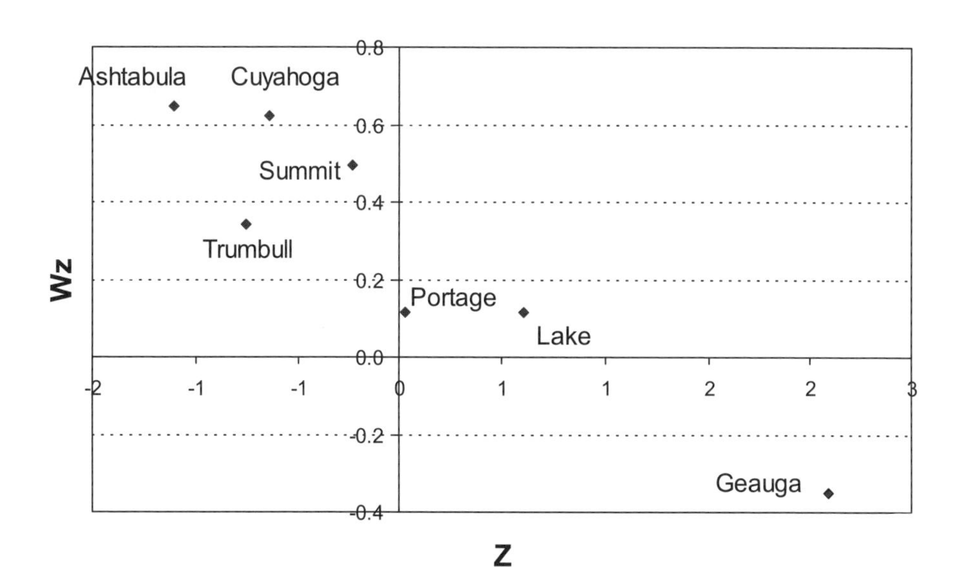

Figure 8.10 Moran scatterplot for the seven Ohio counties using median household income.

any obvious outliers except Trumbull County, which deviates from the downward trend slightly.

Figures 8.11 and 8.12 provide another example using 49 states and Washington, DC, in the continental United States and the foreign-born population, according to Census 2000 Summary File 3 data. The map in Figure 8.11 shows that high percentages of the foreign-born population are found in the West, Southwest, and East Coast regions, while the Midwest and South-Central states have relatively low foreign-born populations. Executing the Moran's scatterplot routine, a regression is developed for z (percent foreign born) and Wz (a neighboring average), with an R-square of 0.2856, much lower than that for the Ohio case. But the slope or Moran's I is 0.3162, a moderate level of positive spatial autocorrelation. This moderately positive slope is also depicted by the upward trend of data shown in Figure 8.12. In this case, the percentage of foreign born population of a state is positively and moderately associated with the percentage of foreign born population of the surrounding states.

A useful way to interpret the scatterplot is to focus on two quadrants: the upper right, where observations with high values are next to high values, and the lower left, where observations with low values are next to low values. Selected states are labeled in the plot. In the upper right quadrant, states with high foreign-born percentages, such as California and New York, are next to

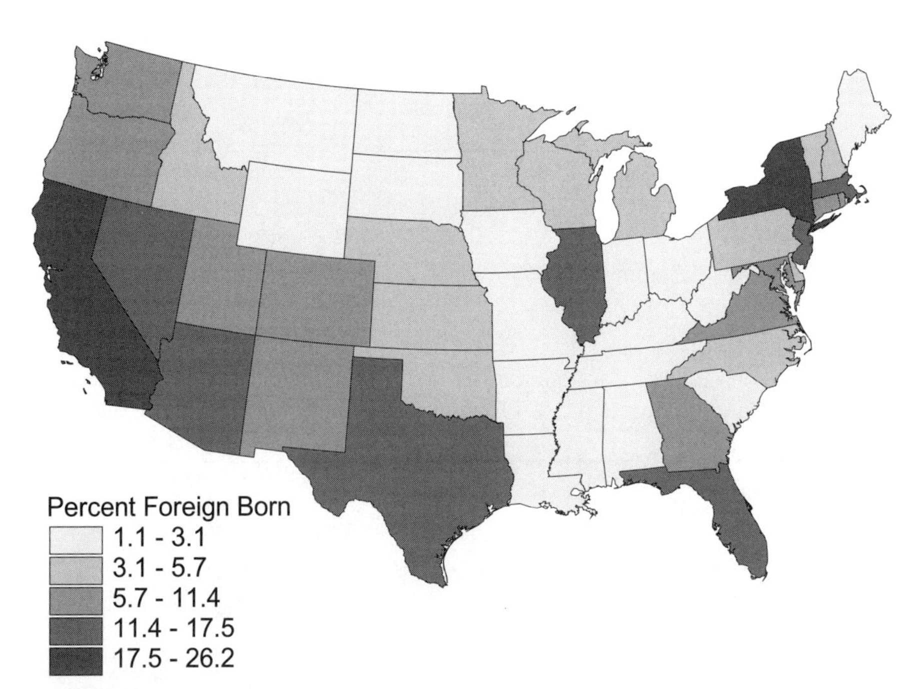

Figure 8.11 Map showing the percentage of foreign-born population 2000 Census, Summary File 3.

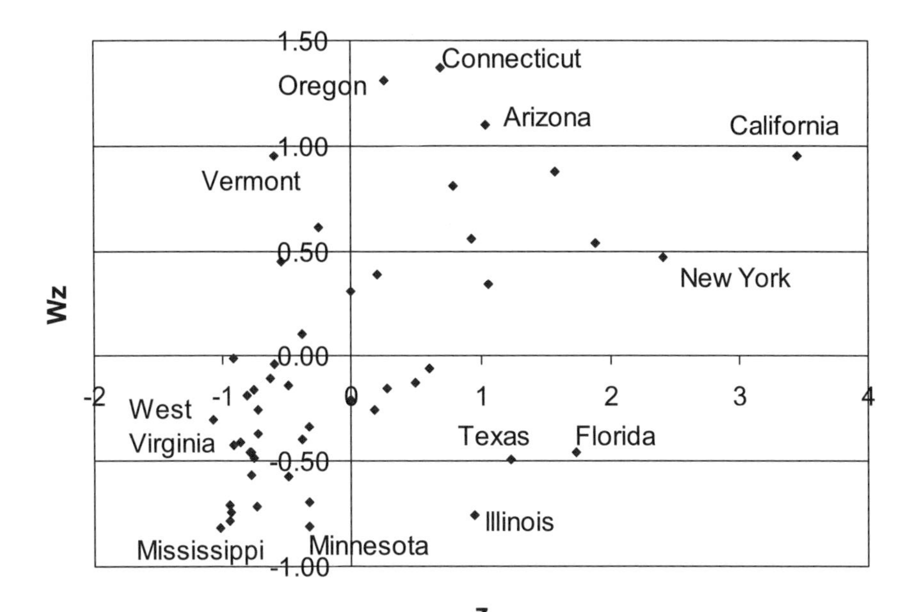

Figure 8.12 Moran scatterplot for 49 states and Washington, DC, percentage of foreign born population.

states with high values. In the lower left quadrant, states such as Mississippi and Minnesota, with low foreign born percentages, are next to states with low values. States in the other quadrants either have high values next to low-value states (such as Illinois and Texas) or low values next to high-value states (such as Vermont).

8.9.1 ArcView Example: Local Spatial Autocorrelation Statistics and Moran Scatterplot

We have discussed three local statistics for spatial autocorrelation and the Moran scatterplot in the past few sections. ArcView Example 8.4 demonstrates the procedures for calculating such statistics. Similar to the other ArcView examples, we will use Ohcounties.shp and the Ch8.avx extension.

ArcView Example 8.4: Local Spatial Autocorrelation Statistics and Moran Scatterplot

Step 1 Data and extension
As with other ArcView examples, Ohcounties.shp and Ch8.avx are used for this example. Refer to ArcView Example 8.1 for steps for adding the data theme and loading the Ch8.avx extension.

In addition, we will continue using the spatial weights matrix created in ArcView Example 8.3, which was created using the stochastic weights as the spatial weights.

Step 2 Local indicators of spatial association
The first set of local statistics to be calculated are LISA.
From the **Ch.8** menu, choose **Local Indicators of Spatial Association** (LISA).
In the Local Indicators of Spatial Association (LISA) dialog box:

- Click to select `Ohcounties.shp` as the data theme.
- Click the **Choose the Theme** button to list the attributes in the two list boxes.
- Check the box to indicate that the stochastic spatial weights matrix has been created.
- Check the **All** option to include all records for calculation.
- From the left list box, choose `Cnty_fips` as the ID field.
- From the right list box, choose `SeniorPCT` as the variable.
- Click the **Calculate** button to proceed.

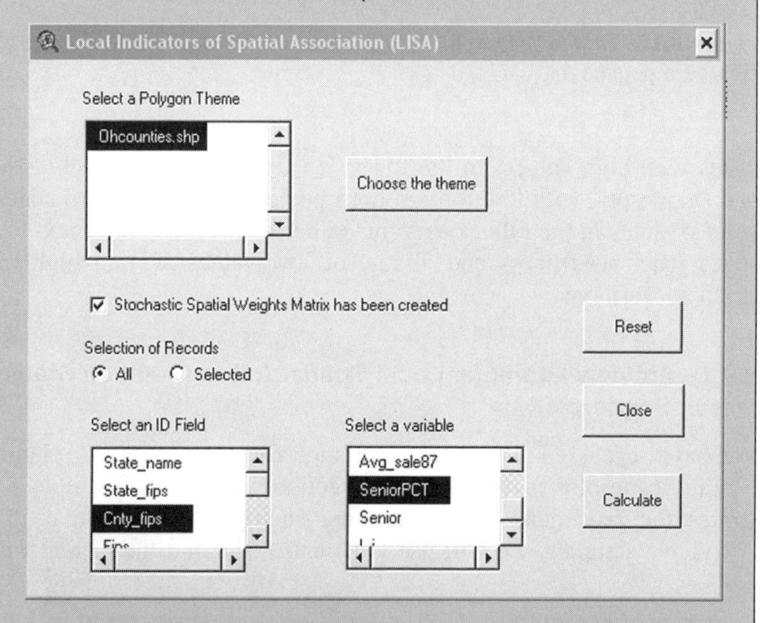

- When prompted, navigate to the folder where the weight matrix was created and stored. Select it for the calculation.
- When the calculation is completed, the Local Indicators of Spatial Association (LISA) dialog box will be closed.

Step 3 Map LISA
After the calculation is completed, use the **Open Theme Table** to open the attribute table of Ohcounties.shp.

Scroll to the right. You will see that several attribute fields have been added to the attribute table: the local Moran statistics (*I_i*), its expected values (*Exp_I_i*), variances (*Var_I_i*), Z-scores (*Z_I_i*), and the local Geary (*C_i*).

Attributes of Ohcounties.shp

Avg_sale8	SeniorPCT	Senior	I_i	Exp_I_i	Var_I_i	Z_I_i	C_i
70618	13.4133	1	-0.00753	-0.01149	0.23501	0.00817	0.91715
76303	11.4772	2	-0.13014	-0.01149	0.31695	-0.21076	0.52392
21984	9.5702	2	-0.31743	-0.01150	0.15307	-0.78196	2.98837
50128	13.0773	1	-0.03373	-0.01149	0.31695	-0.03949	0.27242
96680	16.1065	1	-1.27335	-0.01150	0.15307	-3.22529	6.11081
33595	15.0853	1	-0.12116	-0.01149	0.31695	-0.19480	2.06557
49469	9.5274	2	-0.39323	-0.01150	0.12965	-1.06015	2.42797
61130	11.0517	2	0.03479	-0.01149	0.18581	0.10737	1.40826
56440	13.3674	1	0.01648	-0.01149	0.18581	0.06491	0.78001
25518	14.5574	1	0.06543	-0.01149	0.23501	0.15868	2.91081
46041	13.2046	1	-0.07177	-0.01150	0.12965	-0.16738	0.54190

To display the distribution of the statistics, we suggest using the **Legend Editor** to create the display of *z*-scores with the following classes:

<-1.96
$-19.6 \sim -1.645$
$-1.645 \sim 0$
$0 \sim 1.645$
$1.645 \sim 1.96$
> 1.96

For example, the map below highlights the clusters of low and high *z*-scores.

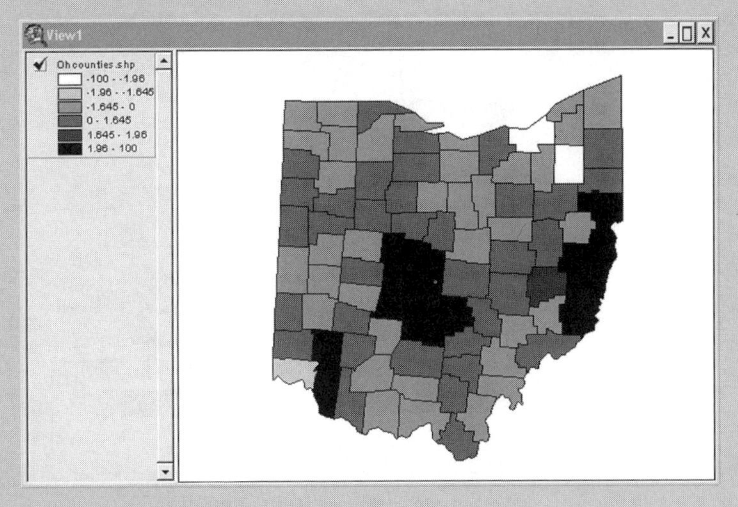

Note that the dark clusters indicate local clusters of counties that have high senior population percentages. The white clusters indicate clusters of counties with lower senior population percentages.

Step 4 Local *G*-statistics

Another statistic for measuring local spatial autocorrelation is the local $G_i(d)$ statistic. To compute this statistic, we must decide on a distance threshold to define the neighborhood relationship between the polygons. In this example, we will use 60 miles.
To compute the $G_i(d)$ statistic:

- From the **Ch.8** menu, choose **Local *Gi*(*d*) Statistics.**

In the Local $G_i(d)$ Statistic dialog box:

- Click `Ohcounties.shp` as the polygon theme.
- Click the **Choose the Theme** button to list attributes in the two list boxes.
- Check the box to indicate that the distance matrix has been created (in previous ArcView examples).
- Click the **All** option to indicate the inclusion of all records for calculation.
- In the left list box, scroll down and click `Cnty_fips` as the ID field.
- In the right list box, scroll down and click `SeniorPCT` as the variable.
- For the threshold distance defining the neighborhood, enter `60`.
- Click the **Calculate** button to proceed.

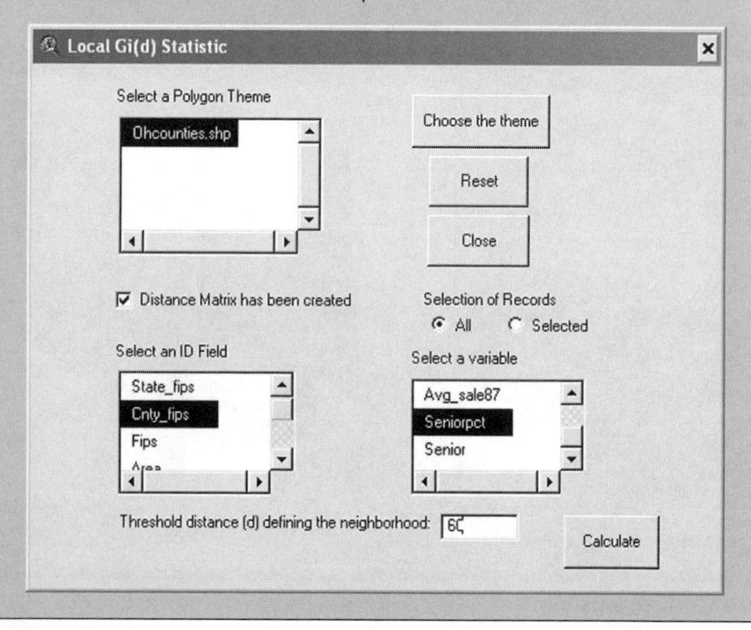

In the **Choose the Distance Matrix File** dialog box, navigate to the folder where the weight matrix was created and stored.

- Click the centroid distance matrix file to proceed.

Similar to calculating LISA, the calculated G_i-statistic will be added to the data theme's attribute table. In turn, it can be used to map the distribution of the local G-statistic so that hot spots and cold spots can be identified.

Step 5 Moran's scatterplot
The last statistical indication of local spatial autocorrelation is Moran's scatterplot. This option produces a chart that allows us to detect any outliers along the regression trend line.
For this data theme of 88 Ohio counties:

- From the **Ch.8** menu, choose **Moran Scatterplot.**

In the Moran Scatterplot dialog box, specify the following:

- Click to select `Ohcounties.shp` as the polygon theme.
- Click the **Choose the Theme** button to generate the lists of attributes in the two list boxes.
- Check to indicate that the stochastic spatial weights matrix has been created.
- Check to indicate that all records are to be included in the calculation.
- In the left list box, scroll down to click/select `Cnty_fips` as the ID field.
- In the right list box, scroll down to click/select `SeniorPCT` as the variable.
- Click the **Calculate** button.

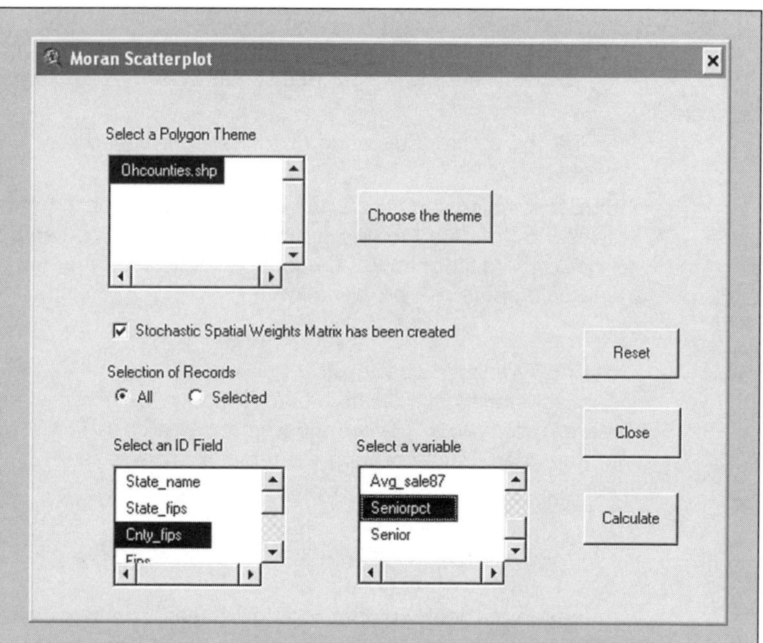

When prompted, navigate to the folder where the stochastic weights matrix was created and stored. Click the dbf file to use it. When the calculation is completed, the first set of reports lists regression parameters for the $z = f(Wz)$ function.

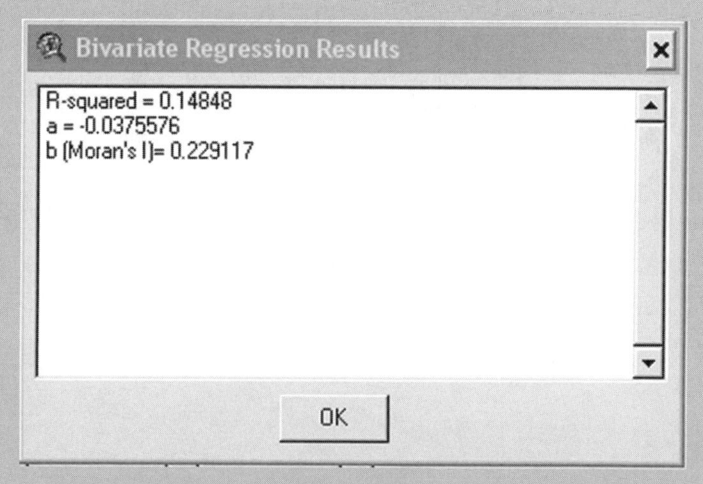

According to the report, we should expect outliers on the scatterplot because R-square is a mere 0.1485, indicating that most observations did not follow the trend line closely. The next step would be to generate the scatterplot. When generated, it may be necessary to resize the plot by extending it horizontally or vertically to obtain a better display of the trend.

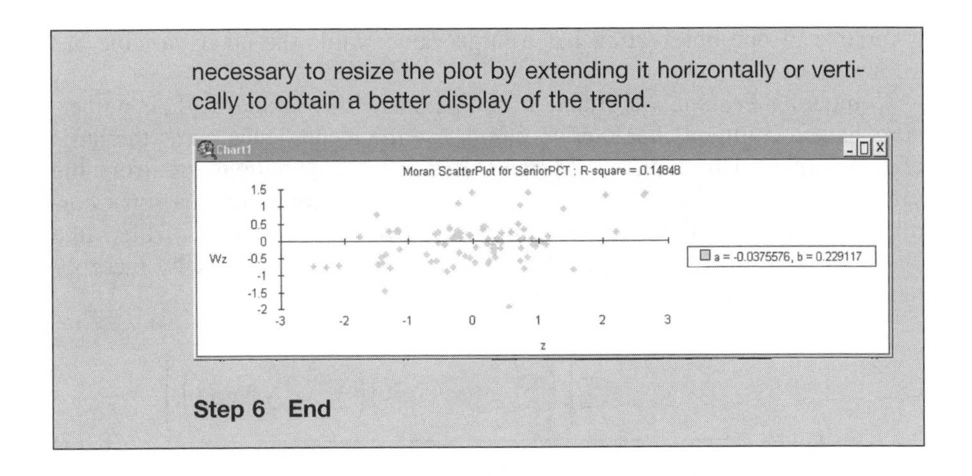

Step 6 End

8.10 BIVARIATE SPATIAL AUTOCORRELATION

All of the spatial autocorrelation measures discussed so far evaluate how values of one variable are spatially related or are interdependent over space. The First Law of Geography basically invalidates the fundamental assumption of independence in classical statistics that spatial dependency or correlation is the norm rather than the exception. As all measures of spatial dependency we have discussed so far are for the univariate situation, development of measures for evaluating spatial dependency for multiple variables (the multivariate situation) has not experienced much progress. Wartenberg (1985) suggested a framework to model spatial autocorrelation in a multivariate setting. Unfortunately, it was difficult to implement. Recently, Lee (2001) suggested a measure combining the characteristics of Pearson's correlation coefficient and Moran's I to evaluate the spatial autocorrelation of two variables—the bivariate situation.

This section reviews the essence of measuring spatial autocorrelation of two variables. Recall that Pearson's correlation coefficient is defined as

$$r_{X,Y} = \frac{\sum_i (x_i - \bar{x})(y_i - \bar{y})}{\sqrt{\sum_i (x_i - \bar{x})^2}\sqrt{\sum_i (y_i - \bar{y})^2}}. \tag{8.74}$$

The numerator is the covariance: how much x and y vary together. To capture the covariation, each x_i is compared with the mean of x, and each y_i is compared with the mean of y. If x's and y's correlate strongly, then large x values will be associated with large y values and vice versa. If this is the case, the numerator will be a large positive number. The situation is the opposite when a variable of one observation has a large value, while the other variable has a low value.

Comparing Pearson's correlation coefficient with Moran's I, their numerators have a common characteristic: both use the mean deviation as the basis of comparison. However, for Moran's I, the mean deviations are from the same variable, but for neighboring observations, whereas for Pearson's correlation coefficient, the mean deviations are from two variables of the same observation. The bivariate spatial association statistic is formed by incorporating these two correlation measures:

$$L_{x,y} = \frac{n}{\sum_i \left(\sum_j w_{ij}\right)^2} \frac{\sum_i \left[\left(\sum_j w_{ij}(x_j - \bar{x})\right)\left(\sum_j w_{ij}(y_j - \bar{y})\right)\right]}{\sqrt{\sum_i (x_i - \bar{x})} \sqrt{\sum_i (y_i - \bar{y})}}. \qquad (8.75)$$

In Equation 8.75, the first part can be regarded as a scaling factor. In the second part, the denominator is composed of the sum of deviations of x and the sum of deviations of y, analogous to the standard deviations of the two variables. In the numerator, the mean deviations of x and y in areal unit j will be multiplied together first through the spatial relationship between i and j as defined by w_{ij}, a spatial weight. If w_{ij} is a binary weight, then the numerator will be the sum of the products of the sums of deviations from x and y.

In this case, the numerator in the second term of $L_{x,y}$ is used to evaluate how similar the values of x and y are in the neighborhood of i, but not actually the values of x and y in areal unit i. In other words, the similarity of the two variables is based on a neighborhood comparison, and the neighborhood is defined through the specification of the spatial weights, w_{ij}. Because the bivariate spatial association statistic is relatively new, its statistical properties have not been fully explored. Therefore, it is still difficult to test for its significance.

Using the hypothetical landscapes and values shown in Figure 8.13, we can compare and contrast Pearson's correlation coefficient with the bivariate spatial association measure, $L_{x,y}$. In the figure, the first set of values gives a correlation coefficient of 1, and the other two sets a correlation of -1. In both cases the x and y values are compared in each areal unit i, disregarding the values in the neighborhood of i. When the spatial dimension is included, the bivariate spatial association measure indicates a strong positive association on the first one and a strong negative association on the second one. These results are reasonable, as there are high levels of spatial autocorrelation in both sets of configurations. However, the two variables in the first set have the same spatial trend, resulting in a high positive $L_{x,y}$ value. By contrast, the two variables in the second set have opposite trends, resulting in a highly negative $L_{x,y}$ value. The third set of configurations has negative spatial auto-

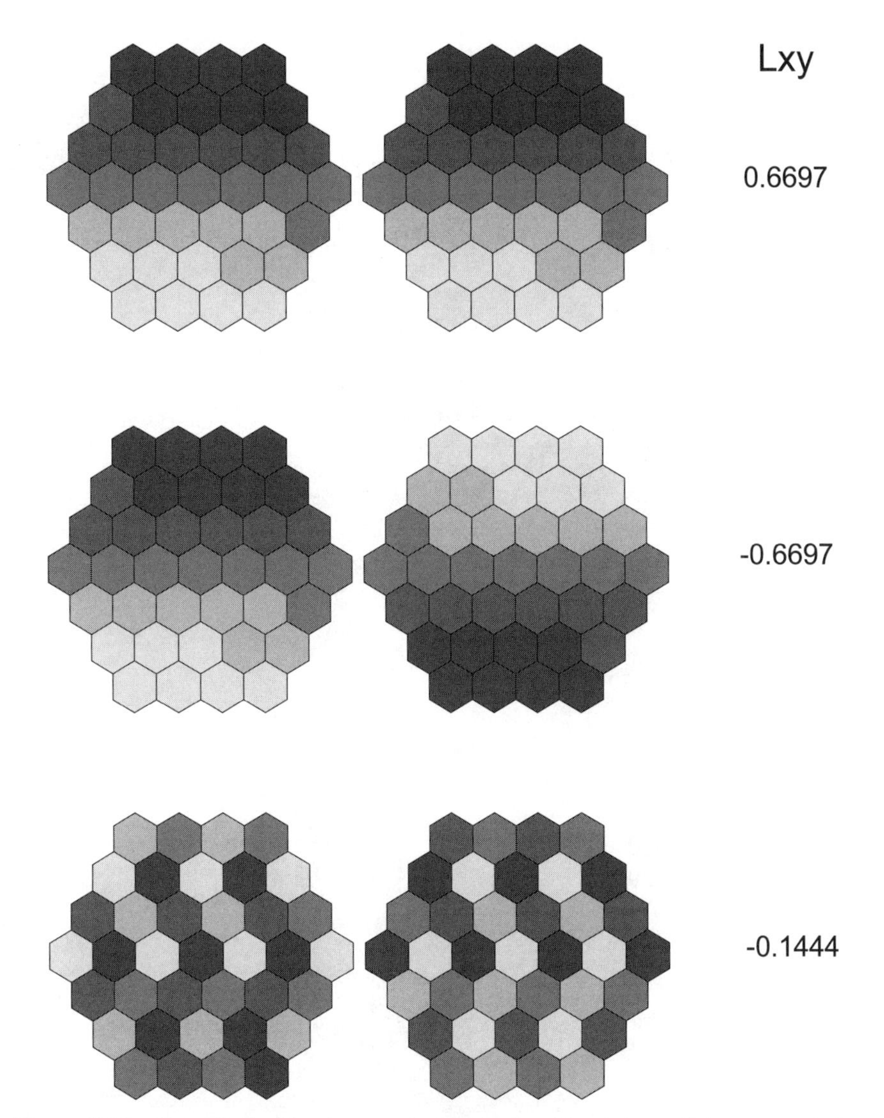

Figure 8.13 Hypothetical landscapes demonstrating bivariate spatial autocorrelation.

correlation in each variable, and the spatial patterns of the two variables are shifted slightly. As a result, the bivariate spatial association, the $L_{x,y}$ value, is slightly negative.

Using the seven Ohio counties data again, in addition to the median household income we have used before, we now include the median house value derived from Census 2000 Summary File 3 as the second variable. Figure 8.14 shows two maps with the two variables, which have almost identical patterns with only minor differences. Pearson's correlation coefficient of the

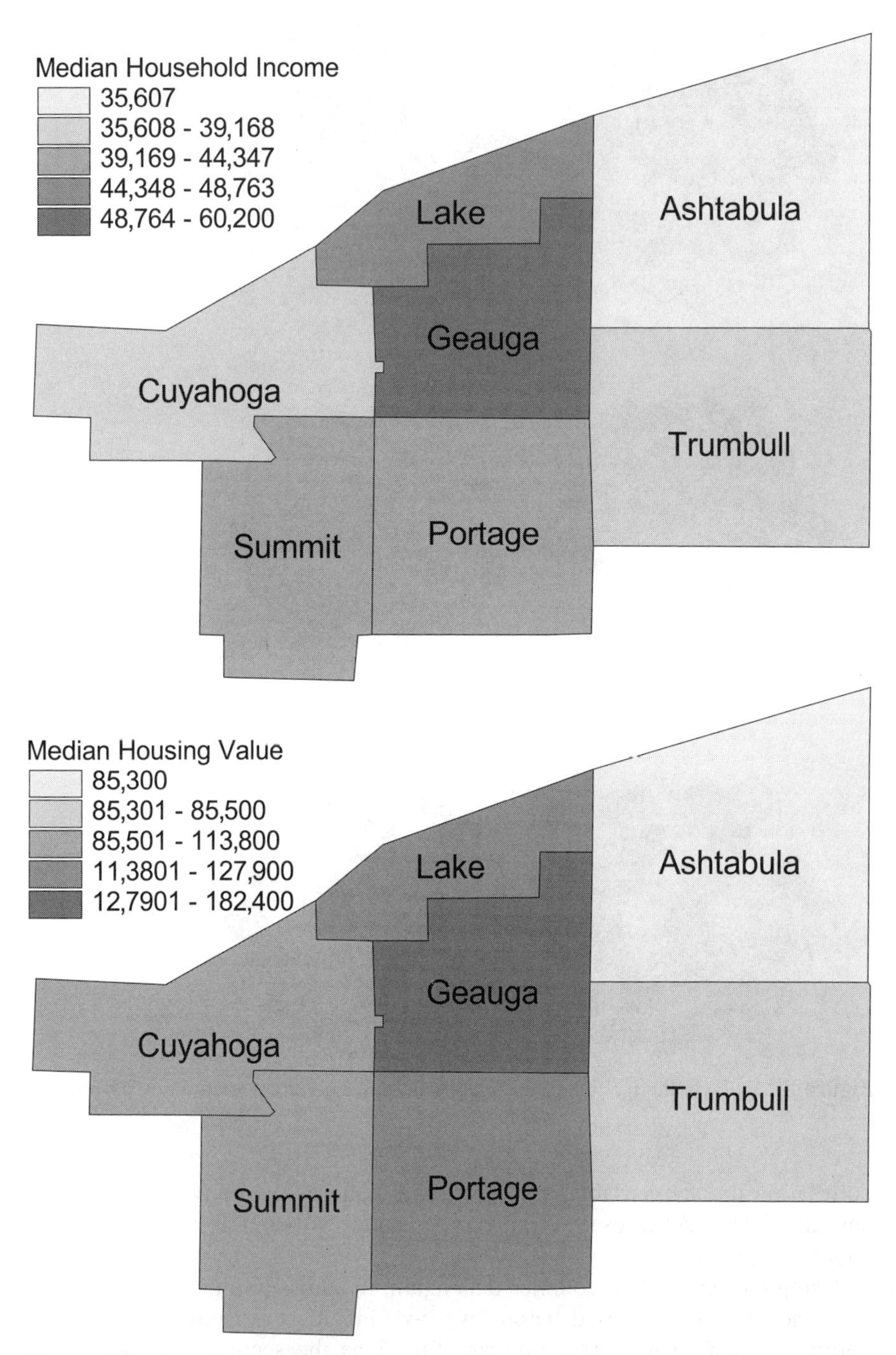

Figure 8.14 Median household income and median house value of the seven Ohio counties, 2000 Census, Summary File 3.

two variables is very high, 0.9677, as expected. But because of the moderate level of negative spatial autocorrelation in median household income (−03468), as demonstrated in the previous section, and the moderate level of negative spatial autocorrelation in median house value (−0.2989), the bivariate spatial association measure, $L_{x,y}$, gives us a slightly positive value of 0.1707. Though the two variables have a very strong aspatial positive correlation, this positive relationship was not expressed in the spatial dimension to overcome the negative spatial autocorrelation embedded in the two variables. As a result, the spatial association of these two variables was just slightly positive.

8.11 APPLICATION EXAMPLES

As discussed in this chapter, we now have a set of measures that can be used to assess the global or local trends of spatial autocorrelation that may exist among the attribute values of a set of polygons. These measures are very useful in helping us understand how geographic phenomena or events distribute or occur over space.

We would like to remind readers that any given pattern of a geographic phenomenon or a type of event is just a geographic pattern existing at a particular time, and we can understand the pattern better if we can identify the spatial process by studying the geographic patterns over time. When multiple spatial patterns of the same phenomenon or type of event are available over different periods for the same area, we can better examine how the events or distributions changed in the past in order to project or predict what change may occur in the future.

The power of the measures discussed in this chapter depends on how they are used. If only the statistics are used to measure the degree of spatial autocorrelation as exhibited in the polygons' attribute values, we can compare such statistics calculated from the same set of polygon attributes from different time periods or similar polygon attributes from different geographic areas at the same time periods. Similar configurations of study can be easily designed and implemented.

As an example, we present a study that uses joint count statistics, the simplest form of statistics that measure spatial autocorrelation, to examine how residential development occurred in urbanized areas. In this study, the temporal changes of the measured joint count statistics were captured to form a model that showed how development of residential areas helped to explain various stages of urbanization, suburbanization, urban sprawl, and saturated development. A detailed discussion of this study can be found in Lee (2002).

In this study, parcel-level residential polygons of Geauga County, Ohio, were used as the primary data. Over 40,000 parcels were recorded with information regarding when each parcel was first built on. In other words, given any year from before 1800 to recent years, we could classify all parcels into either a vacant lot, V, or a built lot, B. This V-B distinction is similar to the black-white attribute of the polygons discussed in Section 8.6. The parcels in Geauga County were further divided into 20+ groups, according to boundaries from its cities, villages, and townships.

For each city/village/township, a series of years were chosen to integrate this building process of converting V lots to B lots in pre-1900, 1910, 1920, 1930, 1940, 1950, 1960, 1970, 1980, 1990, and 2000. At the end of each decade, we calculated the joint count statistics and the associated z-scores for

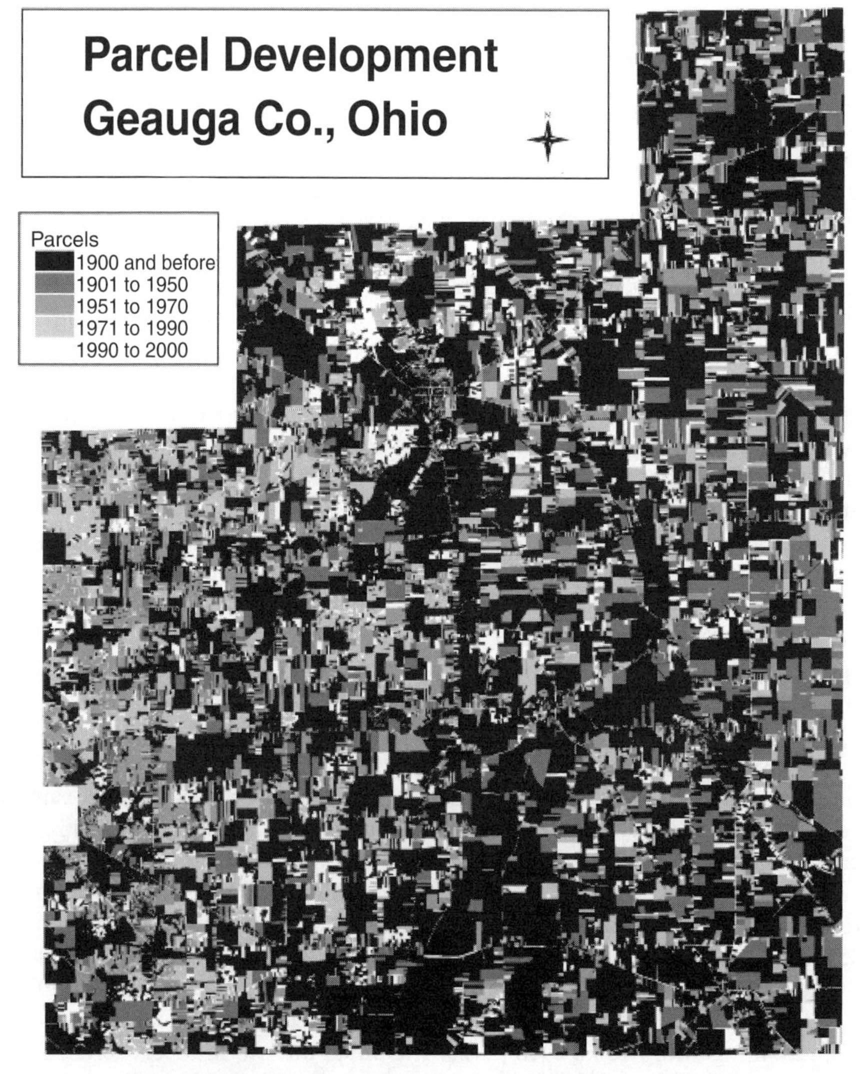

Figure 8.15 Parcel development in Geauga County, Ohio.

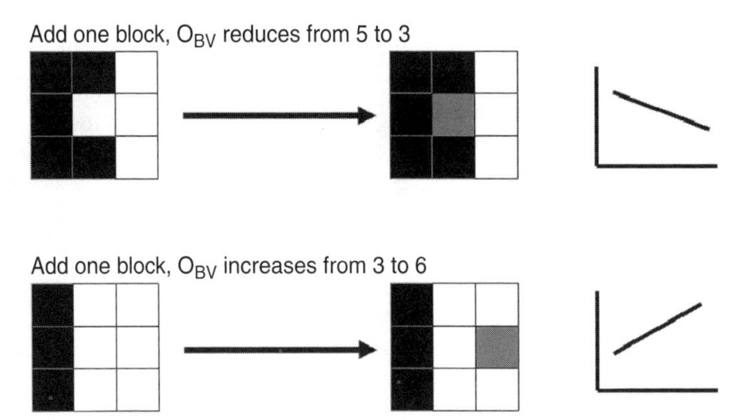

Figure 8.16 Changes in joint count statistics due to new residential developments.

each subarea of Geauga County. In this fashion, we were able to trace, for any city/village/township, how its z-scores changed over time.

Geauga County has experienced significant growth in recent years, possibly under the influence of the greater Cleveland metropolitan area on Geauga's west side. Figure 8.15 shows parcel-level development over time, with darker shades for parcels that were developed earlier and lighter shades for parcels that were developed recently.

First, let's look at how new developments affect the joint count statistics. In Figure 8.16, adding a new residential development to existing residential areas clearly affects the observed counts of BV joints. These, in turn, would affect the values of joint count statistics.

Next, in Figure 8.17, we can see that the slopes of the trend lines between times and the z-scores of the joint count statistics do affect the rate of residential development. The same three patterns are further explained in Figure 8.18, where the different rates at which the number of observed BV joints increase clearly show their association with the three patterns.

Finally, Figure 8.19 shows the quantitative model that consists of various temporal trends, constructed by plotting the z-scores for the joint count statistics (as the vertical axis) against time (as the horizontal axis). As is often

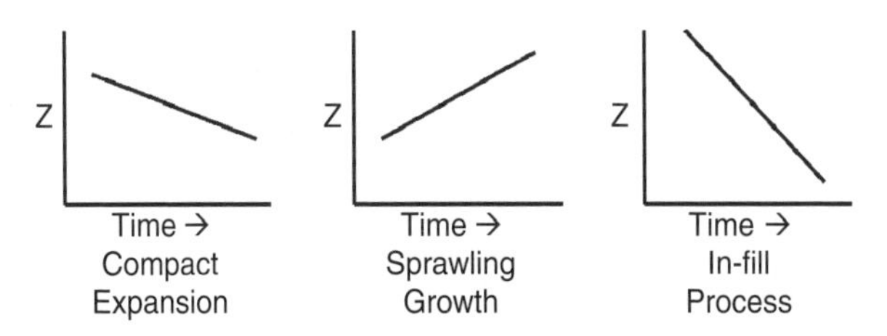

Figure 8.17 Three temporal trends in residential development.

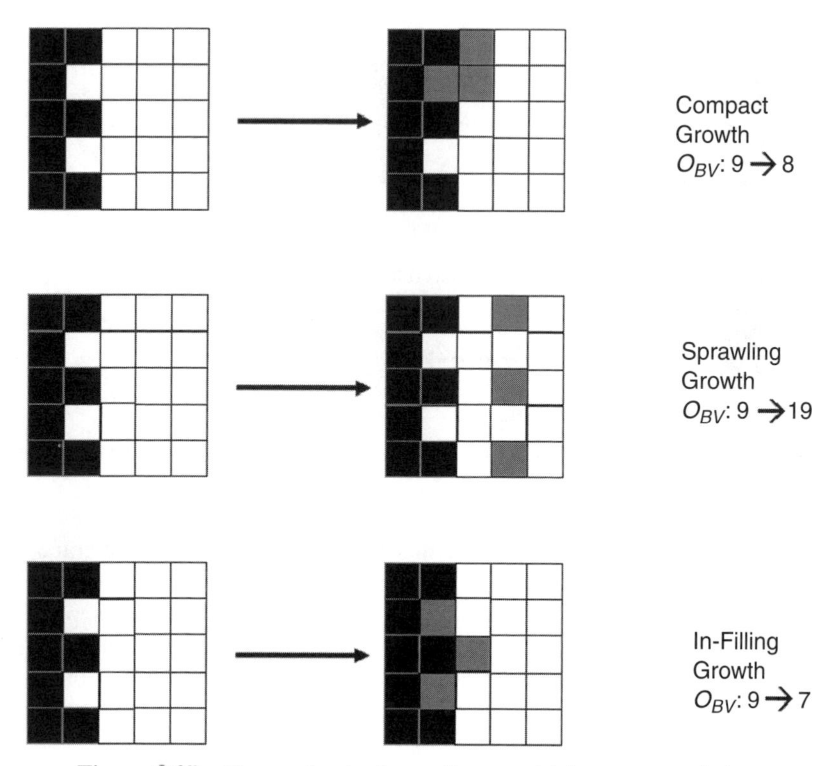

Figure 8.18 Three urbanization patterns and joint count statistics.

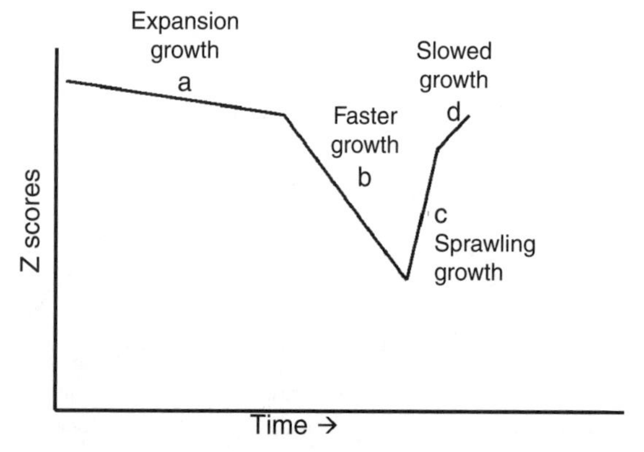

Figure 8.19 A quantitative model of residential development for urban areas.

the case for almost all population settlements, a city may start as a small population settlement that slowly grows, expanding in size. This period is described by section a in Figure 8.19. Following that period, a city may further expand with much faster growth once it surpasses the economy of scale in terms of city size, given enough locational advantages or resource endowments. This period is denoted section b in Figure 8.19. If urban sprawl occurrs in the area surrounding the city, it is called section c. The last section, d, describes the period of slowing growth after a period of significant sprawl.

The empirical data from Geauga County are shown in Figure 8.20. Every city/village/township in Geauga has displayed a development trend closely corresponding to the quantitative model in Figure 8.19. Almost without exception, each trend line shows an initial stage of slow, expansive growth followed by a faster growth period. Since the 1970s, urban sprawl has appeared in many townships and cities, as is evident in these trend lines. It should be noted, however, that each city/village/township began its urban sprawl at a different time. They achieved the lowest statistical significance of joint count statistics at different times and different speeds (slopes).

As indicated in this application example, statistical methods are tools that can assist us in better understanding spatial patterns and processes of geographic phenomena. Users of these methods must clearly understand how to use them, when to use them, and what their limitations are. But most importantly, users should keep in mind that innovative uses of these tools, such as

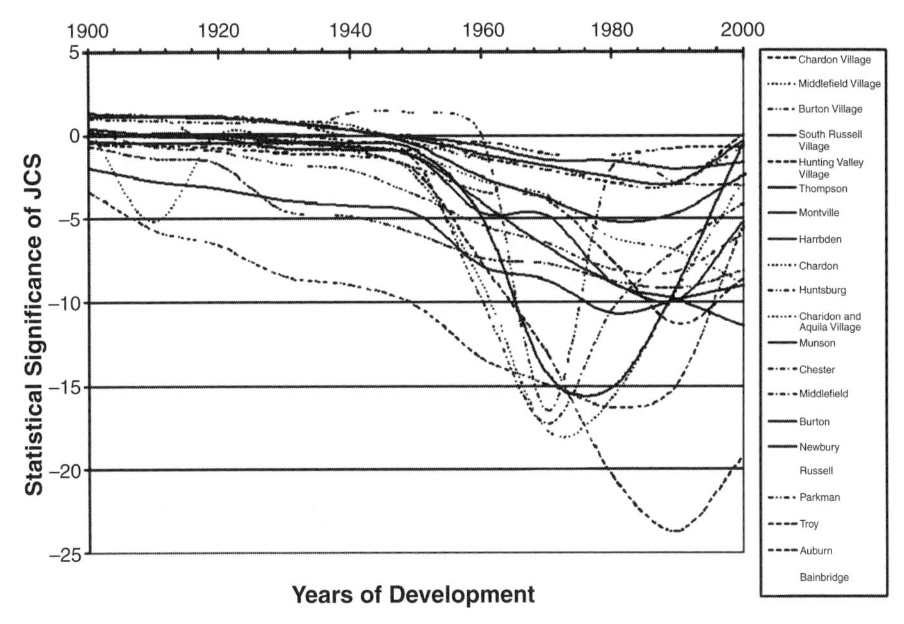

Figure 8.20 Empirical models of urban residential development for Geauga County, Ohio

the approach described in this section, often lead us to new findings, new models, or new research questions.

8.12 SUMMARY

In real-world GIS applications and spatial analyses, spatial data are often treated as data without a spatial dimension. Classical statistical techniques, which often assume that the observations are independent of each other, are used indiscriminately on spatial data as if they are ordinary data. The unique characteristics of spatial data are ignored, and the analytical results may be biased. In this chapter, we argue that a unique characteristic of spatial data is (spatial) dependency, and evaluating spatial dependency or autocorrelation is an important step in correctly analyzing spatial data. If the assumption of independence is violated, classical statistical techniques for drawing inferences will be inappropriate.

Different types of spatial autocorrelation and measures have been discussed in this chapter. Global measures are summary measures for the entire region, while local measures depict the situation of each areal unit. Some measures are effective in identifying spatial trends, while others are useful in distinguishing hot spots from cold spots. All of these measures and tools are descriptive and exploratory in nature. Some of these tools are also for diagnostic purposes. For instance, a special version of Moran's I, $I_e = (N/S_0)(e'We/e'e)$, is to evaluate the spatial autocorrelation of residuals of a regression (Cliff and Ord, 1981). In this formulation, N is the number of areal unit, S_0 is the sum of the spatial weights matrix, W is the weights matrix and e is the vector of the residuals from the regression. If the residuals have significant spatial autocorrelation, using classical regression models will not be appropriate as the spatial autocorrelation has to be modeled.

To model spatial autocorrelation, more advanced techniques and models are needed, but they are beyond the scope of this introductory book. Numerous spatial regressions (Griffith, 1988; 2003) and spatial econometric models (Anselin, 1988) have been developed. While these monographs mainly address global spatial regressions, assuming that the relationships among variables are stationary over the study region, local regression models, such as the Geographically Weighted Regression (GWR), have been developed recently to accommodate the nonstationary nature of parameters in a regression framework (Fotheringham et al., 2002).

Another area that is in the general area, but beyond the scope of this book is advanced point pattern analysis or geostatistics, specifically in estimating statistical surfaces. These techniques include kernel density estimation, deterministic spatial interpolators (such as the inverse distance weight method), and the stochastic spatial interpolators, including the family of kriging techniques (Bailey and Gatrell, 1995). Interested readers can pursue these topics for more advanced techniques.

REFERENCES

Anselin, L. 1988. *Spatial Econometrics: Methods and Models.* Boston: Kluwer Academic Publishers.

Anselin, L. 1995. Local Indicators of Spatial Association—LISA. *Geographical Analysis* 27(2): 93–116.

Anselin, L. and D. A Griffith. 1988. Do spatial effects really matter in regression analysis? *Papers of the Regional Science Association* 65: 11–34.

Bailey, T. C. and A. C. Gatrell. 1995. *Interactive Spatial Data Analysis.* Essex, England: Longman.

Batchelor, B. G. 1978. *Pattern Recognition: Ideas in Practice.* New York: Plenum Press.

Cliff, A. D. and J. K. Ord. 1973. *Spatial Autocorrelation.* London: Pion.

Cliff, A. D. and J. K. Ord. 1981. *Spatial Processes: Models and Applications.* London: Pion.

Costanzo, C. M. 1983. Statistical inference in geography: modern approaches spell better times ahead. *The Professional Geographer* 35(2): 158–164.

Fotheringham, A. S., C. Brunsdon, and M. E. Charlton. 2002, *Geographically Weighted Regression: The Analysis of Spatially Varying Relationships.* Chichester, UK: John Wiley & Sons.

Fotheringham, A. S. and M. E. O'Kelly. 1989. *Spatial Interaction Models: Formulations and Applications.* Kluwer Academic.

Getis, A. 1991. Spatial interaction and spatial autocorrelation: a cross-product approach. *Environment and Planning A* 23: 1269–1277.

Getis, A. and J. Aldstadt. 2004, On the Specification of the Spatial Weights Matrix. *Geographical Analysis,* in press.

Getis, A. and J. K. Ord. 1992. The analysis of spatial association by use of distance statistics. *Geographical Analysis* 24(3): 189–207.

Griffith, D. A. 1987. *Spatial Autocorrelation: A Primer.* Washington, DC: Association of American Geographers.

Griffith, D. A. 1988. *Advanced Spatial Statistics.* Boston: Kluwer Academic.

Griffith, D. A. 1996. Some guidelines for specifying the geographic weights matrix contained in spatial statistical models. In S. L Arlinghaus and D. A. Griffith (eds.), *Practical Handbook of Spatial Statistics.* Boca Raton, FL: CRC Press, pp. 65–82.

Griffith, D. A. 2003. *Spatial Autocorrelation and Spatial Filtering.* Berlin: Springer-Verlag.

Griffith, D. A., D. W. S. Wong, and T. Whitfield. 2003. Exploring relationships between the global and regional measures of spatial autocorrelation. *Journal of Regional Science* 43(4): 683–710.

Lee, J. 2002. Geographic patterns of urban residential devlopment. In B. Boots, A. Okabe, and R. Thomas (eds.), *Modelling Geographical Systems: Statistical and Computational Applications.* Boston: Kluwer Academic, pp. 13–30.

Lee, S. 2001. Developing a bivariate spatial association measure: an integration of Pearson's r and Moran's I. *Journal of Geographical Systems* 3: 369–385.

Odland, J. 1988. *Spatial Autocorrelation.* Newbury Park, CA: Sage Publications.

Summerfield, M. A. 1983. Populations, samples, and statistical inference in geography. *The Professional Geographer* 35(2): 143–148.

Tobler, W. A computer movie simulating urban growth in the Detroit region. *Economic Geography* 46 (Supplement): 234–240.

Wartenberg, D. 1985. Multivariate spatial correlation: a method for exploratory geographical analysis. *Geographical Analysis* 17: 263–283.

Wong, D. W. S. 2002. Modeling local segregation: a spatial interaction approach. *Geographical and Environmental Modelling* 6(1): 81–97.

EXERCISES

1. Construct the binary and stochastic weights matrices for the six New England states. Based upon the binary matrix, also construct the sparse matrix. How much storage space can be saved by using the sparse matrix instead of the square binary matrix?

The table below shows the median housing value of the six New England states according to Summary File 3 of the 2000 Census.

States	Median Housing Value (1999)
Connecticut	166,900
Maine	98,700
Massachusetts	185,700
New Hampshire	133,300
Rhode Island	133,000
Vermont	111,500

2. Compute the mean of the median housing value. Use the mean to form the classification variable ExpensiveHouse. States with housing values below the mean will be assigned a 0, and states with housing values above the mean will be assigned a 1. Compute the joint count statistics to test for spatial autocorrelation.

3. With the shapefile file for the New England state boundaries (C:\Temp\Data\ch8_data\Newengland.shp), the median housing value data are included. Use the extension for Chapter 8 to repeat the joint count statistics calculation to verify if the answers from Question 2 are correct.

4. Using the median housing values in the table, compute Moran's I and Geary's *C*. Determine if the results from the two spatial autocorrelation statistics are consistent or not.

5. Construct a spatial weights matrix using the centroid distance.

6. In order to compute the general *G*-statistic, the distance for the neighborhood definition has to be determined. Review the distance-based matrix in Question 5. What is the appropriate distance to use as the neighborhood definition for the general *G*-statistic? Why?

7. Using that distance from Question 6, compute the general *G*-statistic using the median housing value variable.

8. Compute the LISA for each state and map the results. Do the results shed any light on the location variability of spatial autocorrelation?

9. Compute the local *G*-statistic using the median housing value variable. Use the distance derived from Question 6 as the neighborhood definition. How do the results from the local *G*-statistic complement the results from the global *G*-statistic in Question 7?

10. First, create a stochastic weights matrix. Using this matrix, develop a Moran scatterplot. Is there any outlier in terms of median housing value?

States	Per Capita Income (1999)
Connecticut	28,766
Maine	19,533
Massachusetts	25,952
New Hampshire	23,844
Rhode Island	21,688
Vermont	20,625

11. The table above provides the 1999 per capita income of the six New England states. Compute Pearson's correlation coefficient between the per capita income and the median house value. Also, compute the bivariate spatial association index between the two variables.

APPENDIX

ArcGIS SPATIAL STATISTICS TOOLS

1 SPATIAL STATISTICS IN ArcGIS

With ArcGIS 9.X, a suite of spatial statistics wizards is available for performing spatial analytical and statistics procedures that are similar to some of the methods discussed in this book. This is a significant upgrade of the analytical capability of commercial GIS software packages. While they do not offer functions that are as complete as the ArcView extensions accompanying this book, the spatial statistics wizards in ArcGIS 9.X do provide a variety of useful tools, as shown in Figure A.1. Most, if not all, of these functions were discussed in our first book (*Statistical Analysis with ArcView GIS*) and were incorporated into ArcView 3.X in 2001.

Recognizing the availability of the spatial statistics wizards in ArcGIS 9.X, we are pleased to see the implementation of many spatial statistical tools in ArcGIS 9.X, as this process allows practical uses for academic work and enables many GIS users to apply these tools in their work. We believe this is an encouraging trend. We hope the developers of other GIS software packages will follow this trend and add these functions to their products.

Since all spatial statistics wizards use concepts and procedures similar to those discussed in this book, detailed discussions of them here would be redundant. Instead, we highlight several of these wizards to demonstrate their functions. When appropriate, we offer brief discussions of their uses and comparisons to our extensions.

2 AVERAGE NEAREST NEIGHBOR DISTANCE

The wizard tool for Average Nearest Neighbor performs an analysis similar to the Nearest Neighbor Analysis discussed in Section 6.3 of this book. After

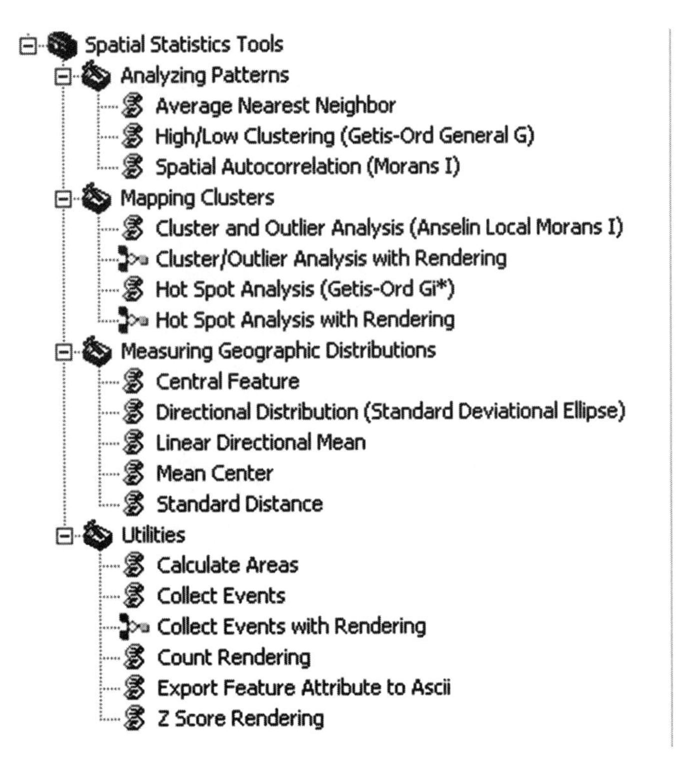

Figure A.1 Spatial statistics wizards in ArcGIS.

launching the average nearest neighbor function, the program displays a ratio between observed mean distance and expected mean distance along with a z-score. In addition, a nice graphic display shows how the calculated statistic is compared with critical values at various levels of significance.

Figures A.2a and A.2b show results of using the Average Nearest Neighbor Distance wizard with the Dip-In and EMAP data sets discussed in Chapter 6. For example, the Dip-In dataset yields a z-score of -15.2 standard deviations. Consequently, one may conclude that the observed pattern resembles a clustered pattern. Since the test result is significant, the wizard suggests that the clustering did not occur by chance. The results of the wizard also suggest that the EMAP dataset also exhibits a clustering pattern, but with a z-score of -22 standard deviations.

The wizard is easy to use. It provides essential information for users to quickly judge if the point pattern has any statistical significance. When comparing two or more patterns objectively, users need to rely on the z-scores in addition to the average nearest neighbor distance.

Note that the results of any Nearest Neighbor Analysis are highly sensitive to the delineation of the study area and how distance is calculated. To ensure that the analysis yields meaningful results, users of this book's ArcView ex-

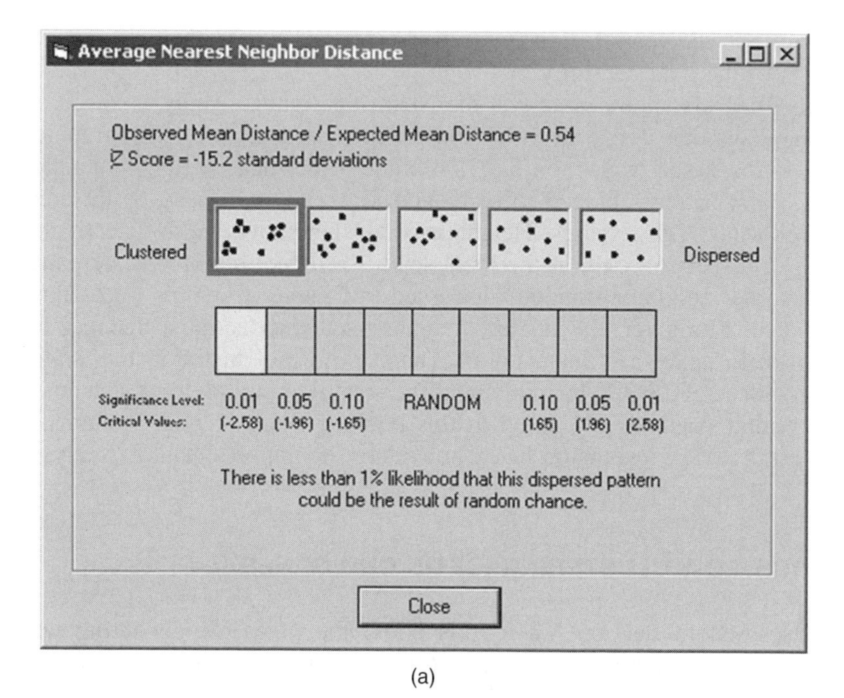

(a)

Figure A.2a Average Nearest Neighbor Analysis of Dip-In data.

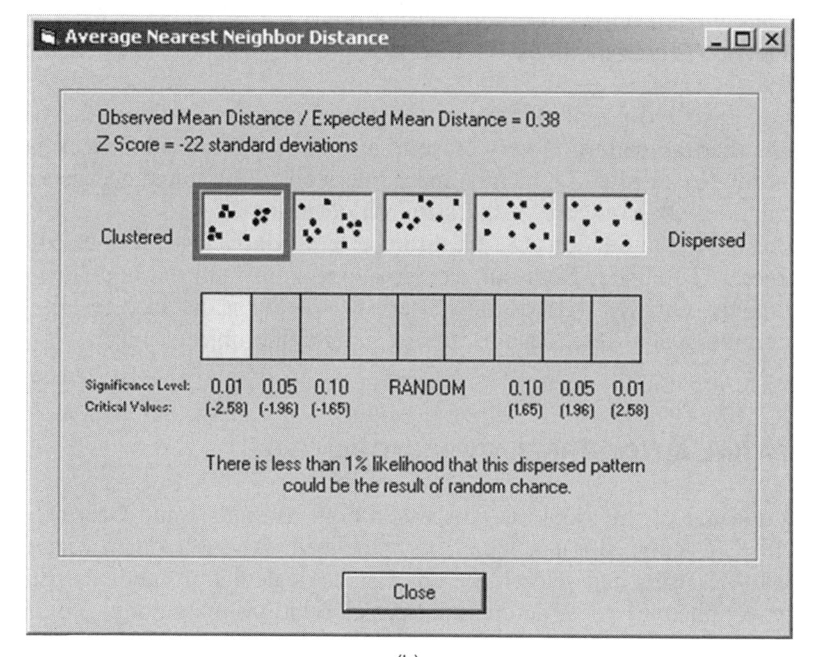

(b)

Figure A.2b Average Nearest Neighbor Analysis of EMAP data.

tensions and the spatial statistics wizards in ArcGIS 9.X must consider these aspects of the data carefully.

Finally, users should note that the Nearest Neighbor Analysis is subject to distance lags that define neighborhood relationships between geographic objects, as discussed in Section 6.3.3 on higher-order nearest neighbor relationships. Point patterns judged to be dispersed at one scale may be different at another scale. Also, Nearest Neighbor Analysis is effective in detecting clustering or dispersion only at the local scale. To detect regional-scale patterns, higher-order neighbor statistics discussed in Chapter 6 of this book must be employed. Moreover, the nearest neighbor statistics adopt a distance-based approach to analyze point patterns. There is no tool in the spatial statistics wizards in ArcGIS 9.X to implement the spatial sampling approach, such as the Quadrat Analysis discussed in this book, to analyze point patterns. The Ch6.avx ArcView extension has a procedure to support Quadrat Analysis.

3 HIGH/LOW CLUSTERING (GETIS-ORD GENERAL G)

As discussed in Section 8.8 of this book, the high/low clustering wizard allows users to calculate the general G-statistic for polygon data. The results of calculation include the general G Index value and a z-score for its statistical significance. A distance threshold can be specified and a weight matrix can be requested, similar to the ArcView extension developed in this book. Finally, this wizard allows users to select from a list of methods for distance calculation. These methods include Euclidean distance and Manhattan distance.

Using the 1990 population counts of the 88 counties in Ohio, the wizard suggests that the pattern is very close to a random pattern. Figures A.3a and A.3b show the results graphically and numerically. The calculated general G Index is 0.8 with a z-score of 0.7 standard deviations.

As discussed in Section 8.8, the distance threshold is critical in calculating the general G-statistic. Both our ArcView extensions and the high/low clustering (Getis-Ord G) wizard allow users to specify a distance threshold to structure the spatial relationships among geographic objects.

4 SPATIAL AUTOCORRELATION (MORAN'S I)

In Section 8.7 of this book, we discussed both Moran's I and Geary's C. In ArcGIS 9.X, only Moran's I can be calculated. As with other wizards in ArcGIS 9.X, users can select different distance calculation methods (Euclidean or Manhattan) to structure the spatial relationship among geographic features.

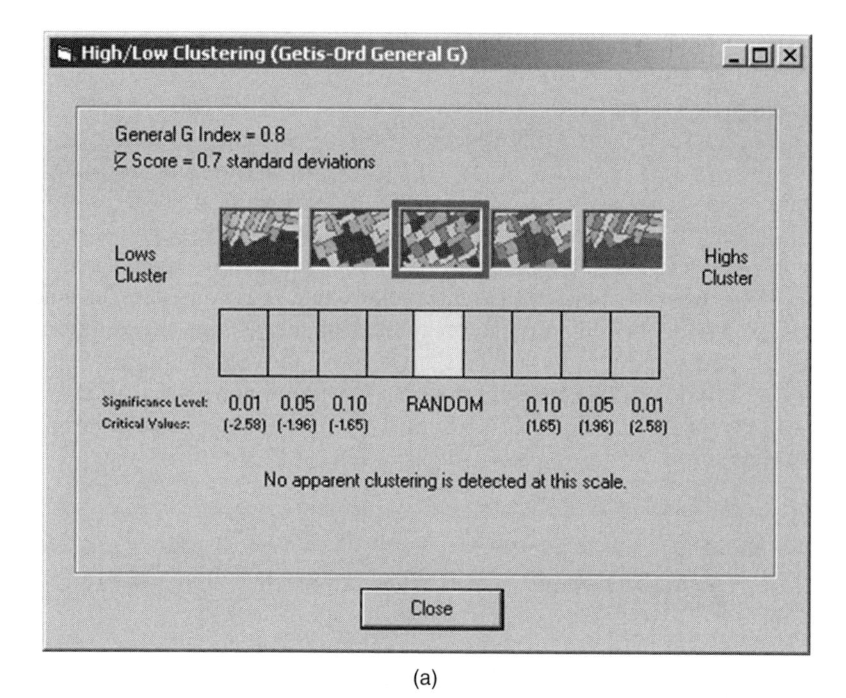

(a)

Figure A.3a Graphic results of the general G index for 1990 Ohio county population counts.

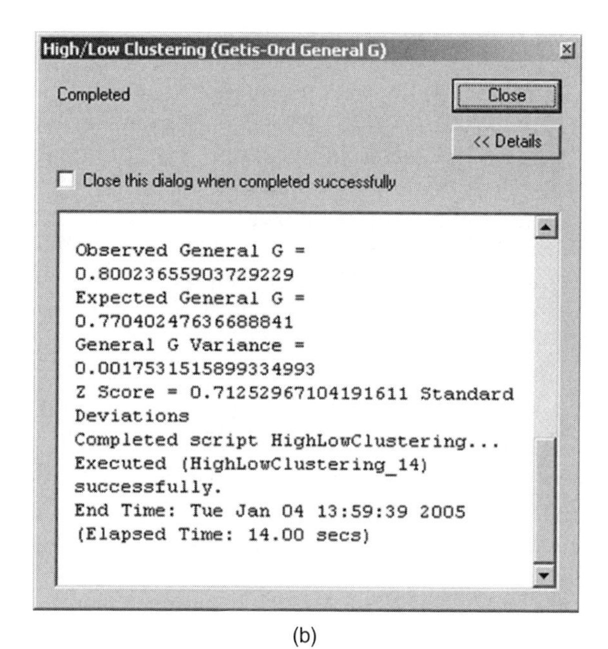

(b)

Figure A.3b Numerical results of the general G index for 1990 Ohio county population counts.

Also, using Ohio's 1990 population counts at the county level, the results of calculating Moran's I are displayed in graphic form in Figure A.4a and numerically in Figure A.4b. The index value suggests that the pattern is a clustering pattern and probably did not occur by chance.

With this book's ArcView extension, Ch8.avx, users can calculate both Moran's I and Geary's *C* with ease. In addition, the extension offers different sampling assumptions (randomization vs. normality) and more methods for distance calculation. Both indices are popular measures of spatial autocorrelation. They differ in how the neighboring values are compared. In general, Moran's I has better statistical properties than Geary's *C*. In addition, the numerical range of Moran's I corresponds more closely to our conventional notion of the correlation coefficient. However, empirical values of Moran's I may go beyond the theoretical range. Still, using both spatial autocorrelation measures is more desirable than using just one.

5 CLUSTER AND OUTLIER ANALYSIS (ANSELIN'S LOCAL MORAN'S I)

Section 8.9.1 of this book provides the necessary background to understand the proper use of local Moran's I. Essentially, the cluster and outlier analysis wizard calculates the local Moran for each polygon using a specified attribute. Afterward, the calculated index values can be mapped to show clusters of high index values or low index values if those clusters exist.

Using the 88 population counts of Ohio counties, Figure A.5a shows one cluster of very high index values around Cuyahoga County in northeast Ohio, another cluster of high index values around the Cincinnati area in southwest Ohio, and a cluster of low index value around the Columbus area (Franklin County) in central Ohio. Figure A.5b shows the list of calculated index values added to the attribute table of the data layer.

The cluster and outlier analysis wizard works very well in producing a graphical display of how the local Moran I values distribute. Users can identify areas with clusters of high or low I values. Note that the ArcView Ch8.avx extension also includes tools to calculate the local Geary ratio in addition to the local Moran I; both indices are LISA indices. In the Ch8.avx extension, the calculated statistics will also be added to the attribute table for thematic map rendering. When these local spatial autocorrelation indices are computed in GIS and stored in the attribute table, they will facilitate the exploration of geographic patterns, and can assist in identifying outliers. This approach is often referred to as *exploratory spatial data analysis* (ESDA). Besides LISA, the Ch8.avx extension in this book has an additional graphic tool, the Moran scatterplot, to explore the pattern of spatial autocorrelation and to identify outliers.

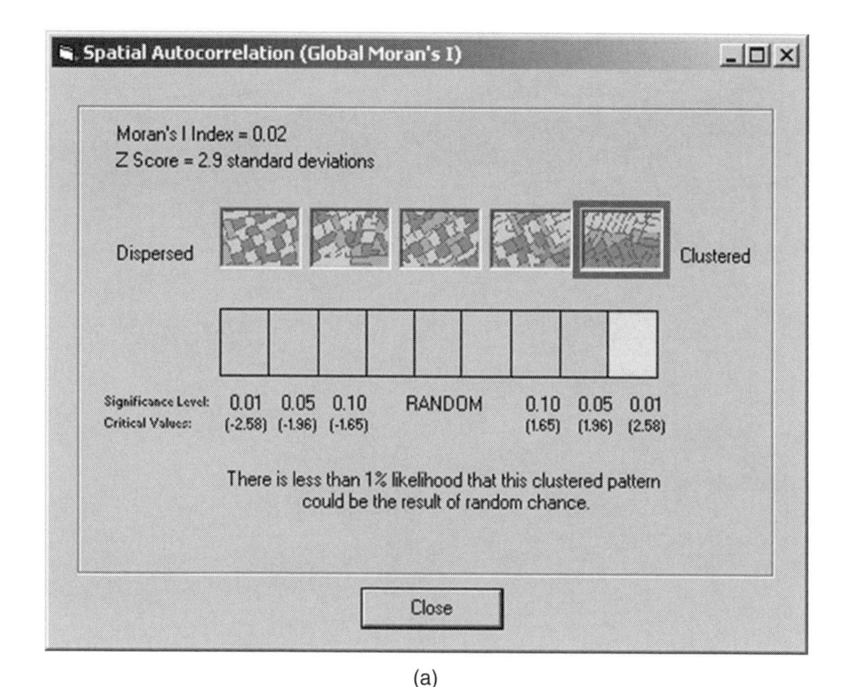

(a)

Figure A.4a Moran's I and Z-scores in a graphic display.

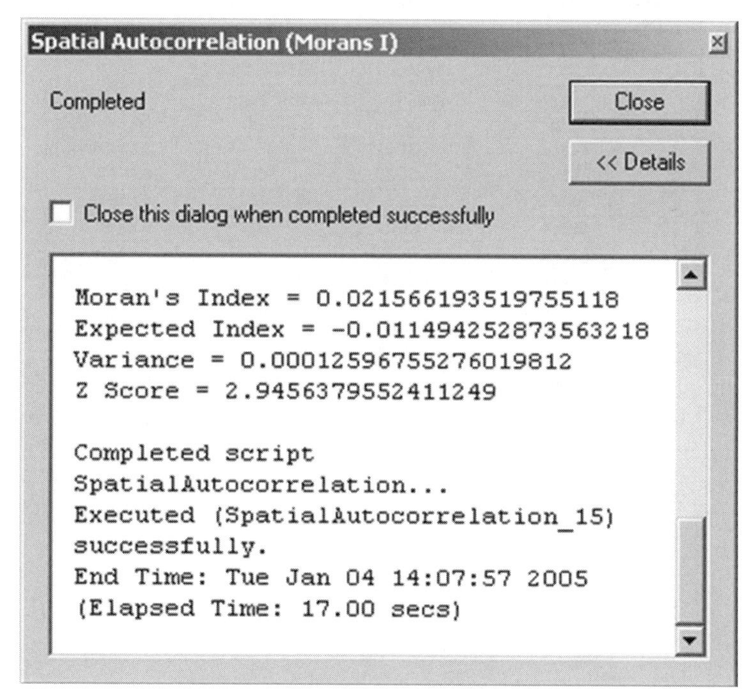

(b)

Figure A.4b Moran's I and Z-scores in a numeral display.

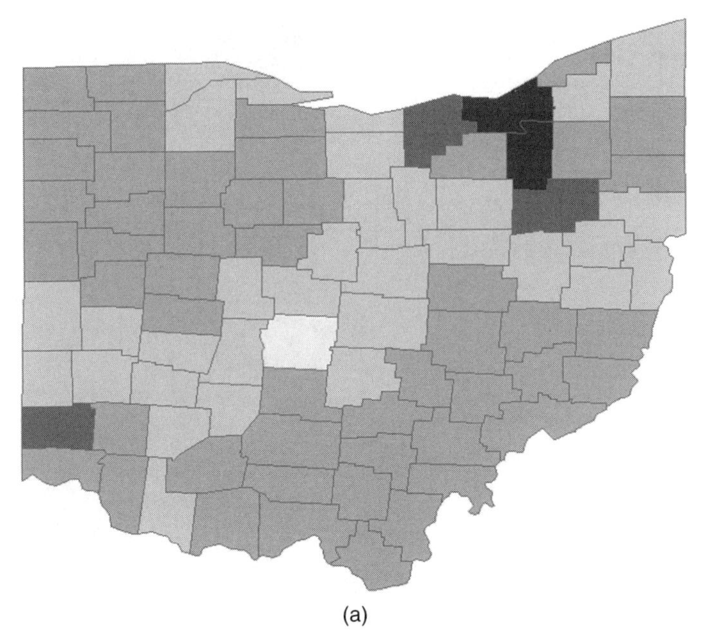

(a)

Figure A.5a Clusters of local Moran's I values.

Attributes of OhioClusters

LMiInvDst	LMzInvDst	ExpectedI	Variance
-7.78776	-1.08766	-0.652672	43.034
1.73186	0.346083	-0.637276	46.8616
-2.7824	-0.275772	-0.655743	59.4698
2.20781	0.416665	-0.582777	44.8558
23.5566	3.31076	-0.69882	53.6739
0.353685	0.131294	-0.741104	69.5296
0.090440	0.110522	-0.745276	57.1766
7.33481	1.06883	-0.741752	57.1
1.07417	0.203799	-0.811729	85.6318
3.0047	0.544326	-0.585254	43.4971
2.83086	0.450135	-0.727867	62.5034
-0.212051	0.069729	-0.761798	62.1589
3.15667	0.472821	-0.671102	65.5388
32.9721	4.15981	-0.759809	65.7558
1.66344	0.305483	-0.707847	60.2548
-0.196223	0.073787	-0.826905	73.0569
2.06882	0.341841	-0.794969	70.1831
1.45944	0.26131	-0.852143	78.2536
3.66428	0.542036	-0.687424	64.4559

Record: 1 Show: All Selected Records (0 out of 88 Se

(b)

Figure A.5b Calculated local Moran's I values.

6 HOT SPOT ANALYSIS (GETIS-ORD G_i^*)

Similar to local Moran's I, the Getis-Ord G_i^* statistics allow us to identify clusters of high index values (hot spots) and clusters of low index values (cold spots). In Section 8.9.2 of this book, we discuss the G_i statistic, which is the local version of the general G-statistic. However, we did not discuss G_i^* statistics in this book. The major difference between G_i and G_i^* is subtle. In computing several components of G_i, unit i cannot equal unit j, but in G_i^*, it can.

Using the wizard in ArcGIS 9.X, the results of the calculated G_i^* statistics can be mapped to show the spatial distribution of the index values. In turns, hot/cold spots may be identified easily on the map. Using the Dip-In and EMAP data sets, Figures A.6a and A.6b show the distributions of G_i^* values with different shades of gray. The two data sets seem to display similar patterns of the G_i^* index values. Massachusetts and Connecticut seem to show some clusters of high index values, while northern parts of Vermont and New Hampshire show few clusters of low index values.

Finally for spatial autocorrelation statistics, we discussed the bivariate spatial autocorrelation index in Section 8.11. The ArcView extension accompanying this book provides the corresponding tool to compute this index.

(a)

Figure A.6a Distribution of G_i^* index values in Dip-In data.

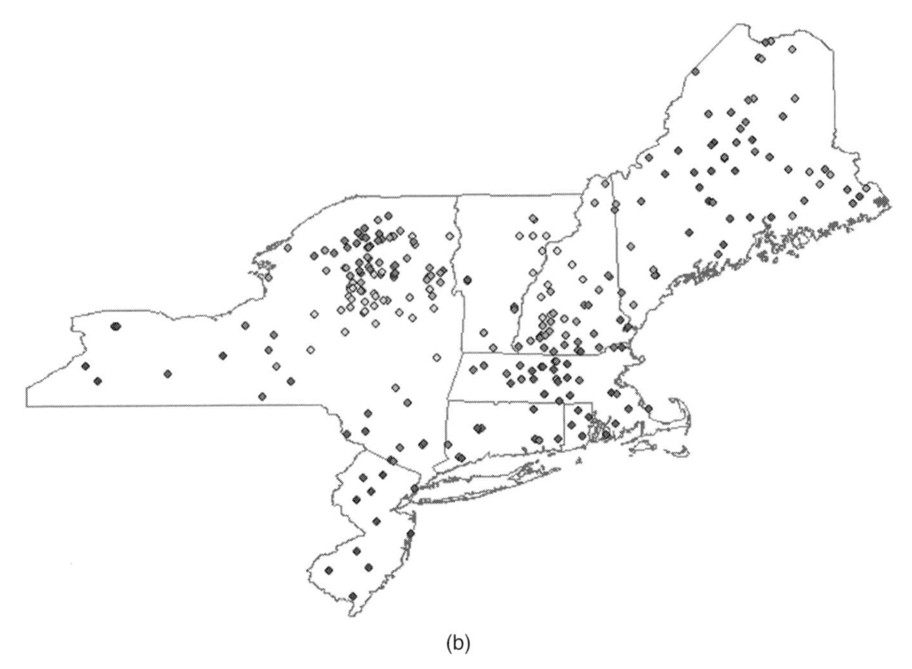

(b)

Figure A.6b Distribution of G_i^* index values in EMAP data.

7 MEASURING GEOGRAPHIC DISTRIBUTION

In ArcGIS's spatial statistics wizards, there are tools for calculating mean centers, standard distance, directional distribution, and linear directional means. These measures are covered in Chapters 5 and 7, and tools to calculate these measures were developed and incorporated in the corresponding extensions. These measures provide effective ways to summarize a distribution of points and directional information on a set of lines.

In short, the spatial statistics wizards in ArcGIS 9.X include some of the spatial statistical procedures and tools that we have discussed in this book and incorporated into several of the accompanying ArcView 3.X extensions. As mentioned in the Preface of this book, the major motivation of our development effort was education, and we believe that our approach in this book and the implementation methods in ArcView's extensions have achieved this objective very well. Our intention is to provide broad-based coverage of fundamental spatial statistical concepts. Therefore, our ArcView extensions include tools to perform or support ordinary statistical sampling and spatial sampling (Chapter 1), classical descriptive statistics (Chapter 2), correlation analyses (Chapter 3), and hypothesis testing using several statistical distributions (Chapter 4). We also have discussed and implemented spatial analytical techniques for network features (Chapter 7). Therefore, our book provides not just a comprehensive treatment of spatial analysis and spatial statistics at the introductory to intermediate levels, but also tools to support those analyses.

Without doubt, the spatial statistics wizards in ArcGIS 9.X are powerful and user-friendly tools. Our book provides thorough discussions of these tools and can help users to use them more effectively and intelligently.

ABOUT THE CD-ROM

Introduction

This appendix provides you with information on the contents of the CD that accompanies this book. For the latest and greatest information, please refer to the ReadMe file located at the root of the CD.

System Requirements

- A computer with a processor running at 120 Mhz or faster
- At least 32 MB of total RAM installed on your computer; for best performance, we recommend at least 64 MB
- A CD-ROM drive
- ArcView® GIS 3.x

Using the CD with Windows

If the opening screen of the CD-ROM does not appear automatically, follow these steps to access the CD:

1. Click the Start button on the left end of the taskbar and then choose Run from the menu that pops up.
2. In the dialog box that appears, type d:\Start.exe. (If your CD-ROM drive is not drive d, fill in the appropriate letter in place of d.) This brings up a listing of the files and local links on the CD.

What's on the CD

This companion CD-ROM contains a complete set of ArcView Extensions used in the book along with accompanying datasets.

Customer Care

If you have trouble with the CD-ROM, please call the Wiley Product Technical Support phone number at (800) 762-2974. Outside the United States, call 1(317) 572-3994. You can also contact Wiley Product Technical Support at **http://support.wiley.com**. John Wiley & Sons will provide technical support only for installation and other general quality control items. For technical support on the applications themselves, consult the program's vendor or author.

To place additional orders or to request information about other Wiley products, please call (877) 762-2974.

INDEX

D

Q

R

S

CUSTOMER NOTE: IF THIS BOOK IS ACCOMPANIED BY SOFTWARE, PLEASE READ THE FOLLOWING BEFORE OPENING THE PACKAGE.